Dark Matter

Theories on its Origin & Substance

Edited by Paul F. Kisak

Contents

1 Dark matter   **1**
    1.1 Overview 1
    1.2 Baryonic and nonbaryonic dark matter 3
    1.3 Observational evidence 4
       1.3.1 Galaxy rotation curves 4
       1.3.2 Velocity dispersions of galaxies 5
       1.3.3 Galaxy clusters and gravitational lensing 6
       1.3.4 Cosmic microwave background 7
       1.3.5 Sky surveys and baryon acoustic oscillations 8
       1.3.6 Type Ia supernovae distance measurements 8
       1.3.7 Lyman-alpha forest 8
       1.3.8 Structure formation 9
    1.4 History of the search for its composition 9
       1.4.1 Cold dark matter 10
       1.4.2 Warm dark matter 11
       1.4.3 Hot dark matter 11
       1.4.4 Mixed dark matter 12
    1.5 Detection 12
       1.5.1 Direct detection experiments 12
       1.5.2 Indirect detection experiments 13
    1.6 Alternative theories 14
       1.6.1 Mass in extra dimensions 14
       1.6.2 Topological defects 14
       1.6.3 Modified gravity 14
       1.6.4 Quantised Inertia 15
    1.7 Popular culture 15
    1.8 See also 15
    1.9 References 15
    1.10 External links 21

2 Dark energy — 31

2.1 Nature of dark energy — 31
2.1.1 Effect of dark energy: a small constant negative pressure of vacuum — 32
2.2 Evidence of existence — 33
2.2.1 Supernovae — 33
2.2.2 Cosmic microwave background — 34
2.2.3 Large-scale structure — 34
2.2.4 Late-time integrated Sachs-Wolfe effect — 34
2.2.5 Observational Hubble constant data — 35
2.3 Theories of explanation — 36
2.3.1 Cosmological constant — 36
2.3.2 Quintessence — 37
2.4 Alternative ideas — 38
2.4.1 Variable Dark Energy models — 38
2.5 Implications for the fate of the universe — 39
2.6 History of discovery and previous speculation — 40
2.7 See also — 41
2.8 References — 41
2.9 External links — 45

3 Big Bang nucleosynthesis — 46

3.1 Characteristics — 46
3.2 Important parameters — 47
3.2.1 Neutron-proton ratio — 47
3.2.2 Baryon-photon ratio — 47
3.3 Sequence — 47
3.3.1 History of theory — 49
3.3.2 Heavy elements — 49
3.3.3 Helium-4 — 50
3.3.4 Deuterium — 50
3.4 Measurements and status of theory — 51
3.5 Non-standard scenarios — 51
3.6 See also — 52
3.7 References — 52
3.8 External links — 53
3.8.1 For a general audience — 53
3.8.2 Technical articles — 53

4 Abundance of the chemical elements — 54

- 4.1 Abundance of elements in the Universe 54
  - 4.1.1 Elemental abundance and nuclear binding energy 55
- 4.2 Abundance of elements in the Earth 55
  - 4.2.1 Earth's detailed bulk (total) elemental abundance in table form 56
  - 4.2.2 Earth's crustal elemental abundance 56
  - 4.2.3 Earth's mantle elemental abundance 57
  - 4.2.4 Earth's core elemental abundance 57
  - 4.2.5 Oceanic elemental abundance 57
  - 4.2.6 Atmospheric elemental abundance 57
  - 4.2.7 Abundances of elements in urban soils 58
- 4.3 Human body elemental abundance 59
- 4.4 See also 59
- 4.5 References 59
  - 4.5.1 Footnotes 59
  - 4.5.2 Notes 60
  - 4.5.3 Notations 60
- 4.6 External links 60

5 Observable universe 65
- 5.1 The Universe versus the observable universe 65
- 5.2 Size 67
  - 5.2.1 Misconceptions on its size 68
- 5.3 Large-scale structure 70
  - 5.3.1 Walls, filaments, nodes, and voids 70
  - 5.3.2 End of Greatness 72
  - 5.3.3 Observations 72
  - 5.3.4 Cosmography of our cosmic neighborhood 73
- 5.4 Mass of ordinary matter 73
  - 5.4.1 Estimates based on critical density 73
  - 5.4.2 Extrapolation from number of stars 74
  - 5.4.3 Estimates based on steady-state universe 74
  - 5.4.4 Comparison of results 75
- 5.5 Matter content — number of atoms 75
- 5.6 Most distant objects 75
- 5.7 Horizons 75
- 5.8 See also 76
- 5.9 References 76
- 5.10 Further reading 80
- 5.11 External links 80

6 Lambda-CDM model — 81
- 6.1 Overview — 81
- 6.2 Cosmic expansion history — 83
- 6.3 Historical development — 84
- 6.4 Successes — 85
- 6.5 Challenges — 85
- 6.6 Parameters — 85
- 6.7 Extended models — 86
- 6.8 See also — 86
- 6.9 References — 86
- 6.10 Further reading — 87
- 6.11 External links — 87

7 Mass–energy equivalence — 88
- 7.1 Nomenclature — 89
- 7.2 Conservation of mass and energy — 90
  - 7.2.1 Fast-moving objects and systems of objects — 91
- 7.3 Applicability of the strict mass–energy equivalence formula, $E = mc^2$ — 92
- 7.4 Meanings of the strict mass–energy equivalence formula, $E = mc^2$ — 92
  - 7.4.1 Binding energy and the "mass defect" — 94
  - 7.4.2 Massless particles — 95
  - 7.4.3 Massless particles contribute rest mass and invariant mass to systems — 95
  - 7.4.4 Relation to gravity — 96
- 7.5 Application to nuclear physics — 96
- 7.6 Practical examples — 97
- 7.7 Efficiency — 98
- 7.8 Background — 99
  - 7.8.1 Mass–velocity relationship — 99
  - 7.8.2 Relativistic mass — 100
  - 7.8.3 Low speed expansion — 101
- 7.9 History — 101
  - 7.9.1 Newton: matter and light — 101
  - 7.9.2 Swedenborg: matter composed of "pure and total motion" — 102
  - 7.9.3 Electromagnetic mass — 102
  - 7.9.4 Radiation pressure and inertia — 102
  - 7.9.5 Einstein: mass–energy equivalence — 102
  - 7.9.6 Others — 105
  - 7.9.7 Radioactivity and nuclear energy — 105
- 7.10 See also — 107

7.11	References	108
7.12	External links	111

8 Gravitational lens — 112

8.1	Description	112
8.2	History	114
8.3	Explanation in terms of space–time curvature	115
8.4	Search for gravitational lenses	116
8.5	Solar gravitational lens	120
8.6	Measuring weak lensing	120
8.7	Gallery	120
8.8	See also	121
8.9	Historical papers and references	121
8.10	References	121
8.11	External links	123
8.12	Featured in science-fiction works	123

9 Physics beyond the Standard Model — 124

9.1	Problems with the Standard Model		124
	9.1.1	Phenomena not explained	124
	9.1.2	Experimental results not explained	125
	9.1.3	Theoretical predictions not observed	126
	9.1.4	Theoretical problems	126
9.2	Grand unified theories		128
9.3	Supersymmetry		128
9.4	Neutrinos		128
9.5	Preon Models		129
9.6	Theories of everything		129
	9.6.1	Theory of everything	129
	9.6.2	String theory	130
9.7	See also		130
9.8	References		131
9.9	Further reading		132
9.10	External resources		132

10 Structure formation — 133

10.1	Overview		133
10.2	Very early Universe		134
	10.2.1	The horizon problem	134

10.3 Primordial plasma 135

  10.3.1 Acoustic oscillations 136

10.4 Linear structure 136

10.5 Nonlinear structure 137

10.6 Gas evolution 138

10.7 Modelling structure formation 138

  10.7.1 Cosmological perturbations 138

  10.7.2 Inflation and initial conditions 138

10.8 See also 140

10.9 References 140

11 Gravitational binding energy   141

11.1 Derivation for a uniform sphere 141

11.2 Non-uniform spheres 142

11.3 See also 142

11.4 References 143

12 Galaxy formation and evolution   144

12.1 Commonly observed properties of galaxies 144

12.2 Formation of disk galaxies 145

12.3 Galaxy mergers and the formation of elliptical galaxies 147

12.4 See also 149

12.5 Further reading 151

12.6 References 151

12.7 External links 151

13 Anisotropy   152

13.1 Fields of interest 152

  13.1.1 Computer graphics 152

  13.1.2 Chemistry 153

  13.1.3 Real-world imagery 153

  13.1.4 Physics 153

  13.1.5 Geology and geophysics 153

  13.1.6 Medical acoustics 155

  13.1.7 Material science and engineering 155

  13.1.8 Microfabrication 155

  13.1.9 Neuroscience 155

  13.1.10 Atmospheric Radiative Transfer 155

13.2 References 156

14 Baryon — 157

- 13.3 External links . . . 156
- 14.1 Background . . . 157
- 14.2 Baryonic matter . . . 158
- 14.3 Baryogenesis . . . 158
- 14.4 Properties . . . 158
  - 14.4.1 Isospin and charge . . . 158
  - 14.4.2 Flavour quantum numbers . . . 159
  - 14.4.3 Spin, orbital angular momentum, and total angular momentum . . . 161
  - 14.4.4 Parity . . . 161
- 14.5 Nomenclature . . . 162
- 14.6 See also . . . 163
- 14.7 Notes . . . 163
- 14.8 References . . . 163
- 14.9 External links . . . 164

15 Weakly interacting massive particles — 165

- 15.1 Theoretical framework and properties . . . 165
- 15.2 WIMPs as dark matter . . . 166
- 15.3 Experimental detection . . . 166
  - 15.3.1 Cryogenic Crystal Detectors . . . 167
  - 15.3.2 Noble Gas Scintillators . . . 167
  - 15.3.3 Crystal Scintillators . . . 168
  - 15.3.4 Bubble Chambers . . . 168
  - 15.3.5 Other . . . 168
- 15.4 Recent Limits . . . 168
- 15.5 See also . . . 169
  - 15.5.1 Theoretical candidates . . . 169
- 15.6 References . . . 169
- 15.7 Further reading . . . 170
- 15.8 External links . . . 170

16 Cold dark matter — 171

- 16.1 Composition . . . 171
- 16.2 Challenges . . . 172
- 16.3 See also . . . 172
- 16.4 References . . . 172
- 16.5 Further reading . . . 173

17 Warm dark matter — 174
- 17.1 keVins and GeVins . . . 174
- 17.2 See also . . . 174
- 17.3 References . . . 175
- 17.4 Further reading . . . 175

18 Hot dark matter — 176
- 18.1 Neutrinos . . . 176
- 18.2 See also . . . 176
- 18.3 References . . . 176
- 18.4 Further reading . . . 176
- 18.5 External links . . . 177

19 Baryonic dark matter — 178
- 19.1 See also . . . 178
- 19.2 References . . . 178

20 Massive compact halo object — 180
- 20.1 Detection . . . 180
- 20.2 Types of MACHOs . . . 180
- 20.3 Theoretical considerations . . . 181
- 20.4 See also . . . 181
- 20.5 References . . . 181

21 Neutralino — 182
- 21.1 Origins in supersymmetric theories . . . 182
- 21.2 Phenomenology . . . 182
- 21.3 Relationship to dark matter . . . 183
- 21.4 See also . . . 183
- 21.5 Notes . . . 183
- 21.6 References . . . 184

22 Axion — 185
- 22.1 History . . . 185
  - 22.1.1 Prediction . . . 185
  - 22.1.2 Searches . . . 186
- 22.2 Experiments . . . 186
- 22.3 Possible detection . . . 186
- 22.4 Properties . . . 187
  - 22.4.1 Predictions . . . 187

22.4.2 Cosmological implications	187
22.5 References	187
22.5.1 Notes	188
22.5.2 Journal entries	190
22.6 External links	190

23 Mixed dark matter — 191

23.1 References	191
23.2 Text and image sources, contributors, and licenses	192
23.2.1 Text	192
23.2.2 Images	198
23.2.3 Content license	202

Chapter 1

Dark matter

Not to be confused with antimatter, dark energy, dark fluid, or dark flow. For other uses, see Dark Matter (disambiguation)
Dark matter is a hypothetical kind of matter that cannot be seen with telescopes but would account for most of the matter in the universe. The existence and properties of dark matter are inferred from its gravitational effects on visible matter, radiation, and the large-scale structure of the universe. Other than neutrinos, a form of hot dark matter, it has not been detected directly, making it one of the greatest mysteries in modern astrophysics.

Dark matter neither emits nor absorbs light or any other electromagnetic radiation at any significant level. According to the Planck mission team, and based on the standard model of cosmology, the total mass–energy of the known universe contains 4.9% ordinary matter, 26.8% dark matter and 68.3% dark energy.[2][3] Thus, dark matter is estimated to constitute 84.5% of the total matter in the universe, while dark energy plus dark matter constitute 95.1% of the total mass–energy content of the universe.[4][5][6]

Astrophysicists hypothesized dark matter because of discrepancies between the mass of large astronomical objects determined from their gravitational effects and the mass calculated from the observable matter (stars, gas, and dust) that they can be seen to contain. Dark matter was postulated by Jan Oort in 1932, albeit based upon flawed or inadequate evidence, to account for the orbital velocities of stars in the Milky Way and by Fritz Zwicky in 1933 to account for evidence of "missing mass" in the orbital velocities of galaxies in clusters. Adequate evidence from galaxy rotation curves was discovered by Horace W. Babcock in 1939, but was not attributed to dark matter. The first to postulate dark matter based upon robust evidence was Vera Rubin in the 1960s–1970s, using galaxy rotation curves.[7][8] Subsequently many other observations have indicated the presence of dark matter in the universe, including gravitational lensing of background objects by galaxy clusters such as the Bullet Cluster, the temperature distribution of hot gas in galaxies and clusters of galaxies and, more recently, the pattern of anisotropies in the cosmic microwave background. According to consensus among cosmologists, dark matter is composed primarily of a not yet characterized type of subatomic particle.[9][10] The search for this particle, by a variety of means, is one of the major efforts in particle physics today.[11]

Although the existence of dark matter is generally accepted by the mainstream scientific community, some alternative theories of gravity have been proposed, such as MOND and TeVeS, which try to account for the anomalous observations without requiring additional matter. However, these theories cannot account for the properties of galaxy clusters.[12]

1.1 Overview

Dark matter's existence is inferred from gravitational effects on visible matter and gravitational lensing of background radiation, and was originally hypothesized to account for discrepancies between calculations of the mass of galaxies, clusters of galaxies and the entire universe made through dynamical and general relativistic means, and calculations based on the mass of the visible "luminous" matter these objects contain: stars and the gas and dust of the interstellar and intergalactic medium.[13]

The most widely accepted explanation for these phenomena is that dark matter exists and that it is most probably[9] composed of weakly interacting massive particles (WIMPs) that interact only through gravity and the weak force. Al-

Dark matter *is invisible. Based on the effect of gravitational lensing, a ring of* dark matter *has been inferred in this image of a galaxy cluster (CL0024+17) and has been represented in blue.*[1]

ternative explanations have been proposed, and there is not yet sufficient experimental evidence to determine whether any of them are correct. Many experiments to detect proposed dark matter particles through non-gravitational means are under way.[11]

One other theory suggests the existence of a "Hidden Valley", a parallel world made of dark matter having very little in common with matter we know,[14] and that could only interact with our visible universe through gravity.[15][16]

According to observations of structures larger than star systems, as well as Big Bang cosmology interpreted under the Friedmann equations and the Friedmann–Lemaître–Robertson–Walker metric, dark matter accounts for 26.8% of the mass-energy content of the observable universe. In comparison, ordinary (baryonic) matter accounts for only 4.9% of the mass-energy content of the observable universe, with the remainder being attributable to dark energy.[3] From these figures, matter accounts for 31.7% of the mass-energy content of the universe, and 84.5% of the matter is dark matter.

Dark matter plays a central role in state-of-the-art modeling of cosmic structure formation and galaxy formation and evolution and has measurable effects on the anisotropies observed in the cosmic microwave background. All these lines of evidence suggest that galaxies, clusters of galaxies, and the universe as a whole contain far more matter than that which is easily visible with electromagnetic radiation.[15]

Important as dark matter is thought to be in the cosmos, direct evidence of its existence and a concrete understanding of its nature have remained elusive. Though the theory of dark matter remains the most widely accepted theory to explain the anomalies in observed galactic rotation, some alternative theoretical approaches have been developed which broadly fall into the categories of modified gravitational laws and quantum gravitational laws.[17]

1.2 Baryonic and nonbaryonic dark matter

There are three separate lines of evidence that the majority of dark matter is not made of baryons (ordinary matter including protons and neutrons):

- The theory of Big Bang nucleosynthesis, which predicts the observed abundance of the chemical elements,[18] predicts that baryonic matter accounts for around 4–5 percent of the critical density of the universe. In contrast, evidence from large-scale structure and other observations indicates that the total matter density is about 30% of the critical density.

- Large astronomical searches for gravitational microlensing, including the MACHO, EROS and OGLE projects, have shown that only a small fraction of the dark matter in the Milky Way can be hiding in dark compact objects; the excluded range covers objects above half the Earth's mass up to 30 solar masses, excluding nearly all the plausible candidates.

- Detailed analysis of the small irregularities (anisotropies) in the cosmic microwave background observed by WMAP and Planck shows that around five-sixths of the total matter is in a form which does not interact significantly with ordinary matter or photons except through gravitational effects.

A small proportion of dark matter may be baryonic dark matter: astronomical bodies, such as massive compact halo objects, which are composed of ordinary matter but emit little or no electromagnetic radiation. Study of nucleosynthesis in the Big Bang produces an upper bound on the amount of baryonic matter in the universe,[19] which indicates that the vast majority of dark matter in the universe cannot be baryons, and thus does not form atoms. It also cannot interact with ordinary matter via electromagnetic forces; in particular, dark matter particles do not carry any electric charge.

Candidates for nonbaryonic dark matter are hypothetical particles such as axions, or supersymmetric particles; neutrinos can only form a small fraction of the dark matter, due to limits from large-scale structure and high-redshift galaxies. Unlike baryonic dark matter, nonbaryonic dark matter does not contribute to the formation of the elements in the early universe ("Big Bang nucleosynthesis")[9] and so its presence is revealed only via its gravitational attraction. In addition, if the particles of which it is composed are supersymmetric, they can undergo annihilation interactions with themselves, possibly resulting in observable by-products such as gamma rays and neutrinos ("indirect detection").[20]

Nonbaryonic dark matter is classified in terms of the mass of the particle(s) that is assumed to make it up, and/or the typical velocity dispersion of those particles (since more massive particles move more slowly). There are three prominent hypotheses on nonbaryonic dark matter, called cold dark matter (CDM), warm dark matter (WDM), and hot dark matter (HDM); some combination of these is also possible. The most widely discussed models for nonbaryonic dark matter are based on the cold dark matter hypothesis, and the corresponding particle is most commonly assumed to be a weakly interacting massive particle (WIMP). Hot dark matter may include (massive) neutrinos, but observations imply that only a small fraction of dark matter can be hot. Cold dark matter leads to a "bottom-up" formation of structure in the universe while hot dark matter would result in a "top-down" formation scenario; since the late 1990s, the latter has been ruled out by observations of high-redshift galaxies such as the Hubble Ultra-Deep Field.[11]

1.3 Observational evidence

The first person to interpret evidence and infer the presence of dark matter was Dutch astronomer Jan Oort, a pioneer in radio astronomy, in 1932.[22] Oort was studying stellar motions in the local galactic neighbourhood and found that the mass in the galactic plane must be more than the material that could be seen, but this measurement was later determined to be essentially erroneous.[23] In 1933, the Swiss astrophysicist Fritz Zwicky, who studied clusters of galaxies while working at the California Institute of Technology, made a similar inference.[24][25] Zwicky applied the virial theorem to the Coma cluster of galaxies and obtained evidence of unseen mass. Zwicky estimated the cluster's total mass based on the motions of galaxies near its edge and compared that estimate to one based on the number of galaxies and total brightness of the cluster. He found that there was about 400 times more estimated mass than was visually observable. The gravity of the visible galaxies in the cluster would be far too small for such fast orbits, so something extra was required. This is known as the "missing mass problem". Based on these conclusions, Zwicky inferred that there must be some non-visible form of matter which would provide enough of the mass and gravity to hold the cluster together. Zwicky's estimates are off by more than an order of magnitude. Had he erred in the opposite direction by as much, he would have had to try explain the opposite – why there was too much visible matter relative to the gravitational observations – and his observations would have indicated dark energy rather than dark matter.[26]

Much of the evidence for dark matter comes from the study of the motions of galaxies.[28] Many of these appear to be fairly uniform, so by the virial theorem, the total kinetic energy should be half the total gravitational binding energy of the galaxies. Observationally, however, the total kinetic energy is found to be much greater: in particular, assuming the gravitational mass is due to only the visible matter of the galaxy, stars far from the center of galaxies have much higher velocities than predicted by the virial theorem. Galactic rotation curves, which illustrate the velocity of rotation versus the distance from the galactic center, show the well known phenomenology that cannot be explained by only the visible matter. Assuming that the visible material makes up only a small part of the cluster is the most straightforward way of accounting for this. Galaxies show signs of being composed largely of a roughly spherically symmetric, centrally concentrated halo of dark matter with the visible matter concentrated in a disc at the center. Low surface brightness dwarf galaxies are important sources of information for studying dark matter, as they have an uncommonly low ratio of visible matter to dark matter, and have few bright stars at the center which would otherwise impair observations of the rotation curve of outlying stars.

Gravitational lensing observations of galaxy clusters allow direct estimates of the gravitational mass based on its effect on light from background galaxies, since large collections of matter (dark or otherwise) will gravitationally deflect light. In clusters such as Abell 1689, lensing observations confirm the presence of considerably more mass than is indicated by the clusters' light alone. In the Bullet Cluster, lensing observations show that much of the lensing mass is separated from the X-ray-emitting baryonic mass. In July 2012, lensing observations were used to identify a "filament" of dark matter between two clusters of galaxies, as cosmological simulations have predicted.[29]

1.3.1 Galaxy rotation curves

Main article: Galaxy rotation curve

The first robust indications that the mass to light ratio was anything other than unity came from measurements of galaxy rotation curves. In 1939, Horace W. Babcock reported in his PhD thesis measurements of the rotation curve for the Andromeda nebula which suggested that the mass-to-luminosity ratio increases radially.[30] He, however, attributed it to either absorption of light within the galaxy or modified dynamics in the outer portions of the spiral and not to any form of missing matter.

In the late 1960s and early 1970s, Vera Rubin at the Department of Terrestrial Magnetism at the Carnegie Institution of Washington was the first to both make robust measurements indicating the existence of dark matter and attribute them to dark matter. Rubin worked with a new sensitive spectrograph that could measure the velocity curve of edge-on spiral galaxies to a greater degree of accuracy than had ever before been achieved.[8] Together with fellow staff-member Kent Ford, Rubin announced at a 1975 meeting of the American Astronomical Society the discovery that most stars in spiral galaxies orbit at roughly the same speed, which implied that the mass densities of the galaxies were uniform well beyond the regions containing most of the stars (the galactic bulge), a result independently found in 1978.[31] An influential paper presented Rubin's results in 1980.[32] Rubin's observations and calculations showed that most galaxies must contain about six times as much "dark" mass as can be accounted for by the visible stars. Eventually other astronomers began to

1.3. OBSERVATIONAL EVIDENCE

corroborate her work and it soon became well-established that most galaxies were dominated by "dark matter":

- Low-surface-brightness (LSB) galaxies.[33] LSBs are probably everywhere dark matter-dominated, with the observed stellar populations making only a small contribution to rotation curves. Such a property is extremely important because it allows one to avoid the difficulties associated with the deprojection and disentanglement of the dark and visible contributions to the rotation curves.[11]

- Spiral galaxies.[34] Rotation curves of both low and high surface luminosity galaxies appear to suggest a universal rotation curve, which can be expressed as the sum of an exponential thin stellar disk, and a spherical dark matter halo with a flat core of radius r_0 and density $\rho_0 = 4.5 \times 10^{-2}(r_0/\text{kpc})^{-2/3}$ $M\odot\text{pc}^{-3}$.

- Elliptical galaxies. Some elliptical galaxies show evidence for dark matter via strong gravitational lensing,[35] X-ray evidence reveals the presence of extended atmospheres of hot gas that fill the dark haloes of isolated ellipticals and whose hydrostatic support provides evidence for dark matter. Other ellipticals have low velocities in their outskirts (tracked for example by planetary nebulae) and were interpreted as not having dark matter haloes.[11] However, simulations of disk-galaxy mergers indicate that stars were torn by tidal forces from their original galaxies during the first close passage and put on outgoing trajectories, explaining the low velocities even with a DM halo.[36] More research is needed to clarify this situation.

Simulated dark matter haloes have significantly steeper density profiles (having central cusps) than are inferred from observations, which is a problem for cosmological models with dark matter at the smallest scale of galaxies as of 2008.[11] This may only be a problem of resolution: star-forming regions which might alter the dark matter distribution via outflows of gas have been too small to resolve and model simultaneously with larger dark matter clumps. A recent simulation[37] of a dwarf galaxy resolving these star-forming regions reported that strong outflows from supernovae remove low-angular-momentum gas, which inhibits the formation of a galactic bulge and decreases the dark matter density to less than half of what it would have been in the central kiloparsec. These simulation predictions—bulgeless and with shallow central dark matter profiles—correspond closely to observations of actual dwarf galaxies. There are no such discrepancies at the larger scales of clusters of galaxies and above, or in the outer regions of haloes of galaxies.

Exceptions to this general picture of dark matter haloes for galaxies appear to be galaxies with mass-to-light ratios close to that of stars. Subsequent to this, numerous observations have been made that do indicate the presence of dark matter in various parts of the cosmos, such as observations of the cosmic microwave background, of supernovas used as distance measures, of gravitational lensing at various scales, and many types of sky survey. Starting with Rubin's findings for spiral galaxies, the robust observational evidence for dark matter has been collecting over the decades to the point that by the 1980s most astrophysicists accepted its existence.[38] As a unifying concept, dark matter is one of the dominant features considered in the analysis of structures on the order of galactic scale and larger.

1.3.2 Velocity dispersions of galaxies

Rubin's pioneering work has stood the test of time. Measurements of velocity curves in spiral galaxies were soon followed up with velocity dispersions of elliptical galaxies.[39] While sometimes appearing with lower mass-to-light ratios, measurements of ellipticals still indicate a relatively high dark matter content. Likewise, measurements of the diffuse interstellar gas found at the edge of galaxies indicate not only dark matter distributions that extend beyond the visible limit of the galaxies, but also that the galaxies are virialized (i.e. gravitationally bound with velocities which appear to disproportionately correspond to predicted orbital velocities of general relativity) up to ten times their visible radii.[40] This has the effect of pushing up the dark matter as a fraction of the total amount of gravitating matter from 50% measured by Rubin to the now accepted value of nearly 95%.

There are places where dark matter seems to be a small component or totally absent. Globular clusters show little evidence that they contain dark matter,[41] though their orbital interactions with galaxies do show evidence for galactic dark matter. For some time, measurements of the velocity profile of stars seemed to indicate concentration of dark matter in the disk of the Milky Way. It now appears, however, that the high concentration of baryonic matter in the disk of the galaxy (especially in the interstellar medium) can account for this motion. Galaxy mass profiles are thought to look very different from the light profiles. The typical model for dark matter galaxies is a smooth, spherical distribution in virialized halos. Such would have to be the case to avoid small-scale (stellar) dynamical effects. Recent research reported in January 2006 from the University of Massachusetts Amherst would explain the previously mysterious warp in the disk of the Milky Way by the interaction of the Large and Small Magellanic Clouds and the predicted 20 fold increase in mass of the Milky Way taking into account dark matter.[42]

In 2005, astronomers from Cardiff University claimed to have discovered a galaxy made almost entirely of dark matter, 50 million light years away in the Virgo Cluster, which was named VIRGOHI21.[43] Unusually, VIRGOHI21 does not appear to contain any visible stars: it was seen with radio frequency observations of hydrogen. Based on rotation profiles, the scientists estimate that this object contains approximately 1000 times more dark matter than hydrogen and has a total mass of about 1/10 that of the Milky Way. For comparison, the Milky Way is estimated to have roughly 10 times as much dark matter as ordinary matter. Models of the Big Bang and structure formation have suggested that such dark galaxies should be very common in the universe, but none had previously been detected.

There are some galaxies whose velocity profile indicates an absence of dark matter, such as NGC 3379.[44]

1.3.3 Galaxy clusters and gravitational lensing

Galaxy clusters are especially important for dark matter studies since their masses can be estimated in three independent ways:

- From the scatter in radial velocities of the galaxies within them (as in Zwicky's early observations, but with accurate measurements and much larger samples).

- From X-rays emitted by very hot gas within the clusters. The temperature and density of the gas can be estimated from the energy and flux of the X-rays, hence the gas pressure; assuming pressure and gravity balance, this enables the mass profile of the cluster to be derived. Many of the experiments of the Chandra X-ray Observatory use this technique to independently determine the mass of clusters. These observations generally indicate a ratio of baryonic to total mass approximately 12–15 percent, in reasonable agreement with the Planck spacecraft cosmic average of 15.5–16 percent.[45]

- From their gravitational lensing effects on background objects, usually more distant galaxies. This is observed as "strong lensing" (multiple images) near the cluster core, and weak lensing (shape distortions) in the outer parts. Several large Hubble projects have used this method to measure cluster masses.

Generally these three methods are in reasonable agreement, that clusters contain much more matter than the visible galaxies and gas.

A gravitational lens is formed when the light from a more distant source (such as a quasar) is "bent" around a massive object (such as a cluster of galaxies) between the source object and the observer. The process is known as gravitational lensing.

The galaxy cluster Abell 2029 is composed of thousands of galaxies enveloped in a cloud of hot gas, and an amount of dark matter equivalent to more than 10^{14} $M\odot$. At the center of this cluster is an enormous, elliptically shaped galaxy that is thought to have been formed from the mergers of many smaller galaxies.[46] The measured orbital velocities of galaxies within galactic clusters have been found to be consistent with dark matter observations.

Another important tool for future dark matter observations is gravitational lensing. Lensing relies on the effects of general relativity to predict masses without relying on dynamics, and so is a completely independent means of measuring the dark matter. Strong lensing, the observed distortion of background galaxies into arcs when the light passes through a gravitational lens, has been observed around a few distant clusters including Abell 1689 (pictured).[47] By measuring the distortion geometry, the mass of the cluster causing the phenomena can be obtained. In the dozens of cases where this has been done, the mass-to-light ratios obtained correspond to the dynamical dark matter measurements of clusters.[48]

Weak gravitational lensing looks at minute distortions of galaxies observed in vast galaxy surveys due to foreground objects through statistical analyses. By examining the apparent shear deformation of the adjacent background galaxies, astrophysicists can characterize the mean distribution of dark matter by statistical means and have found mass-to-light ratios that correspond to dark matter densities predicted by other large-scale structure measurements.[49] The correspondence of the two gravitational lens techniques to other dark matter measurements has convinced almost all astrophysicists that dark matter actually exists as a major component of the universe's composition.

1.3. OBSERVATIONAL EVIDENCE

The most direct observational evidence to date for dark matter is in a system known as the Bullet Cluster. In most regions of the universe, dark matter and visible material are found together,[50] as expected because of their mutual gravitational attraction. In the Bullet Cluster, a collision between two galaxy clusters appears to have caused a separation of dark matter and baryonic matter. X-ray observations show that much of the baryonic matter (in the form of 10^7–10^8 Kelvin[51] gas or plasma) in the system is concentrated in the center of the system. Electromagnetic interactions between passing gas particles caused them to slow down and settle near the point of impact. However, weak gravitational lensing observations of the same system show that much of the mass resides outside of the central region of baryonic gas. Because dark matter does not interact by electromagnetic forces, it would not have been slowed in the same way as the X-ray visible gas, so the dark matter components of the two clusters passed through each other without slowing down substantially. This accounts for the separation. Unlike the galactic rotation curves, this evidence for dark matter is independent of the details of Newtonian gravity, so it is claimed to be direct evidence of the existence of dark matter.[51]

Another galaxy cluster, known as the Train Wreck Cluster/Abell 520, initially appeared to have an unusually massive and dark core containing few of the cluster's galaxies, which presented problems for standard dark matter models.[52] However, more precise observations since this time have shown that the earlier observations were misleading, and that the distribution of dark matter and its ratio to normal matter are very similar to those in galaxies in general, making novel explanations unnecessary.[51]

The observed behavior of dark matter in clusters constrains whether and how much dark matter scatters off other dark matter particles, quantified as its self-interaction cross section. More simply, the question is whether the dark matter has pressure, and thus can be described as a perfect fluid.[53] The distribution of mass (and thus dark matter) in galaxy clusters has been used to argue both for[54] and against[55] the existence of significant self-interaction in dark matter. Specifically, the distribution of dark matter in merging clusters such as the Bullet Cluster shows that dark matter scatters off other dark matter particles only very weakly if at all.[56]

Researchers are conducting a wide-area survey of the distribution of dark matter and changes in distribution over time determine the impact of dark energy causing the expansion of the universe. The survey is using weak lensing to analyze background light, bent by dark matter, to determine how dark matter is distributed in the foreground. The analysis of dark matter and its effects could determine how dark matter assembled over time, which can be related to the history of the expansion of the universe, and could reveal some physical properties of dark energy, its strength and how it has changed over time. The survey is observing galaxies more than a billion light-years away, across an area greater than a thousand square degrees (about one fortieth of the entire sky). The survey is using the Subaru telescope.[57][58]

1.3.4 Cosmic microwave background

Main article: Cosmic microwave background
See also: Wilkinson Microwave Anisotropy Probe

Angular fluctuations in the cosmic microwave background (CMB) spectrum provide evidence for dark matter. Since the 1964 discovery and confirmation of the CMB radiation,[59] many measurements of the CMB have supported and constrained this theory. The NASA Cosmic Background Explorer (COBE) found that the CMB spectrum is a blackbody spectrum with a temperature of 2.726 K. In 1992, COBE detected fluctuations (anisotropies) in the CMB spectrum, at a level of about one part in 10^5.[60] During the following decade, CMB anisotropies were further investigated by a large number of ground-based and balloon experiments. The primary goal of these experiments was to measure the angular scale of the first acoustic peak of the power spectrum of the anisotropies, for which COBE did not have sufficient resolution. In 2000–2001, several experiments, most notably BOOMERanG[61] found the universe to be almost spatially flat by measuring the typical angular size (the size on the sky) of the anisotropies. During the 1990s, the first peak was measured with increasing sensitivity and by 2000 the BOOMERanG experiment reported that the highest power fluctuations occur at scales of approximately one degree. These measurements were able to rule out cosmic strings as the leading theory of cosmic structure formation, and suggested cosmic inflation was the right theory.

A number of ground-based interferometers provided measurements of the fluctuations with higher accuracy over the next three years, including the Very Small Array, the Degree Angular Scale Interferometer (DASI) and the Cosmic Background Imager (CBI). DASI made the first detection of the polarization of the CMB,[62][63] and the CBI provided the first E-mode polarization spectrum with compelling evidence that it is out of phase with the T-mode spectrum.[64] COBE's successor, the Wilkinson Microwave Anisotropy Probe (WMAP) has provided the most detailed measurements of (large-scale) anisotropies in the CMB as of 2009 with ESA's Planck spacecraft returning more detailed results in 2012-2014.[65] WMAP's measurements played the key role in establishing the current Standard Model of Cosmology, namely the Lambda-CDM model, a flat universe dominated by dark energy, supplemented by dark matter and atoms with density fluctuations seeded by a Gaussian, adiabatic, nearly scale invariant process. The basic properties of this universe are determined by five numbers: the density of matter, the density of atoms, the age of the universe (or equivalently, the Hubble constant today), the amplitude of the initial fluctuations, and their scale dependence.

A successful Big Bang cosmology theory must fit with all available astronomical observations, including the CMB. In cosmology, the CMB is explained as relic radiation from shortly after the big bang. The anisotropies in the CMB are explained as acoustic oscillations in the photon-baryon plasma (prior to the emission of the CMB after the photons decouple from the baryons at 379,000 years after the Big Bang) whose restoring force is gravity.[66] Ordinary (baryonic) matter interacts strongly with radiation whereas dark matter particles, such as WIMPs for example, do not. Both affect the oscillations by their gravity, so the two forms of matter will have different effects. The typical angular scales of the oscillations in the CMB, measured as the power spectrum of the CMB anisotropies, thus reveal the different effects of baryonic matter and dark matter. The CMB power spectrum shows a large first peak and smaller successive peaks, with three peaks resolved as of 2009.[65] The first peak tells mostly about the density of baryonic matter and the third peak mostly about the density of dark matter, measuring the density of matter and the density of atoms in the universe.

1.3.5 Sky surveys and baryon acoustic oscillations

Main article: Baryon acoustic oscillations

The acoustic oscillations in the early universe (see the previous section) leave their imprint in the visible matter by Baryon Acoustic Oscillation (BAO) clustering, in a way that can be measured with sky surveys such as the Sloan Digital Sky Survey and the 2dF Galaxy Redshift Survey.[67] These measurements are consistent with those of the CMB derived from the WMAP spacecraft and further constrain the Lambda CDM model and dark matter. Note that the CMB data and the BAO data measure the acoustic oscillations at very different distance scales.[66]

1.3.6 Type Ia supernovae distance measurements

Main article: Type Ia supernova

Type Ia supernovae can be used as "standard candles" to measure extragalactic distances, and extensive data sets of these supernovae can be used to constrain cosmological models.[68] They constrain the dark energy density $\Omega\Lambda$ = ~0.713 for a flat, Lambda CDM universe and the parameter w for a quintessence model. Once again, the values obtained are roughly consistent with those derived from the WMAP observations and further constrain the Lambda CDM model and (indirectly) dark matter.[66]

1.3.7 Lyman-alpha forest

Main article: Lyman-alpha forest

In astronomical spectroscopy, the Lyman-alpha forest is the sum of absorption lines arising from the Lyman-alpha transition of the neutral hydrogen in the spectra of distant galaxies and quasars. Observations of the Lyman-alpha forest can also be used to constrain cosmological models.[69] These constraints are again in agreement with those obtained from WMAP data.

1.3.8 Structure formation

Main article: Structure formation

Dark matter is crucial to the Big Bang model of cosmology as a component which corresponds directly to measurements of the parameters associated with Friedmann cosmology solutions to general relativity. In particular, measurements of the cosmic microwave background anisotropies correspond to a cosmology where much of the matter interacts with photons more weakly than the known forces that couple light interactions to baryonic matter. Likewise, a significant amount of non-baryonic, cold matter is necessary to explain the large-scale structure of the universe.

Observations suggest that structure formation in the universe proceeds hierarchically, with the smallest structures collapsing first and followed by galaxies and then clusters of galaxies. As the structures collapse in the evolving universe, they begin to "light up" as the baryonic matter heats up through gravitational contraction and the object approaches hydrostatic pressure balance. Ordinary baryonic matter had too high a temperature, and too much pressure left over from the Big Bang to collapse and form smaller structures, such as stars, via the Jeans instability. Dark matter acts as a compactor of structure. This model not only corresponds with statistical surveying of the visible structure in the universe but also corresponds precisely to the dark matter predictions of the cosmic microwave background.

This *bottom up* model of structure formation requires something like cold dark matter to succeed. Large computer simulations of billions of dark matter particles have been used[71] to confirm that the cold dark matter model of structure formation is consistent with the structures observed in the universe through galaxy surveys, such as the Sloan Digital Sky Survey and 2dF Galaxy Redshift Survey, as well as observations of the Lyman-alpha forest. These studies have been crucial in constructing the Lambda-CDM model which measures the cosmological parameters, including the fraction of the universe made up of baryons and dark matter.

There are, however, several points of tension between observation and simulations of structure formation driven by dark matter. There is evidence that there are 10 to 100 times fewer small galaxies than permitted by what the dark matter theory of galaxy formation predicts.[72][73] This is known as the dwarf galaxy problem. In addition, the simulations predict dark matter distributions with a very dense cusp near the centers of galaxies, but the observed halos are smoother than predicted.

1.4 History of the search for its composition

Although dark matter had historically been inferred by many astronomical observations, its composition long remained speculative. Early theories of dark matter concentrated on hidden heavy normal objects (such as black holes, neutron stars, faint old white dwarfs, and brown dwarfs) as the possible candidates for dark matter, collectively known as massive compact halo objects or MACHOs. Astronomical surveys for gravitational microlensing, including the MACHO, EROS and OGLE projects, along with Hubble telescope searches for ultra-faint stars, have not found enough of these hidden MACHOs.[74][75][76] Some hard-to-detect baryonic matter, such as MACHOs and some forms of gas, were additionally speculated to make a contribution to the overall dark matter content, but evidence indicated such would constitute only a small portion.[77][78][79]

Furthermore, data from a number of lines of other evidence, including galaxy rotation curves, gravitational lensing, structure formation, and the fraction of baryons in clusters and the cluster abundance combined with independent evidence for the baryon density, indicated that 85–90% of the mass in the universe does not interact with the electromagnetic force. This "nonbaryonic dark matter" is evident through its gravitational effect. Consequently, the most commonly held view was that dark matter is primarily non-baryonic, made of one or more elementary particles other than the usual electrons, protons, neutrons, and known neutrinos. The most commonly proposed particles then became WIMPs (Weakly Interacting Massive Particles, including neutralinos), axions, or sterile neutrinos, though many other possible candidates have been proposed.

Dark matter candidates can be approximately divided into three classes, called *cold*, *warm* and *hot* dark matter.[80] These categories do not correspond to an actual temperature, but instead refer to how fast the particles were moving, thus how far they moved due to random motions in the early universe, before they slowed down due to the expansion of the universe – this is an important distance called the "free streaming length". Primordial density fluctuations smaller than this free-streaming length get washed out as particles move from overdense to underdense regions, while fluctuations larger than the free-streaming length are unaffected; therefore this free-streaming length sets a minimum scale for structure formation.

- Cold dark matter – objects with a free-streaming length much smaller than a protogalaxy.[81]
- Warm dark matter – particles with a free-streaming length similar to a protogalaxy.
- Hot dark matter – particles with a free-streaming length much larger than a protogalaxy.[82]

Though a fourth category had been considered early on, called mixed dark matter, it was quickly eliminated (from the 1990s) since the discovery of dark energy.

As an example, Davis *et al.* wrote in 1985:

> Candidate particles can be grouped into three categories on the basis of their effect on the fluctuation spectrum (Bond *et al.* 1983). If the dark matter is composed of abundant light particles which remain relativistic until shortly before recombination, then it may be termed "hot". The best candidate for hot dark matter is a neutrino ... A second possibility is for the dark matter particles to interact more weakly than neutrinos, to be less abundant, and to have a mass of order 1 keV. Such particles are termed "warm dark matter", because they have lower thermal velocities than massive neutrinos ... there are at present few candidate particles which fit this description. Gravitinos and photinos have been suggested (Pagels and Primack 1982; Bond, Szalay and Turner 1982) ... Any particles which became nonrelativistic very early, and so were able to diffuse a negligible distance, are termed "cold" dark matter (CDM). There are many candidates for CDM including supersymmetric particles.[83]

The full calculations are quite technical, but an approximate dividing line is that "warm" dark matter particles became non-relativistic when the universe was approximately 1 year old and 1 millionth of its present size; standard hot big bang theory implies the universe was then in the radiation-dominated era (photons and neutrinos), with a photon temperature 2.7 million K. Standard physical cosmology gives the particle horizon size as 2ct in the radiation-dominated era, thus 2 light-years, and a region of this size would expand to 2 million light years today (if there were no structure formation). The actual free-streaming length is roughly 5 times larger than the above length, since the free-streaming length continues to grow slowly as particle velocities decrease inversely with the scale factor after they become non-relativistic; therefore, in this example the free-streaming length would correspond to 10 million light-years or 3 Mpc today, which is around the size containing on average the mass of a large galaxy.

The above temperature of 2.7 million K gives a typical photon energy of 250 electron-volts, thereby setting a typical mass scale for "warm" dark matter: particles much more massive than this, such as GeV – TeV mass WIMPs, would become non-relativistic much earlier than 1 year after the Big Bang and thus have a free-streaming length much smaller than a proto-galaxy, making them cold dark matter. Conversely, much lighter particles, such as neutrinos with masses of only a few eV, have a free-streaming length much larger than a proto-galaxy, thus making them hot dark matter.

1.4.1 Cold dark matter

Main article: Cold dark matter

Today, cold dark matter is the simplest explanation for most cosmological observations. "Cold" dark matter is dark matter composed of constituents with a free-streaming length much smaller than the ancestor of a galaxy-scale perturbation. This is currently the area of greatest interest for dark matter research, as hot dark matter does not seem to be viable for galaxy and galaxy cluster formation, and most particle candidates become non-relativistic at very early times, hence are classified as cold.

The composition of the constituents of cold dark matter is currently unknown. Possibilities range from large objects like MACHOs (such as black holes[84]) or RAMBOs, to new particles like WIMPs and axions. Possibilities involving normal baryonic matter include brown dwarfs, other stellar remnants such as white dwarfs, or perhaps small, dense chunks of heavy elements.

1.4. HISTORY OF THE SEARCH FOR ITS COMPOSITION

Studies of big bang nucleosynthesis and gravitational lensing have convinced most scientists[11][85][86][87][88][89] that MACHOs of any type cannot be more than a small fraction of the total dark matter.[9][85] Black holes of nearly any mass are ruled out as a primary dark matter constituent by a variety of searches and constraints.[85][87] According to A. Peter: "...the only *really plausible* dark-matter candidates are new particles."[86]

The DAMA/NaI experiment and its successor DAMA/LIBRA have claimed to directly detect dark matter particles passing through the Earth, but many scientists remain skeptical, as negative results from similar experiments seem incompatible with the DAMA results.

Many supersymmetric models naturally give rise to stable dark matter candidates in the form of the Lightest Supersymmetric Particle (LSP). Separately, heavy sterile neutrinos exist in non-supersymmetric extensions to the standard model that explain the small neutrino mass through the seesaw mechanism.

1.4.2 Warm dark matter

Main article: Warm dark matter

Warm dark matter refers to particles with a free-streaming length comparable to the size of a region which subsequently evolved into a dwarf galaxy. This leads to predictions which are very similar to cold dark matter on large scales, including the CMB, galaxy clustering and large galaxy rotation curves, but with less small-scale density perturbations. This reduces the predicted abundance of dwarf galaxies and may lead to lower density of dark matter in the central parts of large galaxies; some researchers consider this may be a better fit to observations. A challenge for this model is that there are no very well-motivated particle physics candidates with the required mass ~ 300 eV to 3000 eV.

There have been no particles discovered so far that can be categorized as warm dark matter. There is a postulated candidate for the warm dark matter category, which is the sterile neutrino: a heavier, slower form of neutrino which does not even interact through the Weak force unlike regular neutrinos. Interestingly, some modified gravity theories, such as Scalar-tensor-vector gravity, also require that a warm dark matter exist to make their equations work out.

1.4.3 Hot dark matter

Main article: Hot dark matter

Hot dark matter consists of particles that have a free-streaming length much larger than that of a proto-galaxy.

An example of hot dark matter is already known: the neutrino. Neutrinos were discovered quite separately from the search for dark matter, and long before it seriously began: they were first postulated in 1930, and first detected in 1956. Neutrinos have a very small mass: at least 100,000 times less massive than an electron. Other than gravity, neutrinos only interact with normal matter via the weak force making them very difficult to detect (the weak force only works over a small distance, thus a neutrino will only trigger a weak force event if it hits a nucleus directly head-on). This would make them 'weakly interacting light particles' (WILPs), as opposed to cold dark matter's theoretical candidates, the weakly interacting massive particles (WIMPs).

There are three different known flavors of neutrinos (i.e. the *electron*, *muon*, and *tau* neutrinos), and their masses are slightly different. The resolution to the solar neutrino problem demonstrated that these three types of neutrinos actually change and oscillate from one flavor to the others and back as they are in-flight. It's hard to determine an exact upper bound on the collective average mass of the three neutrinos (let alone a mass for any of the three individually). For example, if the average neutrino mass were chosen to be over 50 eV/c^2 (which is still less than 1/10,000th of the mass of an electron), just by the sheer number of them in the universe, the universe would collapse due to their mass. So other observations have served to estimate an upper-bound for the neutrino mass. Using cosmic microwave background data and other methods, the current conclusion is that their average mass probably does not exceed 0.3 eV/c^2 Thus, the normal forms of neutrinos cannot be responsible for the measured dark matter component from cosmology.[90]

Hot dark matter was popular for a time in the early 1980s, but it suffers from a severe problem: because all galaxy-size density fluctuations get washed out by free-streaming, the first objects that can form are huge supercluster-size pancakes, which then were theorised somehow to fragment into galaxies. Deep-field observations clearly show that galaxies formed at early times, with clusters and superclusters forming later as galaxies clump together, so any model dominated by hot dark matter is seriously in conflict with observations.

1.4.4 Mixed dark matter

Main article: Mixed dark matter

Mixed dark matter is a now obsolete model, with a specifically chosen mass ratio of 80% cold dark matter and 20% hot dark matter (neutrinos) content. Though it is presumable that hot dark matter coexists with cold dark matter in any case, there was a very specific reason for choosing this particular ratio of hot to cold dark matter in this model. During the early 1990s it became steadily clear that a universe with critical density of cold dark matter did not fit the COBE and large-scale galaxy clustering observations; either the 80/20 mixed dark matter model, or LambdaCDM, were able to reconcile these. With the discovery of the accelerating universe from supernovae, and more accurate measurements of CMB anisotropy and galaxy clustering, the mixed dark matter model was essentially ruled out while the concordance LambdaCDM model remained a good fit.

1.5 Detection

If the dark matter within the Milky Way is made up of Weakly Interacting Massive Particles (WIMPs), then millions, possibly billions, of WIMPs must pass through every square centimeter of the Earth each second.[91][92] There are many experiments currently running, or planned, aiming to test this hypothesis by searching for WIMPs. Although WIMPs are the historically more popular dark matter candidate for searches,[11] there are experiments searching for other particle candidates; the Axion Dark Matter eXperiment (ADMX) is currently searching for the dark matter axion, a well-motivated and constrained dark matter source. It is also possible that dark matter consists of very heavy hidden sector particles which only interact with ordinary matter via gravity.

These experiments can be divided into two classes: direct detection experiments, which search for the scattering of dark matter particles off atomic nuclei within a detector; and indirect detection, which look for the products of WIMP annihilations.[20]

An alternative approach to the detection of WIMPs in nature is to produce them in the laboratory. Experiments with the Large Hadron Collider (LHC) may be able to detect WIMPs produced in collisions of the LHC proton beams. Because a WIMP has negligible interactions with matter, it may be detected indirectly as (large amounts of) missing energy and momentum which escape the LHC detectors, provided all the other (non-negligible) collision products are detected.[93] These experiments could show that WIMPs can be created, but it would still require a direct detection experiment to show that they exist in sufficient numbers to account for dark matter.

1.5.1 Direct detection experiments

Direct detection experiments usually operate in deep underground laboratories to reduce the background from cosmic rays. These include: the Soudan mine; the SNOLAB underground laboratory at Sudbury, Ontario (Canada); the Gran Sasso National Laboratory (Italy); the Canfranc Underground Laboratory (Spain); the Boulby Underground Laboratory (United Kingdom); the Deep Underground Science and Engineering Laboratory, South Dakota (United States); and the Particle and Astrophysical Xenon Detector (China).

The majority of present experiments use one of two detector technologies: cryogenic detectors, operating at temperatures below 100mK, detect the heat produced when a particle hits an atom in a crystal absorber such as germanium. Noble liquid detectors detect the flash of scintillation light produced by a particle collision in liquid xenon or argon. Cryogenic detector experiments include: CDMS, CRESST, EDELWEISS, EURECA. Noble liquid experiments include ZEPLIN, XENON, DEAP, ArDM, WARP, DarkSide, PandaX, and LUX, the Large Underground Xenon experiment. Both of these detector techniques are capable of distinguishing background particles which scatter off electrons, from dark matter

1.5. DETECTION

particles which scatter off nuclei. Other experiments include SIMPLE and PICASSO.

The DAMA/NaI, DAMA/LIBRA experiments have detected an annual modulation in the event rate,[94] which they claim is due to dark matter particles. (As the Earth orbits the Sun, the velocity of the detector relative to the dark matter halo will vary by a small amount depending on the time of year). This claim is so far unconfirmed and difficult to reconcile with the negative results of other experiments assuming that the WIMP scenario is correct.[95]

Directional detection of dark matter is a search strategy based on the motion of the Solar System around the Galactic Center.[96][97][98][99]

By using a low pressure TPC, it is possible to access information on recoiling tracks (3D reconstruction if possible) and to constrain the WIMP-nucleus kinematics. WIMPs coming from the direction in which the Sun is travelling (roughly in the direction of the Cygnus constellation) may then be separated from background noise, which should be isotropic. Directional dark matter experiments include DMTPC, DRIFT, Newage and MIMAC.

On 17 December 2009 CDMS researchers reported two possible WIMP candidate events. They estimate that the probability that these events are due to a known background (neutrons or misidentified beta or gamma events) is 23%, and conclude "this analysis cannot be interpreted as significant evidence for WIMP interactions, but we cannot reject either event as signal."[100]

More recently, on 4 September 2011, researchers using the CRESST detectors presented evidence[101] of 67 collisions occurring in detector crystals from subatomic particles, calculating there is a less than 1 in 10,000 chance that all were caused by known sources of interference or contamination. It is quite possible then that many of these collisions were caused by WIMPs, and/or other unknown particles.

1.5.2 Indirect detection experiments

Indirect detection experiments search for the products of WIMP annihilation or decay. If WIMPs are Majorana particles (WIMPs are their own antiparticle) then two WIMPs could annihilate to produce gamma rays or Standard Model particle-antiparticle pairs. Additionally, if the WIMP is unstable, WIMPs could decay into standard model particles. These processes could be detected indirectly through an excess of gamma rays, antiprotons or positrons emanating from regions of high dark matter density. The detection of such a signal is not conclusive evidence for dark matter, as the production of gamma rays from other sources is not fully understood.[11][20]

The EGRET gamma ray telescope observed more gamma rays than expected from the Milky Way, but scientists concluded that this was most likely due to a mis-estimation of the telescope's sensitivity.[103]

The Fermi Gamma-ray Space Telescope, launched 11 June 2008, is searching for gamma rays from dark matter annihilation and decay.[104] In April 2012, an analysis[105] of previously available data from its Large Area Telescope instrument produced strong statistical evidence of a 130 GeV line in the gamma radiation coming from the center of the Milky Way. At the time, WIMP annihilation was the most probable explanation for that line.[106]

At higher energies, ground-based gamma-ray telescopes have set limits on the annihilation of dark matter in dwarf spheroidal galaxies[107] and in clusters of galaxies.[108]

The PAMELA experiment (launched 2006) has detected a larger number of positrons than expected. These extra positrons could be produced by dark matter annihilation, but may also come from pulsars. No excess of anti-protons has been observed.[109] The Alpha Magnetic Spectrometer on the International Space Station is designed to directly measure the fraction of cosmic rays which are positrons. The first results, published in April 2013, indicate an excess of high-energy cosmic rays which could potentially be due to annihilation of dark matter.[110][111][112][113][114][115]

A few of the WIMPs passing through the Sun or Earth may scatter off atoms and lose energy. This way a large population of WIMPs may accumulate at the center of these bodies, increasing the chance that two will collide and annihilate. This could produce a distinctive signal in the form of high-energy neutrinos originating from the center of the Sun or Earth.[116] It is generally considered that the detection of such a signal would be the strongest indirect proof of WIMP dark matter.[11] High-energy neutrino telescopes such as AMANDA, IceCube and ANTARES are searching for this signal.

WIMP annihilation from the Milky Way Galaxy as a whole may also be detected in the form of various annihilation products.[117] The Galactic Center is a particularly good place to look because the density of dark matter may be very high there.[118]

1.6 Alternative theories

1.6.1 Mass in extra dimensions

In some multidimensional theories, the force of gravity is the unique force able to have an effect across all the various extra dimensions,[16] which would explain the relative weakness of the force of gravity compared to the other known forces of nature that would not be able to cross into extra dimensions: electromagnetism, strong interaction, and weak interaction.

In that case, dark matter would be a perfect candidate for matter that would exist in other dimensions and that could only interact with the matter on our dimensions through gravity. That dark matter located on different dimensions could potentially aggregate in the same way as the matter in our visible universe does, forming exotic galaxies.[15]

1.6.2 Topological defects

Dark matter could consist of primordial defects (defects originating with the birth of the universe) in the topology of quantum fields, which would contain energy and therefore gravitate. This possibility may be investigated by the use of an orbital network of atomic clocks, which would register the passage of topological defects by monitoring the synchronization of the clocks. The Global Positioning System may be able to operate as such a network.[119]

1.6.3 Modified gravity

Numerous alternative theories have been proposed to explain these observations without the need for a large amount of undetected matter. Most of these theories modify the laws of gravity established by Newton and Einstein.

The earliest modified gravity model to emerge was Mordehai Milgrom's Modified Newtonian Dynamics (MOND) in 1983, which adjusts Newton's laws to create a stronger gravitational field when gravitational acceleration levels become tiny (such as near the rim of a galaxy). It had some success explaining galactic-scale features, such as rotational velocity curves of elliptical galaxies, and dwarf elliptical galaxies, but did not successfully explain galaxy cluster gravitational lensing. However, MOND was not relativistic, since it was just a straight adjustment of the older Newtonian account of gravitation, not of the newer account in Einstein's general relativity. Soon after 1983, attempts were made to bring MOND into conformity with general relativity; this is an ongoing process, and many competing hypotheses have emerged based around the original MOND model—including TeVeS, MOG or STV gravity, and phenomenological covariant approach,[120] among others.

In 2007, John W. Moffat proposed a modified gravity hypothesis based on the nonsymmetric gravitational theory (NGT) that claims to account for the behavior of colliding galaxies.[121] This model requires the presence of non-relativistic neutrinos, or other candidates for (cold) dark matter, to work.

Another proposal uses a gravitational backreaction in an emerging theoretical field that seeks to explain gravity between objects as an action, a reaction, and then a back-reaction. Simply, an object A affects an object B, and the object B then re-affects object A, and so on: creating a sort of feedback loop that strengthens gravity.[122]

Recently, another group has proposed a modification of large-scale gravity in a hypothesis named "dark fluid". In this formulation, the attractive gravitational effects attributed to dark matter are instead a side-effect of dark energy. Dark fluid combines dark matter and dark energy in a single energy field that produces different effects at different scales. This treatment is a simplified approach to a previous fluid-like model called the generalized Chaplygin gas model where the whole of spacetime is a compressible gas.[123] Dark fluid can be compared to an atmospheric system. Atmospheric pressure causes air to expand, but part of the air can collapse to form clouds. In the same way, the dark fluid might generally expand, but it also could collect around galaxies to help hold them together.[123]

Another set of proposals is based on the possibility of a double metric tensor for space-time.[124] It has been argued that time-reversed solutions in general relativity require such double metric for consistency, and that both dark matter and dark energy can be understood in terms of time-reversed solutions of general relativity.[125]

1.6.4 Quantised Inertia

One proposal is that if inertia is assumed to be due to the effect of horizons on Unruh radiation then this predicts galaxy rotation without dark matter.[126]

1.7 Popular culture

Main article: Dark matter in fiction

Mention of dark matter is made in some video games and other works of fiction. In such cases, it is usually attributed extraordinary physical or magical properties. Such descriptions are often inconsistent with the properties of dark matter proposed in physics and cosmology.

1.8 See also

- Chameleon particle
- Conformal gravity
- General Antiparticle Spectrometer
- Illustris project
- Light dark matter
- Mirror matter
- Multidark (research program)
- Scalar field dark matter
- Self-interacting dark matter
- SIMP
- Unparticle physics

1.9 References

[1] "Hubble Finds Dark Matter Ring in Galaxy Cluster".

[2] Ade, P. A. R.; Aghanim, N.; Armitage-Caplan, C.; (Planck Collaboration) et al. (22 March 2013). "Planck 2013 results. I. Overview of products and scientific results – Table 9". *Astronomy and Astrophysics* **1303**: 5062. arXiv:1303.5062. Bibcode:2014A&A...571A...1P.doi:10.1051/0004-6361/201321529.

[3] Francis, Matthew (22 March 2013). "First Planck results: the Universe is still weird and interesting". *Arstechnica*.

[4] "Planck captures portrait of the young Universe, revealing earliest light". University of Cambridge. 21 March 2013. Retrieved 21 March 2013.

[5] Sean Carroll, Ph.D., Cal Tech, 2007, The Teaching Company, *Dark Matter, Dark Energy: The Dark Side of the Universe*, Guidebook Part 2 page 46, Accessed Oct. 7, 2013, "...dark matter: An invisible, essentially collisionless component of matter that makes up about 25 percent of the energy density of the universe... it's a different kind of particle... something not yet observed in the laboratory..."

[6] Ferris, Timothy. "Dark Matter". Retrieved 2015-06-10.

[7] First observational evidence of dark matter. Darkmatterphysics.com. Retrieved on 6 August 2013.

[8] Rubin, Vera C.; Ford, W. Kent, Jr. (February 1970). "Rotation of the Andromeda Nebula from a Spectroscopic Survey of Emission Regions". *The Astrophysical Journal* **159**: 379–403. Bibcode:1970ApJ...159..379R. doi:10.1086/150317.

[9] Copi, C. J.; Schramm, D. N.; Turner, M. S. (1995). "Big-Bang Nucleosynthesis and the Baryon Density of the Universe". *Science* **267** (5195): 192–199. arXiv:astro-ph/9407006. Bibcode:1995Sci...267..192C. doi:10.1126/science.7809624. PMID 7809624.

[10] Bergstrom, L. (2000). "Non-baryonic dark matter: Observational evidence and detection methods". *Reports on Progress in Physics* **63** (5): 793–841. arXiv:hep-ph/0002126. Bibcode:2000RPPh...63..793B. doi:10.1088/0034-4885/63/5/2r3.

[11] Bertone, G.; Hooper, D.; Silk, J. (2005). "Particle dark matter: Evidence, candidates and constraints". *Physics Reports* **405** (5–6): 279–390. arXiv:hep-ph/0404175. Bibcode:2005PhR...405..279B. doi:10.1016/j.physrep.2004.08.031.

[12] Angus, G. (2013). "Cosmological simulations in MOND: the cluster scale halo mass function with light sterile neutrinos". *Monthly Notices of the Royal Astronomical Society* **436**: 202–211. arXiv:1309.6094. Bibcode:2013MNRAS.436..202A.doi:10.1093/mnras/stt1564.

[13] Trimble, V. (1987). "Existence and nature of dark matter in the universe". *Annual Review of Astronomy and Astrophysics* **25**: 425–472. Bibcode:1987ARA&A..25..425T. doi:10.1146/annurev.aa.25.090187.002233.

[14] Dark matter. CERN. Retrieved on 17 November 2014.

[15] Siegfried, T. (5 July 1999). "Hidden Space Dimensions May Permit Parallel Universes, Explain Cosmic Mysteries". *The Dallas Morning News*.

[16] Extra dimensions, gravitons, and tiny black holes. CERN. Retrieved on 17 November 2014.

[17] Kroupa, P. et al. (2010). "Local-Group tests of dark-matter Concordance Cosmology: Towards a new paradigm for structure formation". *Astronomy and Astrophysics* **523**: 32–54. arXiv:1006.1647. Bibcode:2010A&A...523A..32K. doi:10.1051/0004-6361/201014892.

[18] Achim Weiss, "Big Bang Nucleosynthesis: Cooking up the first light elements" in: Einstein Online Vol. 2 (2006), 1017

[19] Raine, D.; Thomas, T. (2001). *An Introduction to the Science of Cosmology*. IOP Publishing. p. 30. ISBN 0-7503-0405-7.

[20] Bertone, G.; Merritt, D. (2005). "Dark Matter Dynamics and Indirect Detection". *Modern Physics Letters A* **20** (14): 1021–1036. arXiv:astro-ph/0504422. Bibcode:2005MPLA...20.1021B. doi:10.1142/S0217732305017391.

[21] "Serious Blow to Dark Matter Theories?" (Press release). European Southern Observatory. 18 April 2012.

[22] "The Hidden Lives of Galaxies: Hidden Mass". *Imagine the Universe!*. NASA/GSFC.

[23] Kuijken K. and Gilmore G. (1989). "The Mass Distribution in the Galactic Disc - Part III - the Local Volume Mass Density". *Monthly Notices of the Royal Astronomical Society* **239**: 651. Bibcode:1989MNRAS.239..651K. doi:10.1093/mnras/239.2.651.

[24] Zwicky, F. (1933). "Die Rotverschiebung von extragalaktischen Nebeln". *Helvetica Physica Acta* **6**: 110–127. Bibcode:1933AcHPh...6..110Z.

[25] Zwicky, F. (1937). "On the Masses of Nebulae and of Clusters of Nebulae".*The Astrophysical Journal* doi:10.1086/143864.

[26] Freese, Katerine (2014). *The Cosmic Cocktail: Three Parts Dark Matter*. Princeton, New Jersey: Princeton University Press. ISBN 978-0691153353.

[27] "First Signs of Self-interacting Dark Matter?". *ESO Press Release*. European Southern Observatory. Retrieved 15 April 2015.

[28] Freeman, K.; McNamara, G. (2006). *In Search of Dark Matter*. Birkhäuser. p. 37. ISBN 0-387-27616-5.

[29] Jörg, D. et al. (2012). "A filament of dark matter between two clusters of galaxies". *Nature* **487** (7406): 202. arXiv:1207.0809. Bibcode:2012Natur.487..202D. doi:10.1038/nature11224.

[30] Babcock, H, 1939, "The rotation of the Andromeda Nebula", Lick Observatory bulletin ; no. 498

[31] Bosma, A. (1978). "The distribution and kinematics of neutral hydrogen in spiral galaxies of various morphological types" (Ph.D. Thesis). Rijksuniversiteit Groningen.

1.9. REFERENCES

[32] Rubin, V.; Thonnard, W. K. Jr.; Ford, N. (1980). "Rotational Properties of 21 Sc Galaxies with a Large Range of Luminosities and Radii from NGC 4605 ($R = 4$kpc) to UGC 2885 ($R = 122$kpc)". *The Astrophysical Journal* **238**: 471. Bibcode: doi:10.1086/158003.

[33] de Blok, W. J. G.; McGaugh, S. S.; Bosma, A.; Rubin, V. C. (2001). "Mass Density Profiles of Low Surface Brightness Galaxies". *The Astrophysical Journal Letters* **552** (1): L23–L26. arXiv:astro-ph/0103102. Bibcode:2001ApJ...552L..23D. doi:10.1086/320262.

[34] Salucci, P.; Borriello, A. (2003). "The Intriguing Distribution of Dark Matter in Galaxies". *Lecture Notes in Physics*. Lecture Notes in Physics **616**: 66–77. arXiv:astro-ph/0203457. Bibcode:2003LNP...616...66S. doi:10.1007/3-540-36539-7_5. ISBN 978-3-540-00711-1.

[35] Koopmans, L. V. E.; Treu, T. (2003). "The Structure and Dynamics of Luminous and Dark Matter in the Early-Type Lens Galaxy of 0047–281 at $z = 0.485$". *The Astrophysical Journal* **583** (2): 606–615. arXiv: doi:10.1086/345423.

[36] Dekel, A. et al. (2005). "Lost and found dark matter in elliptical galaxies". *Nature* **437** (7059): 707–710. arXiv:astro-ph/0501622. Bibcode:2005Natur.437..707D. doi:10.1038/nature03970. PMID 16193046.

[37] Governato, F. et al. (2010). "Bulgeless dwarf galaxies and dark matter cores from supernova-driven outflows". *Nature* **463** (7278): 203–206. arXiv:0911.2237. Bibcode:2010Natur.463..203G. doi:10.1038/nature08640.

[38] Ostriker, J. P.; Steinhardt, P. (2003). "New Light on Dark Matter". *Science* **300** (5627): 1909–1913. doi:10.1126/science.1085976. PMID 12817140.

[39] Faber, S. M.; Jackson, R. E. (1976). "Velocity dispersions and mass-to-light ratios for elliptical galaxies". *The Astrophysical Journal* **204**: 668–683. Bibcode:1976ApJ...204..668F. doi:10.1086/154215.

[40] Collins, G. W. (1978). "The Virial Theorem in Stellar Astrophysics". Pachart Press.

[41] Rejkuba, M.; Dubath, P.; Minniti, D.; Meylan, G. (2008). "Masses and M/L Ratios of Bright Globular Clusters in NGC 5128". *Proceedings of the International Astronomical Union* **246**: 418–422.Bibcode:2008IAUS..246..418R.doi:10.1017/S174392130

[42] Weinberg, M. D.; Blitz, L. (2006). "A Magellanic Origin for the Warp of the Galaxy". *The Astrophysical Journal Letters* **641** (1): L33–L36. arXiv:astro-ph/0601694. Bibcode:2006ApJ...641L..33W. doi:10.1086/503607.

[43] Minchin, R. et al. (2005). "A Dark Hydrogen Cloud in the Virgo Cluster". *The Astrophysical Journal Letters* **622**: L21–L24. arXiv:astro-ph/0502312. Bibcode:2005ApJ...622L..21M. doi:10.1086/429538.

[44] Ciardullo, R.; Jacoby, G. H.; Dejonghe, H. B. (1993). "The radial velocities of planetary nebulae in NGC 3379". *The Astrophysical Journal* **414**: 454–462. Bibcode:1993ApJ...414..454C. doi:10.1086/173092.

[45] Vikhlinin, A. et al. (2006). "Chandra Sample of Nearby Relaxed Galaxy Clusters: Mass, Gas Fraction, and Mass–Temperature Relation". *The Astrophysical Journal* **640** (2): 691–709. arXiv:astro-ph/0507092. Bibcode:2006ApJ...640..691V.doi

[46] "Abell 2029: Hot News for Cold Dark Matter". Chandra X-ray Observatory. 11 June 2003.

[47] Taylor, A. N. et al. (1998). "Gravitational Lens Magnification and the Mass of Abell 1689". *The Astrophysical Journal* **501** (2): 539. arXiv:astro-ph/9801158. Bibcode:1998ApJ...501..539T. doi:10.1086/305827.

[48] Wu, X.; Chiueh, T.; Fang, L.; Xue, Y. (1998). "A comparison of different cluster mass estimates: consistency or discrepancy?". *Monthly Notices of the Royal Astronomical Society* **301** (3): 861–871. arXiv:astro-ph/9808179. Bibcode:1998MNRAS.301..861W. doi:10.1046/j.1365-8711.1998.02055.x.

[49] Refregier, A. (2003). "Weak gravitational lensing by large-scale structure". *Annual Review of Astronomy and Astrophysics* **41** (1): 645–668. arXiv:astro-ph/0307212. Bibcode:2003ARA&A..41..645R. doi:10.1146/annurev.astro.41.111302.102207.

[50] Massey, R. et al. (2007). "Dark matter maps reveal cosmic scaffolding". *Nature* **445** (7125): 286–290. arXiv:astro-ph/0701594. Bibcode:2007Natur.445..286M. doi:10.1038/nature05497. PMID 17206154.

[51] Clowe, D. et al. (2006). "A direct empirical proof of the existence of dark matter". *The Astrophysical Journal* **648** (2): 109–113. arXiv:astro-ph/0608407. Bibcode:2006ApJ...648L.109C. doi:10.1086/508162.

[52] "Dark Matter Mystery Deepens in Cosmic "Train Wreck"". Chandra X-Ray Observatory. 16 August 2007.

[53] Tiberiu, H.; Lobo, F. S. N. (2011). "Two-fluid dark matter models". *Physical Review D* **83** (12): 124051. arXiv:1106.2642. Bibcode:2011PhRvD..83l4051H. doi:10.1103/PhysRevD.83.124051.

[54] Spergel, D. N.; Steinhardt, P. J. (2000). "Observational evidence for self-interacting cold dark matter". *Physical Review Letters* **84** (17): 3760–3763. arXiv:astro-ph/9909386. Bibcode:2000PhRvL..84.3760S. doi:10.1103/PhysRevLett.84.3760.

[55] Allen, S. W.; Evrard, A. E.; Mantz, A. B. (2011). "Cosmological Parameters from Observations of Galaxy Clusters". *Annual Review of Astronomy & Astrophysics* **49**: 409–470. arXiv:1103.4829. Bibcode:2011ARA&A..49..409A. doi:10.1146/annurev-astro-081710-102514.

[56] Markevitch, M. et al. (2004). "Direct Constraints on the Dark Matter Self-Interaction Cross Section from the Merging Galaxy Cluster 1E 0657-56". *The Astrophysical Journal* **606** (2): 819–824. arXiv:astro-ph/0309303. Bibcode:2004ApJ...606..819M. doi:10.1086/383178.

[57] "Press Release - Dark Matter Map Begins to Reveal the Universe's Early History - Subaru Telescope". *www.subarutelescope.org*. Retrieved 2015-07-03.

[58] Miyazaki, Satoshi; Oguri, Masamune; Hamana, Takashi; Tanaka, Masayuki; Miller, Lance; Utsumi, Yousuke; Komiyama, Yutaka; Furusawa, Hisanori; Sakurai, Junya (2015-07-01). "Properties of Weak Lensing Clusters Detected on Hyper Suprime-Cam's 2.3 deg2 field". *The Astrophysical Journal* (in English) **807** (1): 22. arXiv:1504.06974. Bibcode:2015ApJ...807...22M. doi:10.1088/0004-637X/807/1/22. ISSN 0004-637X.

[59] Penzias, A.A.; Wilson, R. W. (1965). "A Measurement of Excess Antenna Temperature at 4080 Mc/s". *The Astrophysical Journal* **142**: 419. Bibcode:1965ApJ...142..419P. doi:10.1086/148307.

[60] Boggess, N. W. et al. (1992). "The COBE Mission: Its Design and Performance Two Years after the launch". *The Astrophysical Journal* **397**: 420. Bibcode:1992ApJ...397..420B. doi:10.1086/171797.

[61] Melchiorri, A. et al. (2000). "A Measurement of Ω from the North American Test Flight of Boomerang". *The Astrophysical Journal Letters* **536** (2): L63–L66. arXiv:astro-ph/9911445. Bibcode:2000ApJ...536L..63M. doi:10.1086/312744.

[62] Leitch, E. M. et al. (2002). "Measurement of polarization with the Degree Angular Scale Interferometer". *Nature* **420** (6917): 763–771. arXiv:astro-ph/0209476. Bibcode:2002Natur.420..763L. doi:10.1038/nature01271. PMID 12490940.

[63] Leitch, E. M. et al. (2005). "Degree Angular Scale Interferometer 3 Year Cosmic Microwave Background Polarization Results". *The Astrophysical Journal* **624** (1): 10–20. arXiv:astro-ph/0409357. Bibcode:2005ApJ...624...10L. doi:10.1086/428825.

[64] Readhead, A. C. S. et al. (2004). "Polarization Observations with the Cosmic Background Imager". *Science* **306** (5697): 836–844. arXiv:astro-ph/0409569. Bibcode:2004Sci...306..836R. doi:10.1126/science.1105598. PMID 15472038.

[65] Hinshaw, G. et al. (2009). "Five-Year Wilkinson Microwave Anisotropy Probe Observations: Data Processing, Sky Maps, and Basic Results". *The Astrophysical Journal Supplement* **180** (2): 225–245. arXiv:0803.0732. Bibcode:2009ApJS..180..225H. doi:10.1088/0067-0049/180/2/225.

[66] Komatsu, E. et al. (2009). "Five-Year Wilkinson Microwave Anisotropy Probe Observations: Cosmological Interpretation". *The Astrophysical Journal Supplement* **180** (2): 330–376. arXiv:0803.0547. Bibcode:2009ApJS..180..330K. doi:10.1088/0067-0049/180/2/330.

[67] Percival, W. J. et al. (2007). "Measuring the Baryon Acoustic Oscillation scale using the Sloan Digital Sky Survey and 2dF Galaxy Redshift Survey". *Monthly Notices of the Royal Astronomical Society* **381** (3): 1053–1066. arXiv:0705.3323. Bibcode:2007MNRAS.381.1053P. doi:10.1111/j.1365-2966.2007.12268.x.

[68] Kowalski, M. et al. (2008). "Improved Cosmological Constraints from New, Old, and Combined Supernova Data Sets". *The Astrophysical Journal* **686** (2): 749–778. arXiv:0804.4142. Bibcode:2008ApJ...686..749K. doi:10.1086/589937.

[69] Viel, M.; Bolton, J. S.; Haehnelt, M. G. (2009). "Cosmological and astrophysical constraints from the Lyman α forest flux probability distribution function". *Monthly Notices of the Royal Astronomical Society* **399** (1): L39–L43. arXiv:0907.2927. Bibcode:2009MNRAS.399L..39V. doi:10.1111/j.1745-3933.2009.00720.x.

[70] "Hubble Maps the Cosmic Web of "Clumpy" Dark Matter in 3-D" (Press release). NASA. 7 January 2007.

[71] Springel, V. et al. (2005). "Simulations of the formation, evolution and clustering of galaxies and quasars". *Nature* **435** (7042): 629–636. arXiv:astro-ph/0504097. Bibcode:2005Natur.435..629S. doi:10.1038/nature03597. PMID 15931216.

1.9. REFERENCES

[72] Mateo, M. L. (1998). "Dwarf Galaxies of the Local Group". *Annual Review of Astronomy and Astrophysics* **36** (1): 435–506. arXiv:astro-ph/9810070. Bibcode:1998ARA&A..36..435M. doi:10.1146/annurev.astro.36.1.435.

[73] Moore, B. et al. (1999). "Dark Matter Substructure within Galactic Halos". *The Astrophysical Journal Letters* **524** (1): L19–L22. arXiv:astro-ph/9907411. Bibcode:1999ApJ...524L..19M. doi:10.1086/312287.

[74] Tisserand, P.; Le Guillou, L.; Afonso, C.; Albert, J. N.; Andersen, J.; Ansari, R.; Aubourg, É.; Bareyre, P.; Beaulieu, J. P.; Charlot, X.; Coutures, C.; Ferlet, R.; Fouqué, P.; Glicenstein, J. F.; Goldman, B.; Gould, A.; Graff, D.; Gros, M.; Haissinski, J.; Hamadache, C.; De Kat, J.; Lasserre, T.; Lesquoy, É.; Loup, C.; Magneville, C.; Marquette, J. B.; Maurice, É.; Maury, A.; Milsztajn, A.; Moniez, M. (2007). "Limits on the Macho content of the Galactic Halo from the EROS-2 Survey of the Magellanic Clouds". *Astronomy and Astrophysics* **469** (2): 387. arXiv:astro-ph/0607207. Bibcode:2007A&A...469..387T. doi:10.1051/0004-6361:20066017.

[75] Graff, D. S.; Freese, K. (1996). "Analysis of a *Hubble Space Telescope* Search for Red Dwarfs: Limits on Baryonic Matter in the Galactic Halo". *The Astrophysical Journal* **456**. arXiv:astro-ph/9507097. Bibcode:1996ApJ...456L..49G. doi:10.1086/309850.

[76] Najita, J. R.; Tiede, G. P.; Carr, J. S. (2000). "From Stars to Superplanets: The Low-Mass Initial Mass Function in the Young Cluster IC 348". *The Astrophysical Journal* **541** (2): 977. doi:10.1086/309477.

[77] Wyrzykowski, Lukasz et al. (2011) The OGLE view of microlensing towards the Magellanic Clouds – IV. OGLE-III SMC data and final conclusions on MACHOs, MNRAS, 416, 2949

[78] Freese, Katherine; Fields, Brian; Graff, David (2000). "Death of Stellar Baryonic Dark Matter Candidates". arXiv:astro-ph/0007444 [astro-ph].

[79] Freese, Katherine; Fields, Brian; Graff, David (2000). "Death of Stellar Baryonic Dark Matter". *The First Stars*. ESO Astrophysics Symposia. p. 18. arXiv:astro-ph/0002058. Bibcode:2000fist.conf...18F. doi:10.1007/10719504_3. ISBN 3-540-67222-2.

[80] Silk, Joseph (1980). *The Big Bang* (1989 ed.). San Francisco: Freeman. chapter ix, page 182. ISBN 0-7167-1085-4.

[81] Vittorio, N.; J. Silk (1984). "Fine-scale anisotropy of the cosmic microwave background in a universe dominated by cold dark matter". *Astrophysical Journal, Part 2 – Letters to the Editor* **285**: L39–L43. Bibcode:1984ApJ...285L..39V. doi:10.1086/184361.

[82] Umemura, Masayuki; Satoru Ikeuchi (1985). "Formation of Subgalactic Objects within Two-Component Dark Matter". *Astrophysical Journal* **299**: 583–592. Bibcode:1985ApJ...299..583U. doi:10.1086/163726.

[83] Davis, M.; Efstathiou, G., Frenk, C. S., & White, S. D. M. (May 15, 1985). "The evolution of large-scale structure in a universe dominated by cold dark matter". *Astrophysical Journal* **292**: 371–394. Bibcode:1985ApJ...292..371D. doi:10.1086/163168.

[84] Hawkins, M. R. S. (2011). "The case for primordial black holes as dark matter". *Monthly Notices of the Royal Astronomical Society* **415** (3): 2744–2757. arXiv:1106.3875. Bibcode:2011MNRAS.415.2744H. doi:10.1111/j.1365-2966.2011.18890.x.

[85] Carr, B. J. et al. (May 2010). "New cosmological constraints on primordial black holes". *Physical Review D* **81** (10): 104019. arXiv:0912.5297. Bibcode:2010PhRvD..81j4019C. doi:10.1103/PhysRevD.81.104019.

[86] Peter, A. H. G. (2012). "Dark Matter: A Brief Review". arXiv:1201.3942 [astro-ph.CO].

[87] Garrett, Katherine; Dūda, Gintaras (2011). "Dark Matter: A Primer". *Advances in Astronomy* **2011**: 1. arXiv:1006.2483. Bibcode:2011AdAst2011E...8G. doi:10.1155/2011/968283. MACHOs can only account for a very small percentage of the nonluminous mass in our galaxy, revealing that most dark matter cannot be strongly concentrated or exist in the form of baryonic astrophysical objects. Although microlensing surveys rule out baryonic objects like brown dwarfs, black holes, and neutron stars in our galactic halo, can other forms of baryonic matter make up the bulk of dark matter? The answer, surprisingly, is no...

[88] Bertone, G. (2010). "The moment of truth for WIMP dark matter". *Nature* **468** (7322): 389–393. doi:10.1038/nature09509. PMID 21085174.

[89] Olive, Keith A. (2003). "TASI Lectures on Dark Matter". p. 21

[90] "Neutrinos as Dark Matter". Astro.ucla.edu. 21 September 1998. Retrieved 6 January 2011.

[91] Gaitskell, Richard J. (2004). "Direct Detection of Dark Matter". *"Annual Review of Nuclear and Particle Systems"* **54**: 315–359. Bibcode:2004ARNPS..54..315G. doi:10.1146/annurev.nucl.54.070103.181244.

[92] "NEUTRALINO DARK MATTER". Retrieved 26 December 2011. Griest, Kim. "WIMPs and MACHOs" (PDF). Retrieved 26 December 2011.

[93] Kane, G. and Watson, S. (2008). "Dark Matter and LHC:. what is the Connection?". *Modern Physics Letters A* **23** (26): 2103–2123. arXiv:0807.2244. Bibcode:2008MPLA...23.2103K. doi:10.1142/S0217732308028314.

[94] Drukier, A.; Freese, K. and Spergel, D. (1986). "Detecting Cold Dark Matter Candidates". *Physical Review D* **33** (12): 3495–3508. Bibcode:1986PhRvD..33.3495D. doi:10.1103/PhysRevD.33.3495.

[95] Bernabei, R.; Belli, P.; Cappella, F.; Cerulli, R.; Dai, C. J.; d'Angelo, A.; He, H. L.; Incicchitti, A.; Kuang, H. H.; Ma, J. M.; Montecchia, F.; Nozzoli, F.; Prosperi, D.; Sheng, X. D.; Ye, Z. P. (2008). "First results from DAMA/LIBRA and the combined results with DAMA/NaI". *Eur. Phys. J. C* **56** (3): 333–355. arXiv:0804.2741. doi:10.1140/epjc/s10052-008-0662-y.

[96] Stonebraker, Alan (2014-01-03). "Synopsis: Dark-Matter Wind Sways through the Seasons". *Physics - Synopses* (American Physical Society). Retrieved 6 January 2014.

[97] Lee, Samuel K.; Mariangela Lisanti, Annika H. G. Peter, and Benjamin R. Safdi (2014-01-03). "Effect of Gravitational Focusing on Annual Modulation in Dark-Matter Direct-Detection Experiments". *Phys. Rev. Lett.* (American Physical Society) **112** (1): 011301 (2014) [5 pages]. arXiv:1308.1953. Bibcode:2014PhRvL.112a1301L. doi:10.1103/PhysRevLett.112.011301.

[98] The Dark Matter Group. "An Introduction to Dark Matter". *Dark Matter Research* (Sheffield, UK: University of Sheffield). Retrieved 7 January 2014.

[99] "Blowing in the Wind". *Kavli News* (Sheffield, UK: Kavli Foundation). Retrieved 7 January 2014. Scientists at Kavli MIT are working on...a tool to track the movement of dark matter.

[100] The CDMS II Collaboration; Ahmed, Z.; Akerib, D. S.; Arrenberg, S.; Bailey, C. N.; Balakishiyeva, D.; Baudis, L.; Bauer, D. A.; Brink, P. L.; Bruch, T.; Bunker, R.; Cabrera, B.; Caldwell, D. O.; Cooley, J.; Cushman, P.; Daal, M.; Dejongh, F.; Dragowsky, M. R.; Duong, L.; Fallows, S.; Figueroa-Feliciano, E.; Filippini, J.; Fritts, M.; Golwala, S. R.; Grant, D. R.; Hall, J.; Hennings-Yeomans, R.; Hertel, S. A.; Holmgren, D.; Hsu, L. (2010). "Dark Matter Search Results from the CDMS II Experiment". *Science* **327** (5973): 1619–1621. doi:10.1126/science.1186112. PMID 20150446.

[101] Angloher, G.; Bauer; Bavykina; Bento; Bucci; Ciemniak; Deuter; von Feilitzsch; Hauff et al. (2011). "Results from 730kg days of the CRESST-II Dark Matter Search". arXiv:1109.0702v1 [astro-ph.CO].

[102] "Dark matter even darker than once thought". Retrieved 16 June 2015.

[103] Stecker, F.W.; Hunter, S; Kniffen, D (2008). "The likely cause of the EGRET GeV anomaly and its implications". *Astroparticle Physics* **29** (1): 25–29. arXiv:0705.4311. Bibcode:2008APh....29...25S. doi:10.1016/j.astropartphys.2007.11.002.

[104] Atwood, W.B.; Abdo, A. A.; Ackermann, M.; Althouse, W.; Anderson, B.; Axelsson, M.; Baldini, L.; Ballet, J. et al. (2009). "The large area telescope on the Fermi Gamma-ray Space Telescope Mission". *Astrophysical Journal* **697** (2): 1071–1102. arXiv:0902.1089. Bibcode:2009ApJ...697.1071A. doi:10.1088/0004-637X/697/2/1071.

[105] Weniger, Christoph (2012). "A Tentative Gamma-Ray Line from Dark Matter Annihilation at the Fermi Large Area Telescope". *Journal of Cosmology and Astroparticle Physics* **2012** (8): 7. arXiv:1204.2797v2. Bibcode:2012JCAP...08..007W. doi:10.1088/1475-7516/2012/08/007.

[106] Cartlidge, Edwin (24 April 2012). "Gamma rays hint at dark matter". Institute Of Physics. Retrieved 23 April 2013.

[107] Albert, J.; Aliu, E.; Anderhub, H.; Antoranz, P.; Backes, M.; Baixeras, C.; Barrio, J. A.; Bartko, H.; Bastieri, D.; Becker, J. K.; Bednarek, W.; Berger, K.; Bigongiari, C.; Biland, A.; Bock, R. K.; Bordas, P.; Bosch-Ramon, V.; Bretz, T.; Britvitch, I.; Camara, M.; Carmona, E.; Chilingarian, A.; Commichau, S.; Contreras, J. L.; Cortina, J.; Costado, M. T.; Curtef, V.; Danielyan, V.; Dazzi, F.; De Angelis, A. (2008). "Upper Limit for γ-Ray Emission above 140 GeV from the Dwarf Spheroidal Galaxy Draco". *The Astrophysical Journal* **679**: 428. doi:10.1086/529135.

[108] Aleksić, J.; Antonelli, L. A.; Antoranz, P.; Backes, M.; Baixeras, C.; Balestra, S.; Barrio, J. A.; Bastieri, D.; González, J. B.; Bednarek, W.; Berdyugin, A.; Berger, K.; Bernardini, E.; Biland, A.; Bock, R. K.; Bonnoli, G.; Bordas, P.; Tridon, D. B.; Bosch-Ramon, V.; Bose, D.; Braun, I.; Bretz, T.; Britzger, D.; Camara, M.; Carmona, E.; Carosi, A.; Colin, P.; Commichau, S.; Contreras, J. L.; Cortina, J. (2010). "Magic Gamma-Ray Telescope Observation of the Perseus Cluster of Galaxies: Implications for Cosmic Rays, Dark Matter, and Ngc 1275". *The Astrophysical Journal* **710**: 634. doi:10.1088/0004-637X/710/1/634.

[109] Adriani, O.; Barbarino, G. C.; Bazilevskaya, G. A.; Bellotti, R.; Boezio, M.; Bogomolov, E. A.; Bonechi, L.; Bongi, M.; Bonvicini, V.; Bottai, S.; Bruno, A.; Cafagna, F.; Campana, D.; Carlson, P.; Casolino, M.; Castellini, G.; De Pascale, M. P.; De Rosa, G.; De Simone, N.; Di Felice, V.; Galper, A. M.; Grishantseva, L.; Hofverberg, P.; Koldashov, S. V.; Krutkov, S. Y.; Kvashnin, A. N.; Leonov, A.; Malvezzi, V.; Marcelli, L.; Menn, W. (2009). "An anomalous positron abundance in cosmic rays with energies 1.5–100 GeV". *Nature* **458** (7238): 607–609. doi:10.1038/nature07942. PMID 19340076.

[110] Aguilar, M. (AMS Collaboration) et al. (3 April 2013). "First Result from the Alpha Magnetic Spectrometer on the International Space Station: Precision Measurement of the Positron Fraction in Primary Cosmic Rays of 0.5–350 GeV". *Physical Review Letters*. Bibcode:2013PhRvL.110n1102A. doi:10.1103/PhysRevLett.110.141102. Retrieved 3 April 2013.

[111] "First Result from the Alpha Magnetic Spectrometer Experiment". *AMS Collaboration*. 3 April 2013. Retrieved 3 April 2013.

[112] Heilprin, John; Borenstein, Seth (3 April 2013). "Scientists find hint of dark matter from cosmos". Associated Press. Retrieved 3 April 2013.

[113] Amos, Jonathan (3 April 2013). "Alpha Magnetic Spectrometer zeroes in on dark matter". *BBC*. Retrieved 3 April 2013.

[114] Perrotto, Trent J.; Byerly, Josh (2 April 2013). "NASA TV Briefing Discusses Alpha Magnetic Spectrometer Results". *NASA*. Retrieved 3 April 2013.

[115] Overbye, Dennis (3 April 2013). "New Clues to the Mystery of Dark Matter". *New York Times*. Retrieved 3 April 2013.

[116] Freese, K. (1986). "Can Scalar Neutrinos or Massive Dirac Neutrinos be the Missing Mass?". *Physics Letters B* **167** (3): 295–300. Bibcode:1986PhLB..167..295F. doi:10.1016/0370-2693(86)90349-7.

[117] Ellis, J.; Flores, R. A.; Freese, K.; Ritz, S.; Seckel, D.; Silk, J. (1988). "Cosmic ray constraints on the annihilations of relic particles in the galactic halo". *Physics Letters B* **214** (3): 403. Bibcode:1988PhLB..214..403E. doi:10.1016/0370-2693(88)91385-8.

[118] Merritt, David (1 January 2010). "Dark Matter at the Centers of Galaxies". In Bertone, Gianfranco. *Particle Dark Matter : Observations, Models and Searches*. Cambridge University Press. pp. 83–104. arXiv:1001.3706. ISBN 978-0-521-76368-4.

[119] Rzetelny, Xaq (19 November 2014). "Looking for a different sort of dark matter with GPS satellites". *Ars Technica*. Retrieved 24 November 2014.

[120] Exirifard, Q. (2010). "Phenomenological covariant approach to gravity". *General Relativity and Gravitation* **43** (1): 93–106. arXiv:0808.1962. Bibcode:2011GReGr..43...93E. doi:10.1007/s10714-010-1073-6.

[121] Brownstein, J.R.; Moffat, J. W. (2007). "The Bullet Cluster 1E0657-558 evidence shows modified gravity in the absence of dark matter". *Monthly Notices of the Royal Astronomical Society* **382** (1): 29–47. arXiv: astro-ph/0702146.Bibcode:2007MNRAS.38. doi:10.1111/j.1365-2966.2007.12275.x.

[122] Anastopoulos, C. (2009). "Gravitational backreaction in cosmological spacetimes". *Physical Review D* **79** (8): 084029. arXiv:0902.0159. Bibcode:2009PhRvD..79h4029A. doi:10.1103/PhysRevD.79.084029.

[123] "New Cosmic Theory Unites Dark Forces". SPACE.com. 11 February 2008. Retrieved 6 January 2011.

[124] Hossenfelder, S. (2008). "A Bi-Metric Theory with Exchange Symmetry". *Physical Review D* **78** (4): 044015. arXiv:gr-qc/0603005. Bibcode:2008PhRvD..78d4015H. doi:10.1103/PhysRevD.78.044015.

[125] Ripalda, Jose M. (1999). "Time reversal and negative energies in general relativity". arXiv:gr-qc/9906012 [gr-qc].

[126] McCulloch, M.E. (2012). "Testing quantised inertia on galactic scales".*Astrophysics & Space Science***342**: 575–578.arXiv:1207. Bibcode:2012Ap&SS.342..575M.. doi:10.1007/s10509-012-1197-0.

1.10 External links

- Dark matter at DMOZ
- What is dark matter? at cosmosmagazine.com
- The Dark Matter Crisis
- The European astroparticle physics network

- Helmholtz Alliance for Astroparticle Physics
- "NASA Finds Direct Proof of Dark Matter" (Press release). NASA. 21 August 2006.
- Tuttle, Kelen (22 August 2006). "Dark Matter Observed". SLAC (Stanford Linear Accelerator Center) Today.
- "Astronomers claim first 'dark galaxy' find". New Scientist. 23 February 2005.
- Sample, Ian (17 December 2009). "Dark Matter Detected". London: Guardian. Retrieved 1 May 2010.
- Video lecture on dark matter by Scott Tremaine, IAS professor
- Science Daily story "Astronomers' Doubts About the Dark Side ..."
- Gray, Meghan; Merrifield, Mike; Copeland, Ed (2010). "Dark Matter". *Sixty Symbols*. Brady Haran for the University of Nottingham.

1.10. EXTERNAL LINKS

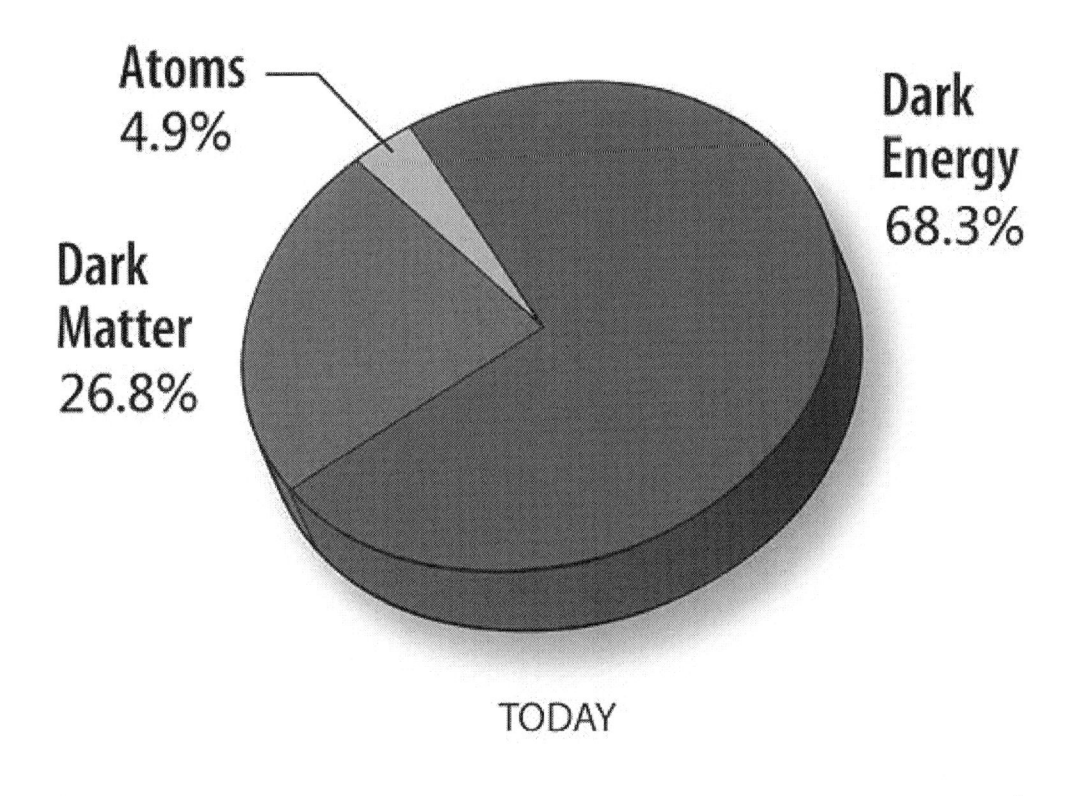

Estimated distribution of matter and energy in the universe, today (top) and when the CMB was released (bottom)

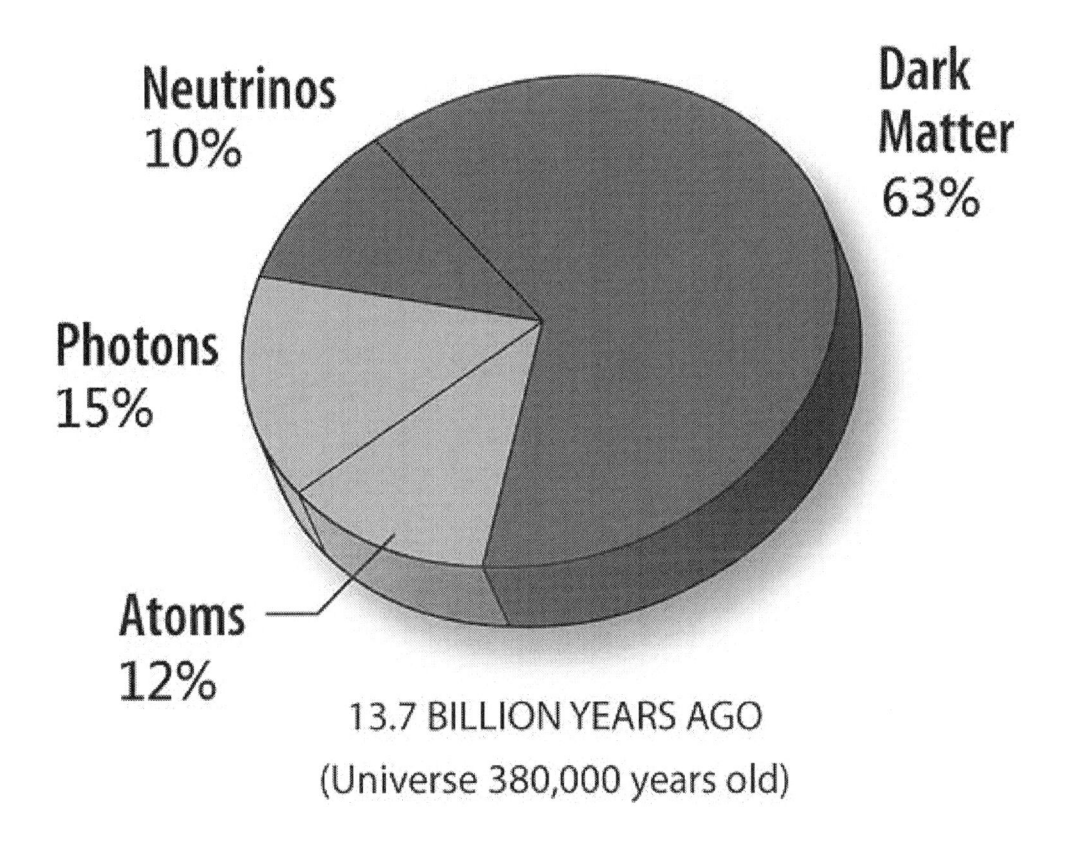

Fermi-LAT observations of dwarf galaxies provide new insights on dark matter.

This artist's impression shows the expected distribution of dark matter in the Milky Way galaxy as a blue halo of material surrounding the galaxy.[21]

1.10. EXTERNAL LINKS   25

Observations have provided hints that the dark matter around one of the central four merging galaxies is not moving with the galaxy itself.[27]

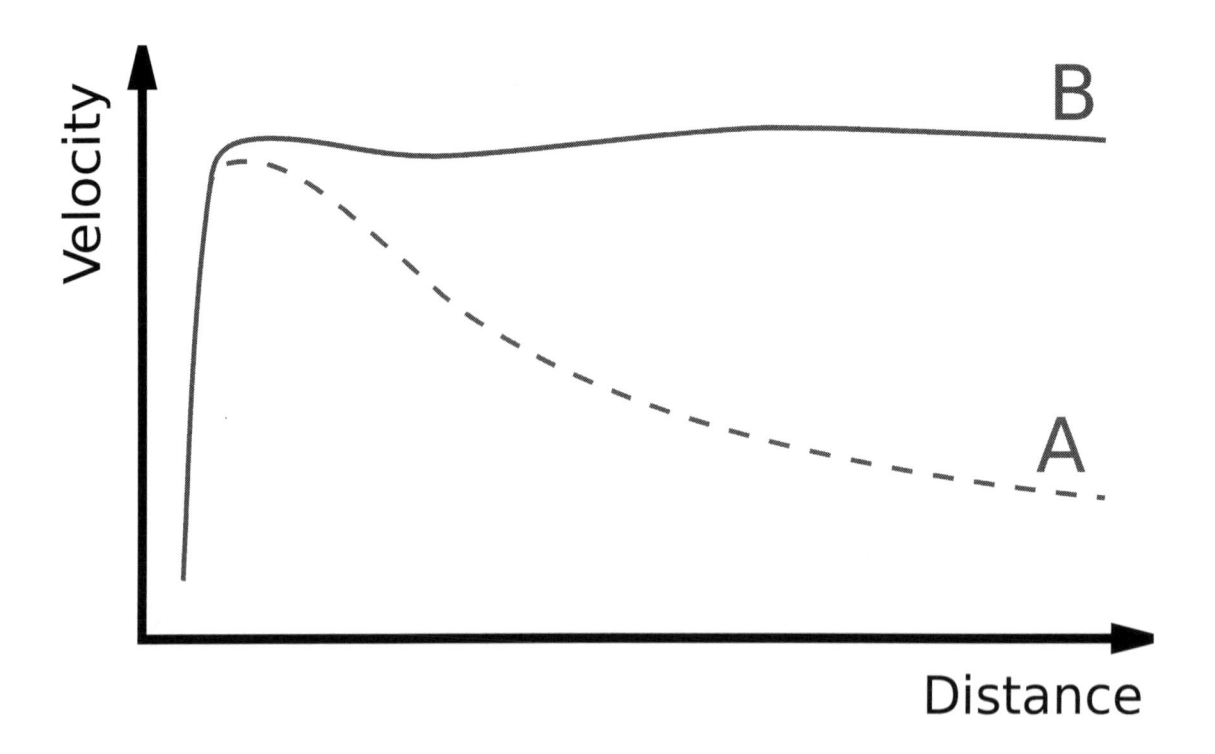

*Rotation curve of a typical spiral galaxy: predicted (**A**) and observed (**B**). Dark matter can explain the 'flat' appearance of the velocity curve out to a large radius*

1.10. EXTERNAL LINKS

Strong gravitational lensing as observed by the Hubble Space Telescope in Abell 1689 indicates the presence of dark matter—enlarge the image to see the lensing arcs.

The Bullet Cluster: HST image with overlays. The total projected mass distribution reconstructed from strong and weak gravitational lensing is shown in blue, while the X-ray emitting hot gas observed with Chandra is shown in red.

The cosmic microwave background by WMAP

1.10. EXTERNAL LINKS 29

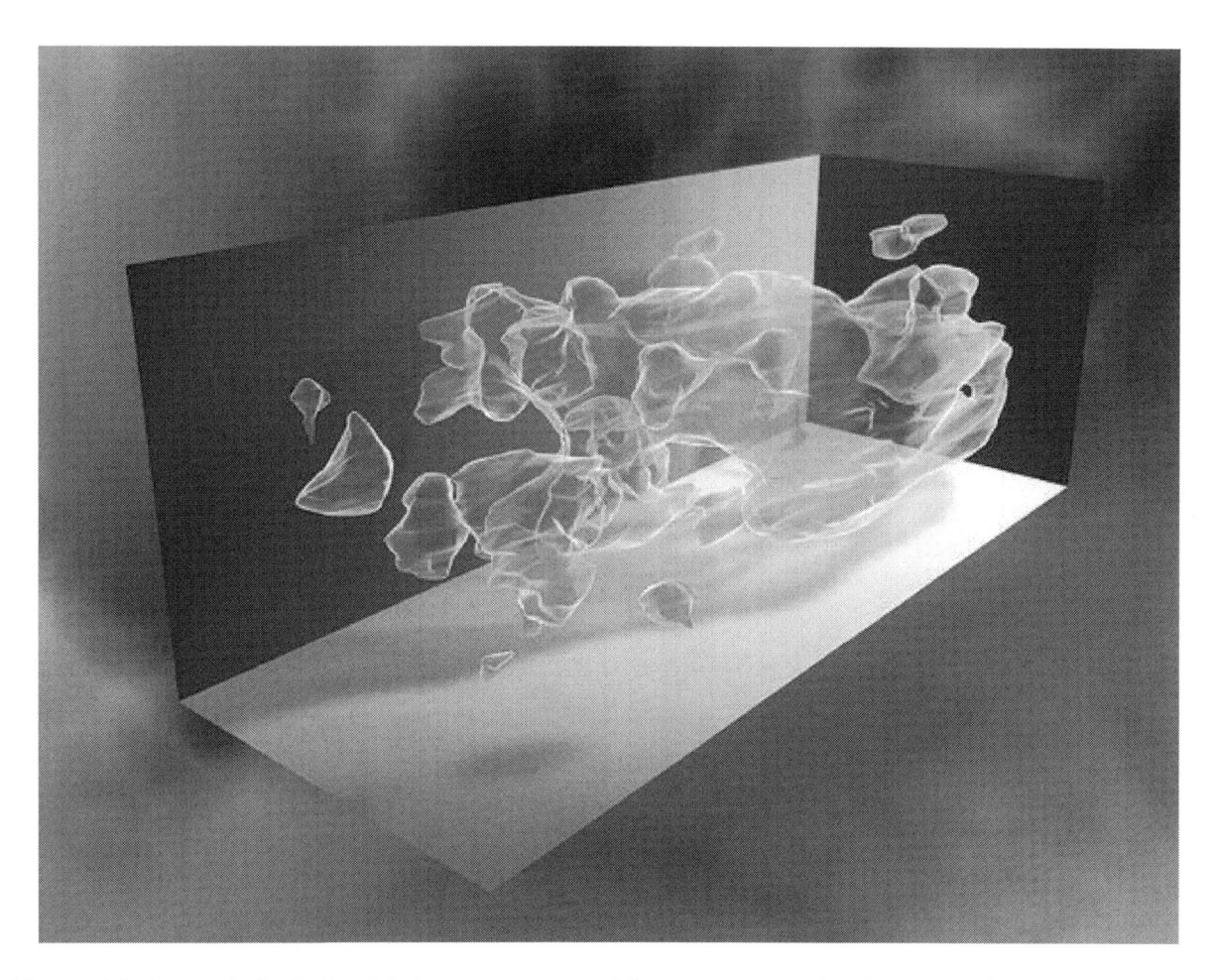

3D map of the large-scale distribution of dark matter, reconstructed from measurements of weak gravitational lensing with the Hubble Space Telescope.[70]

This video introduces a new computer simulation exploring the connection between two of the most elusive phenomena in the universe, black holes and dark matter. In the visualization, dark matter particles are gray spheres attached to shaded trails representing their motion. Redder trails indicate particles more strongly affected by the black hole's gravitation and closer to its event horizon (black sphere at center, mostly hidden by trails). The ergosphere, where all matter and light must follow the black hole's spin, is shown in teal.

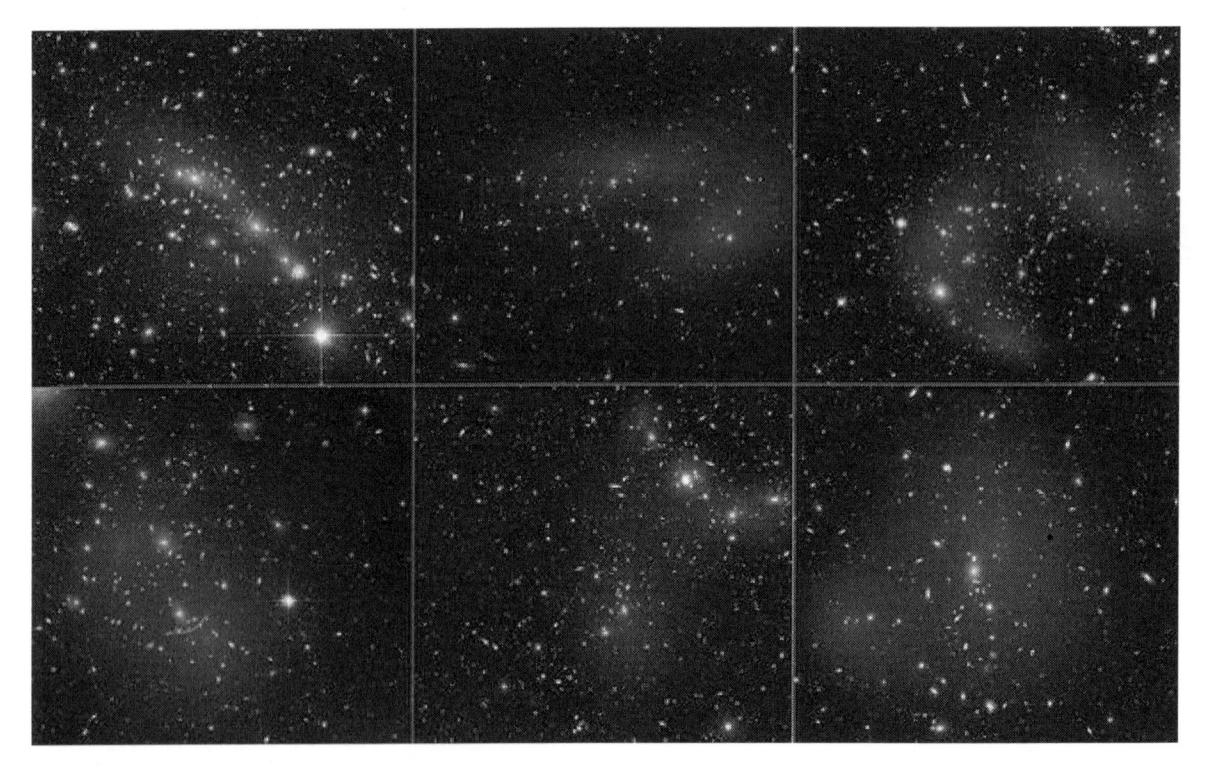

Collage of six cluster collisions with dark matter maps. The clusters were observed in a study of how dark matter in clusters of galaxies behaves when the clusters collide.[102]

Chapter 2

Dark energy

Not to be confused with Dark flow, Dark fluid, or Dark matter.

In physical cosmology and astronomy, **dark energy** is an unknown form of energy which is hypothesized to permeate all of space, tending to accelerate the expansion of the universe.[1] Dark energy is the most accepted hypothesis to explain the observations since the 1990s indicating that the universe is expanding at an accelerating rate. According to the Planck mission team, and based on the standard model of cosmology, on a mass–energy equivalence basis, the observable universe contains 26.8% dark matter, 68.3% dark energy (for a total of 95.1%) and 4.9% ordinary matter.[2][3][4][5] Again on a mass–energy equivalence basis, the density of dark energy (6.91×10^{-27} kg/m^3) is very low, much less than the density of ordinary matter or dark matter within galaxies. However, it comes to dominate the mass–energy of the universe because it is uniform across space.[6][7]

Two proposed forms for dark energy are the cosmological constant, a *constant* energy density filling space homogeneously,[8] and scalar fields such as quintessence or moduli, *dynamic* quantities whose energy density can vary in time and space. Contributions from scalar fields that are constant in space are usually also included in the cosmological constant. The cosmological constant can be formulated to be equivalent to vacuum energy. Scalar fields that do change in space can be difficult to distinguish from a cosmological constant because the change may be extremely slow.

High-precision measurements of the expansion of the universe are required to understand how the expansion rate changes over time and space. In general relativity, the evolution of the expansion rate is parameterized by the cosmological equation of state (the relationship between temperature, pressure, and combined matter, energy, and vacuum energy density for any region of space). Measuring the equation of state for dark energy is one of the biggest efforts in observational cosmology today.

Adding the cosmological constant to cosmology's standard FLRW metric leads to the Lambda-CDM model, which has been referred to as the "*standard model of cosmology*" because of its precise agreement with observations. Dark energy has been used as a crucial ingredient in a recent attempt to formulate a cyclic model for the universe.[9]

2.1 Nature of dark energy

Many things about the nature of dark energy remain matters of speculation. The evidence for dark energy is indirect but comes from three independent sources:

- Distance measurements and their relation to redshift, which suggest the universe has expanded more in the last half of its life.[10]

- The theoretical need for a type of additional energy that is not matter or dark matter to form the observationally flat universe (absence of any detectable global curvature).

- It can be inferred from measures of large scale wave-patterns of mass density in the universe.

Dark energy is thought to be very homogeneous, not very dense and is not known to interact through any of the fundamental forces other than gravity. Since it is quite rarefied—roughly 10^{-30} g/cm^3—it is unlikely to be detectable in laboratory experiments. Dark energy can have such a profound effect on the universe, making up 68% of universal density, only because it uniformly fills otherwise empty space. The two leading models are a cosmological constant and quintessence. Both models include the common characteristic that dark energy must have negative pressure.

2.1.1 Effect of dark energy: a small constant negative pressure of vacuum

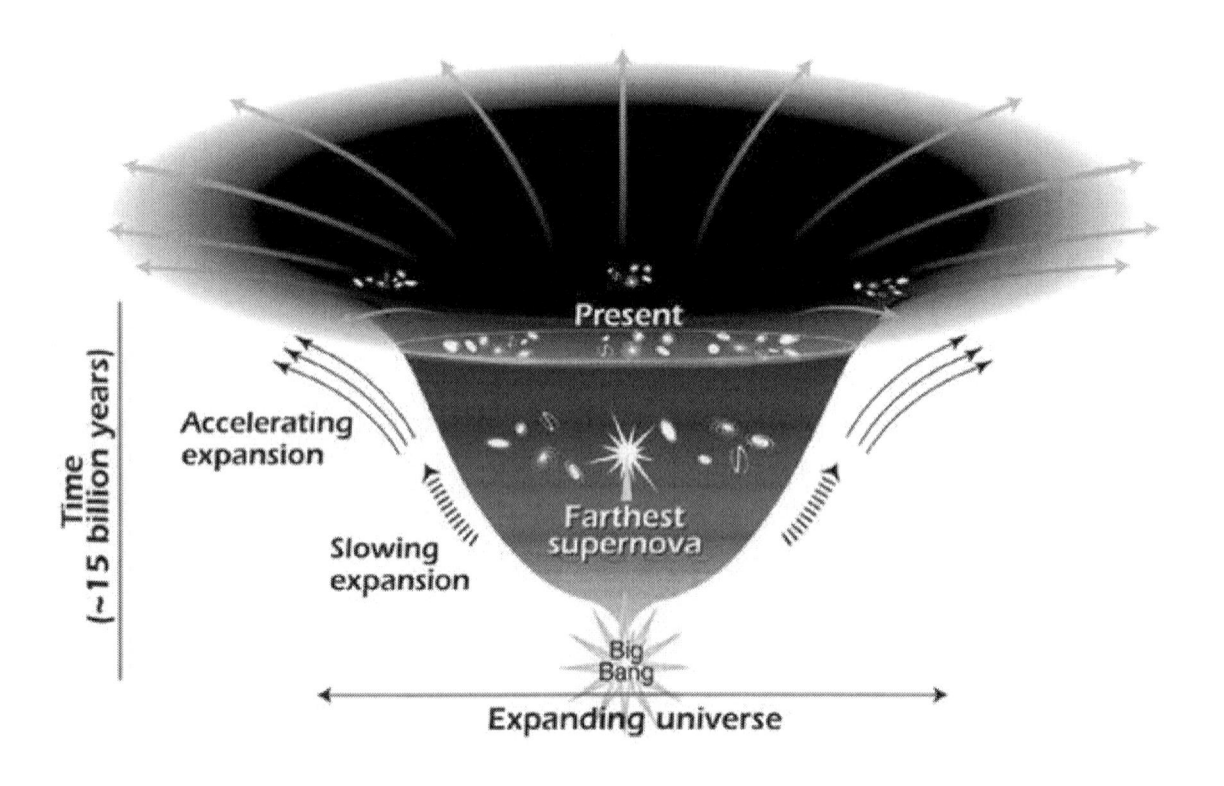

Diagram representing the accelerated expansion of the universe due to dark energy.

Independently of its actual nature, dark energy would need to have a strong negative pressure (acting repulsively) in order to explain the observed acceleration of the expansion of the universe.

According to general relativity, the pressure within a substance contributes to its gravitational attraction for other things just as its mass density does. This happens because the physical quantity that causes matter to generate gravitational effects is the stress–energy tensor, which contains both the energy (or matter) density of a substance and its pressure and viscosity.

In the Friedmann–Lemaître–Robertson–Walker metric, it can be shown that a strong constant negative pressure in all the universe causes an acceleration in universe expansion if the universe is already expanding, or a deceleration in universe contraction if the universe is already contracting. More exactly, the second derivative of the universe scale factor, $\ddot{a}$, is positive if the equation of state of the universe is such that $w < -1/3$ (see Friedmann equations).

This accelerating expansion effect is sometimes labeled "gravitational repulsion", which is a colorful but possibly confusing expression. In fact a negative pressure does not influence the gravitational interaction between masses—which remains attractive—but rather alters the overall evolution of the universe at the cosmological scale, typically resulting in the accelerating expansion of the universe despite the attraction among the masses present in the universe.

The acceleration is simply a function of dark energy density. Dark energy is persistent: its density remains constant (experimentally, within a factor of 1:10), i.e. it does not get diluted when space expands.

2.2 Evidence of existence

2.2.1 Supernovae

A Type Ia supernova (bright spot on the bottom-left) near a galaxy

In 1998, published observations of Type Ia supernovae ("one-A") by the High-Z Supernova Search Team[11] followed in

1999 by the Supernova Cosmology Project[12] suggested that the expansion of the universe is accelerating.[13] The 2011 Nobel Prize in Physics was awarded to Saul Perlmutter, Brian P. Schmidt and Adam G. Riess for their leadership in the discovery.[14][15]

Since then, these observations have been corroborated by several independent sources. Measurements of the cosmic microwave background, gravitational lensing, and the large-scale structure of the cosmos as well as improved measurements of supernovae have been consistent with the Lambda-CDM model.[16] Some people argue that the only indication for the existence of dark energy is observations of distance measurements and associated redshifts. Cosmic microwave background anisotropies and baryon acoustic oscillations are only observations that distances to a given redshift are larger than expected from a "dusty" Friedmann–Lemaître universe and the local measured Hubble constant.[17]

Supernovae are useful for cosmology because they are excellent standard candles across cosmological distances. They allow the expansion history of the universe to be measured by looking at the relationship between the distance to an object and its redshift, which gives how fast it is receding from us. The relationship is roughly linear, according to Hubble's law. It is relatively easy to measure redshift, but finding the distance to an object is more difficult. Usually, astronomers use standard candles: objects for which the intrinsic brightness, the absolute magnitude, is known. This allows the object's distance to be measured from its actual observed brightness, or apparent magnitude. Type Ia supernovae are the best-known standard candles across cosmological distances because of their extreme and consistent luminosity.

Recent observations of supernovae are consistent with a universe made up 71.3% of dark energy and 27.4% of a combination of dark matter and baryonic matter.[18]

2.2.2 Cosmic microwave background

The existence of dark energy, in whatever form, is needed to reconcile the measured geometry of space with the total amount of matter in the universe. Measurements of cosmic microwave background (CMB) anisotropies indicate that the universe is close to flat. For the shape of the universe to be flat, the mass/energy density of the universe must be equal to the critical density. The total amount of matter in the universe (including baryons and dark matter), as measured from the CMB spectrum, accounts for only about 30% of the critical density. This implies the existence of an additional form of energy to account for the remaining 70%.[16] The Wilkinson Microwave Anisotropy Probe (WMAP) spacecraft seven-year analysis estimated a universe made up of 72.8% dark energy, 22.7% dark matter and 4.5% ordinary matter.[4] Work done in 2013 based on the Planck spacecraft observations of the CMB gave a more accurate estimate of 68.3% of dark energy, 26.8% of dark matter and 4.9% of ordinary matter.[20]

2.2.3 Large-scale structure

The theory of large-scale structure, which governs the formation of structures in the universe (stars, quasars, galaxies and galaxy groups and clusters), also suggests that the density of matter in the universe is only 30% of the critical density.

A 2011 survey, the WiggleZ galaxy survey of more than 200,000 galaxies, provided further evidence towards the existence of dark energy, although the exact physics behind it remains unknown.[21][22] The WiggleZ survey from Australian Astronomical Observatory scanned the galaxies to determine their redshift. Then, by exploiting the fact that baryon acoustic oscillations have left voids regularly of ~150 Mpc diameter, surrounded by the galaxies, the voids were used as standard rulers to determine distances to galaxies as far as 2,000 Mpc (redshift 0.6), which allowed astronomers to determine more accurately the speeds of the galaxies from their redshift and distance. The data confirmed cosmic acceleration up to half of the age of the universe (7 billion years) and constrain its inhomogeneity to 1 part in 10.[22] This provides a confirmation to cosmic acceleration independent of supernovae.

2.2.4 Late-time integrated Sachs-Wolfe effect

Accelerated cosmic expansion causes gravitational potential wells and hills to flatten as photons pass through them, producing cold spots and hot spots on the CMB aligned with vast supervoids and superclusters. This so-called late-time Integrated Sachs–Wolfe effect (ISW) is a direct signal of dark energy in a flat universe.[23] It was reported at high significance in 2008 by Ho *et al.*[24] and Giannantonio *et al.*[25]

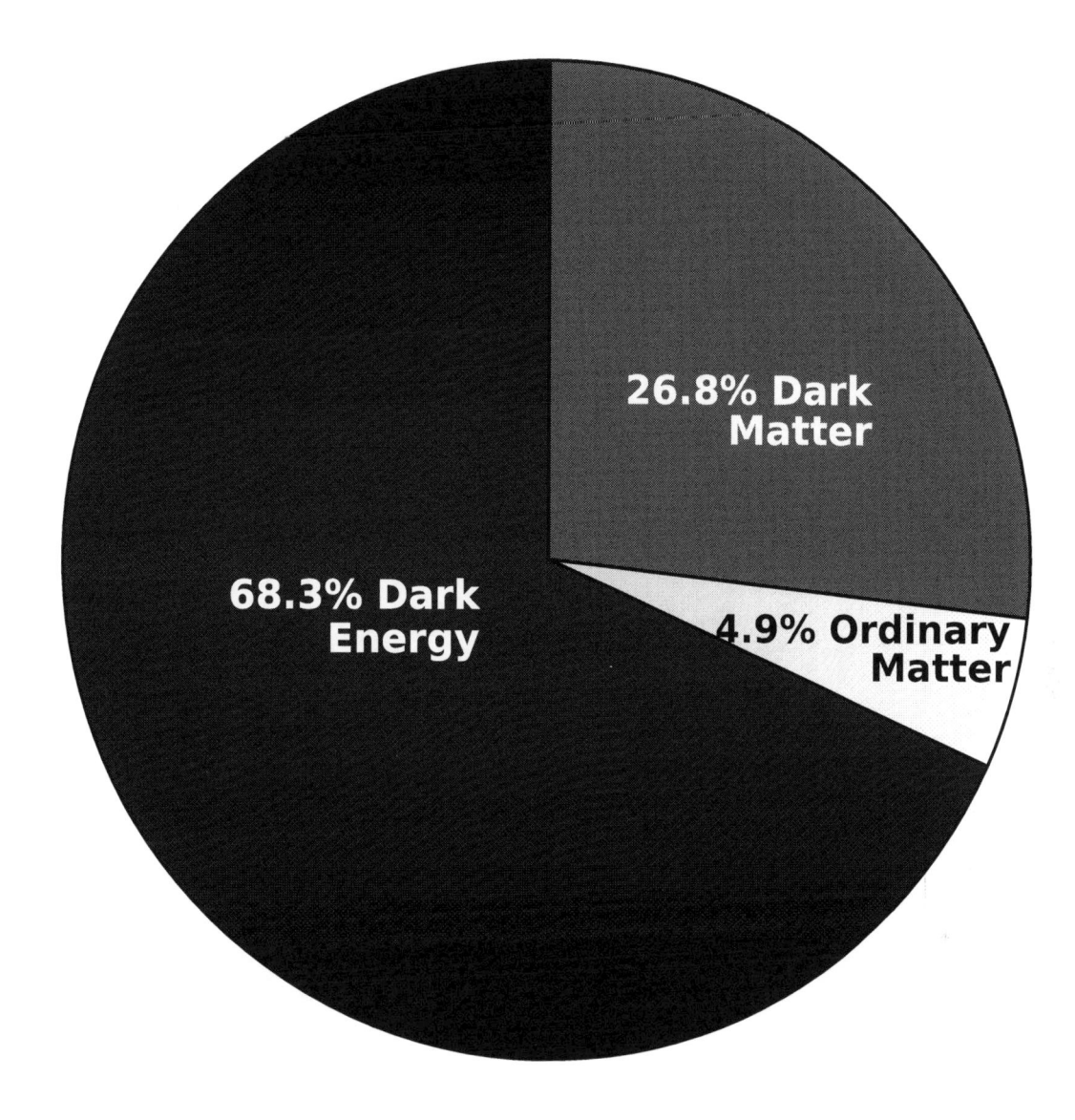

Estimated distribution of matter and energy in the universe[19]

2.2.5 Observational Hubble constant data

A new approach to test evidence of dark energy through observational Hubble constant (H(z)) data (OHD) has gained significant attention in recent years.[26][27][28][29] The Hubble constant is measured as a function of cosmological redshift. OHD directly tracks the expansion history of the universe by taking passively evolving early-type galaxies as "cosmic chronometers".[30] From this point, this approach provides standard clocks in the universe. The core of this idea is the measurement of the differential age evolution as a function of redshift of these cosmic chronometers. Thus, it provides a direct estimate of the Hubble parameter $H(z) = -1/(1+z)dz/dt \approx -1/(1+z)\Delta z/\Delta t$. The merit of this approach is clear: the reliance on a differential quantity, $\Delta z/\Delta t$, can minimize many common issues and systematic effects; and as a direct measurement of the Hubble parameter instead of its integral, like supernovae and baryon acoustic oscillations (BAO), it brings more information and is appealing in computation. For these reasons, it has been widely used to examine the accelerated cosmic expansion and study properties of dark energy.

2.3 Theories of explanation

2.3.1 Cosmological constant

Main article: Cosmological constant
For more details on this topic, see Equation of state (cosmology).

The simplest explanation for dark energy is that it is simply the "cost of having space": that is, a volume of space has

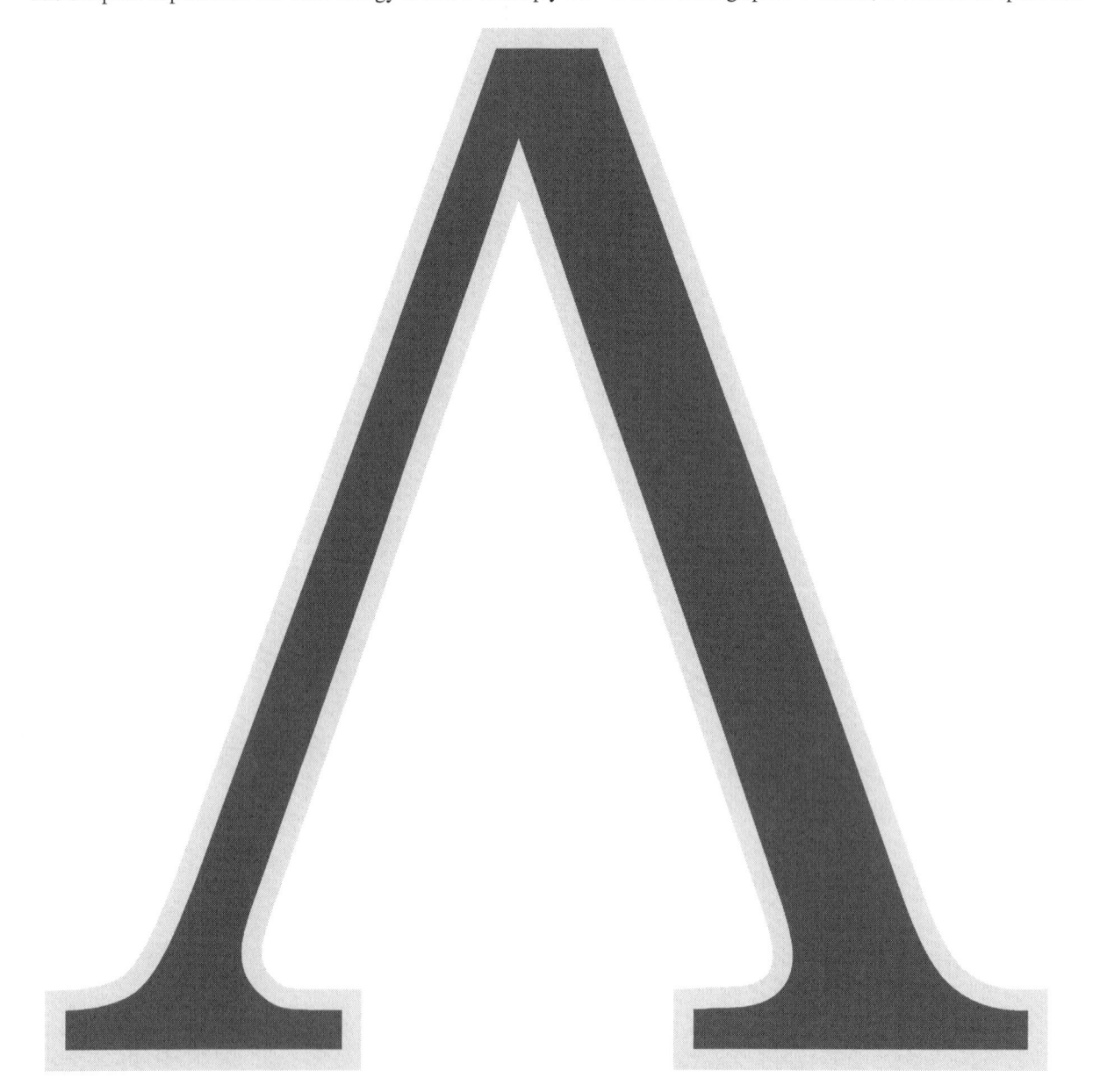

Lambda, the letter that represents the cosmological constant

some intrinsic, fundamental energy. This is the cosmological constant, sometimes called Lambda (hence Lambda-CDM model) after the Greek letter Λ, the symbol used to represent this quantity mathematically. Since energy and mass are related by $E = mc^2$, Einstein's theory of general relativity predicts that this energy will have a gravitational effect. It is sometimes called a vacuum energy because it is the energy density of empty vacuum. In fact, most theories of particle physics predict vacuum fluctuations that would give the vacuum this sort of energy. This is related to the Casimir effect, in which there is a small suction into regions where virtual particles are geometrically inhibited from forming (e.g. between plates with tiny separation). The cosmological constant is estimated by cosmologists to be on the order of 10^{-29} g/cm^3, or about 10^{-120} in reduced Planck units. Particle physics predicts a natural value of 1 in reduced Planck units, leading to a large discrepancy.

The cosmological constant has negative pressure equal to its energy density and so causes the expansion of the universe to accelerate. The reason why a cosmological constant has negative pressure can be seen from classical thermodynamics; Energy must be lost from inside a container to do work on the container. A change in volume dV requires work done equal to a change of energy $-P\,dV$, where P is the pressure. But the amount of energy in a container full of vacuum actually increases when the volume increases (dV is positive), because the energy is equal to ϱV, where ϱ (rho) is the energy density of the cosmological constant. Therefore, P is negative and, in fact, $P = -\varrho$.

A major outstanding problem is that most quantum field theories predict a huge cosmological constant from the energy of the quantum vacuum, more than 100 orders of magnitude too large.[8] This would need to be cancelled almost, but not exactly, by an equally large term of the opposite sign. Some supersymmetric theories require a cosmological constant that is exactly zero,[31] which does not help because supersymmetry must be broken. The present scientific consensus amounts to extrapolating the empirical evidence where it is relevant to predictions, and fine-tuning theories until a more elegant solution is found. Technically, this amounts to checking theories against macroscopic observations. Unfortunately, as the known error-margin in the constant predicts the fate of the universe more than its present state, many such "deeper" questions remain unknown.

In spite of its problems, the cosmological constant is in many respects the most economical solution to the problem of cosmic acceleration. One number successfully explains a multitude of observations. Thus, the current standard model of cosmology, the Lambda-CDM model, includes the cosmological constant as an essential feature.

2.3.2 Quintessence

Main article: Quintessence (physics)

In quintessence models of dark energy, the observed acceleration of the scale factor is caused by the potential energy of a dynamical field, referred to as quintessence field. Quintessence differs from the cosmological constant in that it can vary in space and time. In order for it not to clump and form structure like matter, the field must be very light so that it has a large Compton wavelength.

No evidence of quintessence is yet available, but it has not been ruled out either. It generally predicts a slightly slower acceleration of the expansion of the universe than the cosmological constant. Some scientists think that the best evidence for quintessence would come from violations of Einstein's equivalence principle and variation of the fundamental constants in space or time.[32] Scalar fields are predicted by the *Standard Model of particle physics* and string theory, but an analogous problem to the cosmological constant problem (or the problem of constructing models of cosmic inflation) occurs: renormalization theory predicts that scalar fields should acquire large masses.

The **cosmic coincidence problem** asks why the cosmic acceleration began when it did. If cosmic acceleration began earlier in the universe, structures such as galaxies would never have had time to form and life, at least as we know it, would never have had a chance to exist. Proponents of the anthropic principle view this as support for their arguments. However, many models of quintessence have a so-called **tracker** behavior, which solves this problem. In these models, the quintessence field has a density which closely tracks (but is less than) the radiation density until matter-radiation equality, which triggers quintessence to start behaving as dark energy, eventually dominating the universe. This naturally sets the low energy scale of the dark energy.[33]

In 2004, when scientists fit the evolution of dark energy with the cosmological data, they found that the equation of state had possibly crossed the cosmological constant boundary (w=−1) from above to below. A No-Go theorem has been proved that gives this scenario at least two degrees of freedom as required for dark energy models. This scenario is so-called Quintom scenario.

Some special cases of quintessence are phantom energy, in which the energy density of quintessence actually increases with time, and k-essence (short for kinetic quintessence) which has a non-standard form of kinetic energy. They can have unusual properties: phantom energy, for example, can cause a Big Rip.

2.4 Alternative ideas

Some alternatives to dark energy aim to explain the observational data by a more refined use of established theories, focusing, for example, on the gravitational effects of density inhomogeneities, or on consequences of electroweak symmetry breaking in the early universe. If we are located in an emptier-than-average region of space, the observed cosmic expansion rate could be mistaken for a variation in time, or acceleration.[34][35][36][37] A different approach uses a cosmological extension of the equivalence principle to show how space might appear to be expanding more rapidly in the voids surrounding our local cluster. While weak, such effects considered cumulatively over billions of years could become significant, creating the illusion of cosmic acceleration, and making it appear as if we live in a Hubble bubble.[38][39][40]

Another class of theories attempts to come up with an all-encompassing theory of both dark matter and dark energy as a single phenomenon that modifies the laws of gravity at various scales. An example of this type of theory is the theory of dark fluid. Another class of theories that unifies dark matter and dark energy are suggested to be covariant theories of modified gravities. These theories alter the dynamics of the space-time such that the modified dynamic stems what have been assigned to the presence of dark energy and dark matter.[41]

A 2011 paper in the journal *Physical Review D* by Christos Tsagas, a cosmologist at Aristotle University of Thessaloniki in Greece, argued that it is likely that the accelerated expansion of the universe is an illusion caused by the relative motion of us to the rest of the universe. The paper cites data showing that the 2.5 billion ly wide region of space we are inside of is moving very quickly relative to everything around it. If the theory is confirmed, then dark energy would not exist (but the "dark flow" still might).[42][43]

Some theorists think that dark energy and cosmic acceleration are a failure of general relativity on very large scales, larger than superclusters. However most attempts at modifying general relativity have turned out to be either equivalent to theories of quintessence, or inconsistent with observations. Other ideas for dark energy have come from string theory, brane cosmology and the holographic principle, but have not yet proved as compelling as quintessence and the cosmological constant.

On string theory, an article in the journal *Nature* described:[44]

> *String theories, popular with many particle physicists, make it possible, even desirable, to think that the observable universe is just one of 10^{500} universes in a grander multiverse, says Leonard Susskind, a cosmologist at Stanford University in California. The vacuum energy will have different values in different universes, and in many or most it might indeed be vast. But it must be small in ours because it is only in such a universe that observers such as ourselves can evolve.*

Paul Steinhardt in the same article criticizes string theory's explanation of dark energy stating "...Anthropics and randomness don't explain anything... I am disappointed with what most theorists are willing to accept".[44]

Another set of proposals is based on the possibility of a double metric tensor for space-time.[45][46] It has been argued that time reversed solutions in general relativity require such double metric for consistency, and that both dark matter and dark energy can be understood in terms of time reversed solutions of general relativity.[47]

It has been shown that if inertia is assumed to be due to the effect of horizons on Unruh radiation then this predicts galaxy rotation and a cosmic acceleration similar to that observed.[48]

2.4.1 Variable Dark Energy models

In general, the dark energy can be variable. Modern observational data have determined the density of dark energy in the present. Using baryon acoustic oscillations, we can investigate the effect of dark energy in the history of the Universe and we can constrain parameters of the equation of state of dark energy. One of the proposed solutions to get closer to answering the question of dark energy, is to assume that it is variable. To that end, several models have been proposed. One of their most popular models is Chevallier–Polarski–Linder model (CPL).[50][51] Some other common models are, (Barboza & Alcaniz. 2008),[52] (Jassal et al. 2005),[53] (Wetterich. 2004).[54]

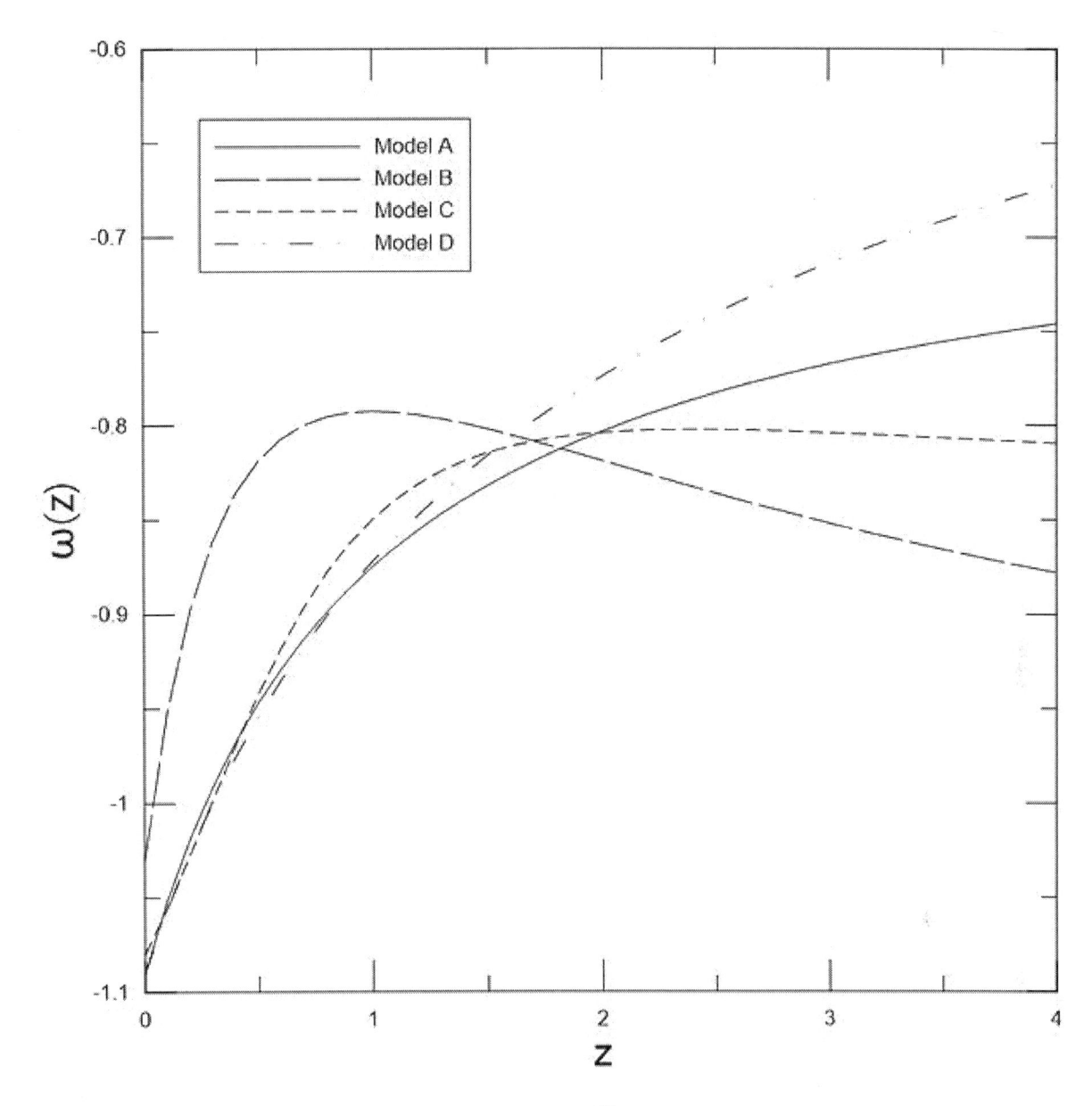

The equation of state of Dark Energy for 4 common models by Redshift.[49]
A: CPL Model,
B: Jassal Model,
C: Barboza & Alcaniz Model,
D: Wetterich Model

2.5 Implications for the fate of the universe

Cosmologists estimate that the acceleration began roughly 5 billion years ago. Before that, it is thought that the expansion was decelerating, due to the attractive influence of dark matter and baryons. The density of dark matter in an expanding universe decreases more quickly than dark energy, and eventually the dark energy dominates. Specifically, when the volume of the universe doubles, the density of dark matter is halved, but the density of dark energy is nearly unchanged (it is exactly constant in the case of a cosmological constant).

If the acceleration continues indefinitely, the ultimate result will be that galaxies outside the local supercluster will have a line-of-sight velocity that continually increases with time, eventually far exceeding the speed of light.[55] This is not a violation of special relativity because the notion of "velocity" used here is different from that of velocity in a local inertial frame of reference, which is still constrained to be less than the speed of light for any massive object (see Uses of the proper distance for a discussion of the subtleties of defining any notion of relative velocity in cosmology). Because the Hubble parameter is decreasing with time, there can actually be cases where a galaxy that is receding from us faster than light does manage to emit a signal which reaches us eventually.[56][57] However, because of the accelerating expansion, it is projected that most galaxies will eventually cross a type of cosmological event horizon where any light they emit past that point will never be able to reach us at any time in the infinite future[58] because the light never reaches a point where its "peculiar velocity" toward us exceeds the expansion velocity away from us (these two notions of velocity are also discussed in Uses of the proper distance). Assuming the dark energy is constant (a cosmological constant), the current distance to this cosmological event horizon is about 16 billion light years, meaning that a signal from an event happening *at present* would eventually be able to reach us in the future if the event were less than 16 billion light years away, but the signal would never reach us if the event were more than 16 billion light years away.[57]

As galaxies approach the point of crossing this cosmological event horizon, the light from them will become more and more redshifted, to the point where the wavelength becomes too large to detect in practice and the galaxies appear to vanish completely[59][60] (*see* Future of an expanding universe). The Earth, the Milky Way, and the Virgo Supercluster, however, would remain virtually undisturbed while the rest of the universe recedes and disappears from view. In this scenario, the local supercluster would ultimately suffer heat death, just as was thought for the flat, matter-dominated universe before measurements of cosmic acceleration.

There are some very speculative ideas about the future of the universe. One suggests that phantom energy causes *divergent* expansion, which would imply that the effective force of dark energy continues growing until it dominates all other forces in the universe. Under this scenario, dark energy would ultimately tear apart all gravitationally bound structures, including galaxies and solar systems, and eventually overcome the electrical and nuclear forces to tear apart atoms themselves, ending the universe in a "Big Rip". On the other hand, dark energy might dissipate with time or even become attractive. Such uncertainties leave open the possibility that gravity might yet rule the day and lead to a universe that contracts in on itself in a "Big Crunch".[61] Some scenarios, such as the cyclic model, suggest this could be the case. It is also possible the universe may never have an end and continue in its present state forever (see The Second Law as a law of disorder). While these ideas are not supported by observations, they are not ruled out.

2.6 History of discovery and previous speculation

The cosmological constant was first proposed by Einstein as a mechanism to obtain a solution of the gravitational field equation that would lead to a static universe, effectively using dark energy to balance gravity.[62] Not only was the mechanism an inelegant example of fine-tuning but it was also later realized that Einstein's static universe would actually be unstable because local inhomogeneities would ultimately lead to either the runaway expansion or contraction of the universe. The equilibrium is unstable: If the universe expands slightly, then the expansion releases vacuum energy, which causes yet more expansion. Likewise, a universe which contracts slightly will continue contracting. These sorts of disturbances are inevitable, due to the uneven distribution of matter throughout the universe. More importantly, observations made by Edwin Hubble in 1929 showed that the universe appears to be expanding and not static at all. Einstein reportedly referred to his failure to predict the idea of a dynamic universe, in contrast to a static universe, as his greatest blunder.[63]

Alan Guth and Alexei Starobinsky proposed in 1980 that a negative pressure field, similar in concept to dark energy, could drive cosmic inflation in the very early universe. Inflation postulates that some repulsive force, qualitatively similar to dark energy, resulted in an enormous and exponential expansion of the universe slightly after the Big Bang. Such expansion is an essential feature of most current models of the Big Bang. However, inflation must have occurred at a much higher energy density than the dark energy we observe today and is thought to have completely ended when the universe was just a fraction of a second old. It is unclear what relation, if any, exists between dark energy and inflation. Even after inflationary models became accepted, the cosmological constant was thought to be irrelevant to the current universe.

Nearly all inflation models predict that the total (matter+energy) density of the universe should be very close to the critical density. During the 1980s, most cosmological research focused on models with critical density in matter only, usually 95% cold dark matter and 5% ordinary matter (baryons). These models were found to be successful at forming realistic galaxies and clusters, but some problems appeared in the late 1980s: notably, the model required a value for the Hubble constant lower than preferred by observations, and the model under-predicted observations of large-scale galaxy clustering. These difficulties became stronger after the discovery of anisotropy in the cosmic microwave background by the COBE spacecraft in 1992, and several modified CDM models came under active study through the mid-1990s: these included the Lambda-CDM model and a mixed cold/hot dark matter model. The first direct evidence for dark energy came from supernova observations in 1998 of accelerated expansion in Riess *et al.*[11] and in Perlmutter *et al.*,[12] and the Lambda-CDM model then became the leading model. Soon after, dark energy was supported by independent observations: in 2000, the BOOMERanG and Maxima cosmic microwave background experiments observed the first acoustic peak in the CMB, showing that the total (matter+energy) density is close to 100% of critical density. Then in 2001, the 2dF Galaxy Redshift Survey gave strong evidence that the matter density is around 30% of critical. The large difference between these two supports a smooth component of dark energy making up the difference. Much more precise measurements from WMAP in 2003–2010 have continued to support the standard model and give more accurate measurements of the key parameters.

The term "dark energy", echoing Fritz Zwicky's "dark matter" from the 1930s, was coined by Michael Turner in 1998.[64]

As of 2013, the Lambda-CDM model is consistent with a series of increasingly rigorous cosmological observations, including the Planck spacecraft and the Supernova Legacy Survey. First results from the SNLS reveal that the average behavior (i.e., equation of state) of dark energy behaves like Einstein's cosmological constant to a precision of 10%.[65] Recent results from the Hubble Space Telescope Higher-Z Team indicate that dark energy has been present for at least 9 billion years and during the period preceding cosmic acceleration.

2.7 See also

- Conformal gravity
- De Sitter relativity
- Illustris project
- The Dark Energy Survey
- Vacuum state

2.8 References

[1] Peebles, P. J. E. and Ratra, Bharat (2003). "The cosmological constant and dark energy". *Reviews of Modern Physics* **75** (2): 559–606. arXiv:astro-ph/0207347. Bibcode:2003RvMP...75..559P. doi:10.1103/RevModPhys.75.559.

[2] Ade, P. A. R.; Aghanim, N.; Armitage-Caplan, C.; et al. (Planck Collaboration), C.; Arnaud, M.; Ashdown, M.; Atrio-Barandela, F.; Aumont, J.; Aussel, H.; Baccigalupi, C.; Banday, A. J.; Barreiro, R. B.; Barrena, R.; Bartelmann, M.; Bartlett, J. G.; Bartolo, N.; Basak, S.; Battaner, E.; Battye, R.; Benabed, K.; Benoît, A.; Benoit-Lévy, A.; Bernard, J.-P.; Bersanelli, M.; Bertincourt, B.; Bethermin, M.; Bielewicz, P.; Bikmaev, I.; Blanchard, A. et al. (22 March 2013). "Planck 2013 results. I. Overview of products and scientific results – Table 9". *Astronomy and Astrophysics* **571**: A1. arXiv:1303.5062. Bibcode:2014A&A...571A...1P. doi:10.1051/0004-6361/201321529.

[3] Ade, P. A. R.; Aghanim, N.; Armitage-Caplan, C.; et al. (Planck Collaboration), C.; Arnaud, M.; Ashdown, M.; Atrio-Barandela, F.; Aumont, J.; Aussel, H.; Baccigalupi, C.; Banday, A. J.; Barreiro, R. B.; Barrena, R.; Bartelmann, M.; Bartlett, J. G.; Bartolo, N.; Basak, S.; Battaner, E.; Battye, R.; Benabed, K.; Benoît, A.; Benoit-Lévy, A.; Bernard, J.-P.; Bersanelli, M.; Bertincourt, B.; Bethermin, M.; Bielewicz, P.; Bikmaev, I.; Blanchard, A. et al. (31 March 2013). "Planck 2013 Results Papers". *Astronomy and Astrophysics* **571**: A1. arXiv:1303.5062. Bibcode:2014A&A...571A...1P. doi:10.1051/0004-6361/201321529.

[4] "First Planck results: the Universe is still weird and interesting".

[5] Sean Carroll, Ph.D., Cal Tech, 2007, The Teaching Company, *Dark Matter, Dark Energy: The Dark Side of the Universe*, Guidebook Part 2 page 46. Retrieved Oct. 7, 2013, "...dark energy: A smooth, persistent component of invisible energy, thought to make up about 70 percent of the current energy density of the universe. Dark energy is known to be smooth because it doesn't accumulate preferentially in galaxies and clusters..."

[6] "Dark Energy". *Hyperphysics*. Retrieved January 4, 2014.

[7] Ferris, Timothy. "Dark Matter(Dark Energy)". Retrieved 2015-06-10.

[8] Carroll, Sean (2001). "The cosmological constant". *Living Reviews in Relativity* **4**. Retrieved 2006-09-28.

[9] Baum, L. and Frampton, P.H. (2007). "Turnaround in Cyclic Cosmology". *Physical Review Letters* **98** (7): 071301. arXiv:hep-th/0610213. Bibcode:2007PhRvL..98g1301B. doi:10.1103/PhysRevLett.98.071301. PMID 17359014.

[10] Durrer, R. (2011). "What do we really know about Dark Energy?". *Philosophical Transactions of the Royal Society A: Mathematical, Physical and Engineering Sciences* **369** (1957): 5102. arXiv:1103.5331. Bibcode:2011RSPTA.369.5102D. doi:10.1098/rsta.2011.0285.

[11] Riess, Adam G.; Filippenko; Challis; Clocchiatti; Diercks; Garnavich; Gilliland; Hogan; Jha; Kirshner; Leibundgut; Phillips; Reiss; Schmidt; Schommer; Smith; Spyromilio; Stubbs; Suntzeff; Tonry (1998). "Observational evidence from supernovae for an accelerating universe and a cosmological constant". *Astronomical J.* **116** (3): 1009–38. arXiv:astro-ph/9805201. Bibcode:1998AJ....116.1009R. doi:10.1086/300499.

[12] Perlmutter, S.; Aldering; Goldhaber; Knop; Nugent; Castro; Deustua; Fabbro; Goobar; Groom; Hook; Kim; Kim; Lee; Nunes; Pain; Pennypacker; Quimby; Lidman; Ellis; Irwin; McMahon; Ruiz-Lapuente; Walton; Schaefer; Boyle; Filippenko; Matheson; Fruchter et al. (1999). "Measurements of Omega and Lambda from 42 high redshift supernovae". *Astrophysical Journal* **517** (2): 565–86. arXiv:astro-ph/9812133. Bibcode:1999ApJ...517..565P. doi:10.1086/307221.

[13] The first paper, using observed data, which claimed a positive Lambda term was Paál, G. et al. (1992). "Inflation and compactification from galaxy redshifts?". *Astrophysics and Space Science* **191**: 107–24. Bibcode:1992Ap&SS.191..107P. doi:10.1007/BF00644200.

[14] "The Nobel Prize in Physics 2011". Nobel Foundation. Retrieved 2011-10-04.

[15] The Nobel Prize in Physics 2011. Perlmutter got half the prize, and the other half was shared between Schmidt and Riess.

[16] Spergel, D. N. (WMAP collaboration) et al. (March 2006). "Wilkinson Microwave Anisotropy Probe (WMAP) three year results: implications for cosmology".

[17] Durrer, R. (2011). "What do we really know about dark energy?". *Philosophical Transactions of the Royal Society A* **369** (1957): 5102–5114. arXiv:1103.5331. Bibcode:2011RSPTA.369.5102D. doi:10.1098/rsta.2011.0285.

[18] Kowalski, Marek; Rubin, David; Aldering, G.; Agostinho, R. J.; Amadon, A.; Amanullah, R.; Balland, C.; Barbary, K.; Blanc, G.; Challis, P. J.; Conley, A.; Connolly, N. V.; Covarrubias, R.; Dawson, K. S.; Deustua, S. E.; Ellis, R.; Fabbro, S.; Fadeyev, V.; Fan, X.; Farris, B.; Folatelli, G.; Frye, B. L.; Garavini, G.; Gates, E. L.; Germany, L.; Goldhaber, G.; Goldman, B.; Goobar, A.; Groom, D. E. et al. (October 27, 2008). "Improved Cosmological Constraints from New, Old and Combined Supernova Datasets". *The Astrophysical Journal* (Chicago: University of Chicago Press) **686** (2): 749–778. arXiv:0804.4142. Bibcode:2008ApJ...686..749K. doi:10.1086/589937.. They find a best fit value of the dark energy density, $\Omega\Lambda$ of 0.713+0.027–0.029(stat)+0.036–0.039(sys), of the total matter density, ΩM, of 0.274+0.016–0.016(stat)+0.013–0.012(sys) with an equation of state parameter w of −0.969+0.059–0.063(stat)+0.063–0.066(sys).

[19] "Planck reveals an almost perfect universe". *Planck*. ESA. 2013-03-21. Retrieved 2013-03-21.

[20] "Big Bang's afterglow shows universe is 80 million years older than scientists first thought". *The Washington Post*. Retrieved 22 March 2013.

[21] "New method 'confirms dark energy'". BBC News. 2011-05-19.

[22] Dark energy is real, Swinburne University of Technology, 19 May 2011

[23] Crittenden; Neil Turok (1995). "Looking for Λ with the Rees-Sciama Effect". *Physical Review Letters* **76** (4): 575–578. arXiv:astro-ph/9510072. Bibcode:1996PhRvL..76..575C. doi:10.1103/PhysRevLett.76.575. PMID 10061494.

[24] Shirley Ho; Hirata; Nikhil Padmanabhan; Uros Seljak; Neta Bahcall (2008). "Correlation of CMB with large-scale structure: I. ISW Tomography and Cosmological Implications". *Physical Review D* **78**(4): 043519.arXiv:0801.0642.Bibcode:2008PhRvdoi: 10.1103/PhysRevD.78.043519.

[25] Tommaso Giannantonio; Ryan Scranton; Crittenden; Nichol; Boughn; Myers; Richards (2008). "Combined analysis of the integrated Sachs-Wolfe effect and cosmological implications".*Physical Review D* **77**(12): 123520.arXiv:0801.4380.Bibcode:.doi: 10.1103/PhysRevD.77.123520.

2.8. REFERENCES

[26] Zelong Yi; Tongjie Zhang (2007). "Constraints on holographic dark energy models using the differential ages of passively evolving galaxies". *Modern Physics Letters A* **22** (1): 41. arXiv:astro-ph/0605596.Bibcode:2007MPLA...22...41Y.doi:10.1142.

[27] Haoyi Wan; Zelong Yi; Tongjie Zhang; Jie Zhou (2007). "Constraints on the DGP Universe Using Observational Hubble parameter". *Physics Letters B* **651** (5): 352. arXiv:0706.2723. Bibcode:2007PhLB..651..352W. doi:10.1016/j.physletb.2007.06.053.

[28] Cong Ma; Tongjie Zhang (2010). "Power of Observational Hubble Parameter Data: a Figure of Merit Exploration". *Astrophysical Journal* **730** (2): 74. arXiv:1007.3787. Bibcode:2011ApJ...730...74M. doi:10.1088/0004-637X/730/2/74.

[29] Tongjie Zhang; Cong Ma; Tian Lan (2010). "Constraints on the Dark Side of the Universe and Observational Hubble Parameter Data". *Advances in Astronomy* **2010** (1): 1. arXiv:1010.1307. Bibcode:2010AdAst2010E..81Z. doi:10.1155/2010/184284.

[30] Joan Simon; Licia Verde; Raul Jimenez (2005). "Constraints on the redshift dependence of the dark energy potential". *Physical Review D* **71** (12): 123001. arXiv:astro-ph/0412269. Bibcode:2005PhRvD..71l3001S. doi:10.1103/PhysRevD.71.123001.

[31] Wess, Julius; Bagger, Jonathan. *Supersymmetry and Supergravity*. ISBN 978-0691025308.

[32] Carroll, Sean M. (1998). "Quintessence and the Rest of the World: Suppressing Long-Range Interactions". *Physical Review Letters* **81** (15): 3067–3070. arXiv:astro-ph/9806099. Bibcode:1998PhRvL..81.3067C. doi:10.1103/PhysRevLett.81.3067. ISSN 0031-9007.

[33] Steinhardt, Paul J.; Wang, Li-Min; Zlatev, Ivaylo. "Cosmological tracking solutions". *Phys. Rev.* **D59**: 123504. arXiv:astro-ph/9812313. Bibcode:1999PhRvD..59l3504S. doi:10.1103/PhysRevD.59.123504.

[34] Wiltshire, David L. (2007). "Exact Solution to the Averaging Problem in Cosmology". *Physical Review Letters* **99** (25): 251101. arXiv:0709.0732. Bibcode:2007PhRvL..99y1101W. doi:10.1103/PhysRevLett.99.251101. PMID 18233512.

[35] Ishak, Mustapha; Richardson, James; Garred, David; Whittington, Delilah; Nwankwo, Anthony; Sussman, Roberto (2007). "Dark Energy or Apparent Acceleration Due to a Relativistic Cosmological Model More Complex than FLRW?". *Physical Review D* **78** (12): 123531. arXiv:0708.2943. Bibcode:2008PhRvD..78l3531I. doi:10.1103/PhysRevD.78.123531.

[36] Mattsson, Teppo (2007). "Dark energy as a mirage".*Gen. Rel. Grav.* **42**(3): 567–599.arXiv:0711.4264.Bibcode:2010GReGr..4. doi:10.1007/s10714-009-0873-z.

[37] Clifton, Timothy; Ferreira, Pedro (April 2009). "Does Dark Energy Really Exist?". *Scientific American* **300** (4): 48–55. doi:10.1038/scientificamerican0409-48. PMID 19363920. Retrieved April 30, 2009.

[38] Wiltshire, D. (2008). "Cosmological equivalence principle and the weak-field limit". *Physical Review D* **78** (8): 084032. arXiv:0809.1183. Bibcode:2008PhRvD..78h4032W. doi:10.1103/PhysRevD.78.084032.

[39] Gray, Stuart. "Dark questions remain over dark energy". ABC Science Australia. Retrieved 27 January 2013.

[40] Merali, Zeeya (March 2012). "Is Einstein's Greatest Work All Wrong—Because He Didn't Go Far Enough?". *Discover magazine*. Retrieved 27 January 2013.

[41] Exirifard, Q. (2010). "Phenomenological covariant approach to gravity". *General Relativity and Gravitation* **43**: 93–106. arXiv:0808.1962. Bibcode:2011GReGr..43...93E. doi:10.1007/s10714-010-1073-6.

[42] Wolchover, Natalie (27 September 2011) 'Accelerating universe' could be just an illusion, MSNBC

[43] Tsagas, Christos G. (2011). "Peculiar motions, accelerated expansion, and the cosmological axis". *Physical Review D* **84** (6): 063503. arXiv:1107.4045. Bibcode:2011PhRvD..84f3503T. doi:10.1103/PhysRevD.84.063503.

[44] Hogan, Jenny (2007). "Unseen Universe: Welcome to the dark side". *Nature* **448** (7151): 240–245. Bibcode:2007Natur.448..240H. doi:10.1038/448240a. PMID 17637630.

[45] Hossenfelder, S. (2008). "A Bi-Metric Theory with Exchange Symmetry". *Physical Review D* **78** (4): 044015. arXiv:0807.2838. Bibcode:2008PhRvD..78d4015H. doi:10.1103/PhysRevD.78.044015.

[46] Henry-Couannier, F. (2005). "Discrete Symmetries and General Relativity, the Dark Side of Gravity". *International Journal of Modern Physics A* **20** (11): 2341. arXiv:gr-qc/0410055. Bibcode:2005IJMPA..20.2341H. doi:10.1142/S0217751X05024602.

[47] Ripalda, Jose M. (1999). "Time reversal and negative energies in general relativity". arXiv:gr-qc/9906012.

[48] McCulloch, M.E. (2010). "Minimum accelerations from quantised inertia".*EPL* **90**(2): 29001.arXiv:1004.3303.Bibcode:2010E. doi:10.1209/0295-5075/90/29001.

[49] by Ehsan Sadri M.A Ap

[50] Chevallier, M; Polarski, D (2001). "Accelerating Universes with Scaling Dark Matter". *International Journal of Modern Physics D* **10**: 213–224. arXiv:gr-qc/0009008. Bibcode:2001IJMPD..10..213C. doi:10.1142/S0218271801000822.

[51] Linder, Eric V. (3 March 2003). "Exploring the Expansion History of the Universe". *Physical Review Letters* **90** (9). arXiv:astro-ph/0208512v1. Bibcode:2003PhRvL..90i1301L. doi:10.1103/PhysRevLett.90.091301.

[52] Alcaniz, E.M.; Alcaniz, J.S. (2008). "A parametric model for dark energy". *Physics Letters B* **666**: 415–419. arXiv:0805.1713. Bibcode:2008PhLB..666..415B. doi:10.1016/j.physletb.2008.08.012.

[53] Jassal, H.K; Bagla, J.S (2010). "Understanding the origin of CMB constraints on Dark Energy". *Monthly Notices of the Royal Astronomical Society* **405**: 2639–2650. arXiv:astro-ph/0601389. Bibcode:2010MNRAS.405.2639J. doi:10.1111/j.1365-2966.2.

[54] Wetterich, C. (2004). "Phenomenological parameterization of quintessence". arXiv:astro-ph/0403289v1.

[55] Krauss, Lawrence M. and Scherrer, Robert J. (March 2008). "The End of Cosmology?". *Scientific American* **82**. Retrieved 2011-01-06.

[56] Is the universe expanding faster than the speed of light? (see the last two paragraphs)

[57] Lineweaver, Charles; Tamara M. Davis (2005). "Misconceptions about the Big Bang" (PDF). *Scientific American*. Retrieved 2008-11-06.

[58] Loeb, Abraham (2002). "The Long-Term Future of Extragalactic Astronomy". *Physical Review D* **65** (4): 047301. arXiv:astro-ph/0107568. Bibcode:2002PhRvD..65d7301L. doi:10.1103/PhysRevD.65.047301.

[59] Krauss, Lawrence M.; Robert J. Scherrer (2007). "The Return of a Static Universe and the End of Cosmology". *General Relativity and Gravitation* **39** (10): 1545–1550. arXiv:0704.0221. Bibcode:2007GReGr..39.1545K. doi:10.1007/s10714-007-0472-9.

[60] Using Tiny Particles To Answer Giant Questions. Science Friday, 3 Apr 2009. According to the transcript, Brian Greene makes the comment "And actually, in the far future, everything we now see, except for our local galaxy and a region of galaxies will have disappeared. The entire universe will disappear before our very eyes, and it's one of my arguments for actually funding cosmology. We've got to do it while we have a chance."

[61] *How the Universe Works 3*. End of the Universe. Discovery Channel. 2014.

[62] Harvey, Alex (2012). "How Einstein Discovered Dark Energy". arXiv:1211.6338.

[63] Gamow, George (1970) *My World Line: An Informal Autobiography*. p. 44: "Much later, when I was discussing cosmological problems with Einstein, he remarked that the introduction of the cosmological term was the biggest blunder he ever made in his life." – Here the "cosmological term" refers to the cosmological constant in the equations of general relativity, whose value Einstein initially picked to ensure that his model of the universe would neither expand nor contract; if he hadn't done this he might have theoretically predicted the universal expansion that was first observed by Edwin Hubble.

[64] The first appearance of the term "dark energy" is in the article with another cosmologist and Turner's student at the time, Dragan Huterer, "Prospects for Probing the Dark Energy via Supernova Distance Measurements", which was posted to the ArXiv.org e-print archive in August 1998 and published in Huterer, D.; Turner, M. (1999). "Prospects for probing the dark energy via supernova distance measurements". *Physical Review D* **60** (8). doi:10.1103/PhysRevD.60.081301., although the manner in which the term is treated there suggests it was already in general use. Cosmologist Saul Perlmutter has credited Turner with coining the term in an article they wrote together with Martin White, where it is introduced in quotation marks as if it were a neologism. Perlmutter, S.; Turner, M.; White, M. (1999). "Constraining Dark Energy with Type Ia Supernovae and Large-Scale Structure". *Physical Review Letters* **83** (4): 670. doi:10.1103/PhysRevLett.83.670.

[65] Astier, Pierre (Supernova Legacy Survey); Guy; Regnault; Pain; Aubourg; Balam; Basa; Carlberg; Fabbro; Fouchez; Hook; Howell; Lafoux; Neill; Palanque-Delabrouille; Perrett; Pritchet; Rich; Sullivan; Taillet; Aldering; Antilogus; Arsenijevic; Balland; Baumont; Bronder; Courtois; Ellis; Filiol et al. (2006). "The Supernova legacy survey: Measurement of Ω_M, Ω_Λ and W from the first year data set". *Astronomy and Astrophysics* **447**: 31–48. arXiv:astro-ph/0510447. Bibcode:2006A&A...447...31A. doi:10.1051/0004-6361:20054185.

2.9 External links

-
- Dark Energy on *In Our Time* at the BBC. (listen now)
- Dark energy: how the paradigm shifted Physicsworld.com
- Dennis Overbye (November 2006). "9 Billion-Year-Old 'Dark Energy' Reported". *The New York Times.*
- "Mysterious force's long presence" BBC News online (2006) More evidence for dark energy being the cosmological constant
- "Astronomy Picture of the Day" one of the images of the Cosmic Microwave Background which confirmed the presence of dark energy and dark matter
- SuperNova Legacy Survey home page The Canada-France-Hawaii Telescope Legacy Survey Supernova Program aims primarily at measuring the equation of state of Dark Energy. It is designed to precisely measure several hundred high-redshift supernovae.
- "Report of the Dark Energy Task Force"
- "HubbleSite.org – Dark Energy Website" Multimedia presentation explores the science of dark energy and Hubble's role in its discovery.
- "Surveying the dark side"
- "Dark energy and 3-manifold topology" Acta Physica Polonica 38 (2007), p. 3633–3639
- The Dark Energy Survey
- The Joint Dark Energy Mission
- Harvard: Dark Energy Found Stifling Growth in Universe, primary source
- April 2010 Smithsonian Magazine Article
- HETDEX Dark energy experiment
- Dark Energy FAQ
- "The Hunt for Dark Energy" George FR Ellis, Peter Cameron and David Tong discuss the presence of dark energy in the Universe

April 2010 Smithsonian Magazine Article]

- Euclid ESA Satellite, a mission to map the geometry of the dark universe

Chapter 3

Big Bang nucleosynthesis

In physical cosmology, **Big Bang nucleosynthesis** (abbreviated BBN, also known as **primordial nucleosynthesis**) refers to the production of nuclei other than those of the lightest isotope of hydrogen (hydrogen-1, ^{1}H, having a single proton as a nucleus) during the early phases of the universe. Primordial nucleosynthesis is believed by most cosmologists to have taken place from 10 seconds to 20 minutes after the Big Bang, and is calculated to be responsible for the formation of most of the universe's helium as the isotope helium-4 (^{4}He), along with small amounts of the hydrogen isotope deuterium (^{2}H or D), the helium isotope helium-3 (^{3}He), and a very small amount of the lithium isotope lithium-7 (^{7}Li). In addition to these stable nuclei, two unstable or radioactive isotopes were also produced: the heavy hydrogen isotope tritium (^{3}H or T); and the beryllium isotope beryllium-7 (^{7}Be); but these unstable isotopes later decayed into ^{3}He and ^{7}Li, as above.

Essentially all of the elements that are heavier than lithium and beryllium were created much later, by stellar nucleosynthesis in evolving and exploding stars.

3.1 Characteristics

There are two important characteristics of Big Bang nucleosynthesis (BBN):

- The era began at temperatures of around 10 MeV (116 gigakelvin) and ended at temperatures below 100 keV (1.16 gigakelvin).[1] The corresponding time interval was from a few tenths of a second to up to 10^3 seconds.[2] The temperature/time relation in this era can be given by the equation:

$$tT^2 = (0.74 \text{ s MeV}^2) \times (10.75/g_*)^{1/2}$$

  where t is time in seconds, T is temperature in MeV and g* is the effective number of particle species.[3] (g* includes contributions of 2 from photons, 7/2 from electron-positron pairs and 7/4 from each neutrino flavor. In the standard model g* is 10.75). This expression also shows how a different number of neutrino flavors will change the rate of cooling of the early universe.

- It was widespread, encompassing the entire observable universe.

The key parameter which allows one to calculate the effects of BBN is the baryon/photon number ratio, which is a small number of order 6×10^{-10}. This parameter corresponds to the baryon density and controls the rate at which nucleons collide and react; from this we can derive elemental abundances. Although the baryon per photon ratio is important in determining elemental abundances, the precise value makes little difference to the overall picture. Without major changes to the Big Bang theory itself, BBN will result in mass abundances of about 75% of hydrogen-1, about 25% helium-4, about 0.01% of deuterium and helium-3, trace amounts (on the order of 10^{-10}) of lithium, and negligible heavier elements. (Traces of boron have been found in some old stars, giving rise to the question whether some boron, not really predicted by the theory, might have been produced in the Big Bang. The question is not presently resolved.[4]) That the observed abundances in the universe are generally consistent with these abundance numbers is considered strong evidence for the Big Bang theory.

In this field, for historical reasons it is customary to quote the Helium-4 fraction *by mass*, symbol Y, so that 25% helium-4 means that helium-4 atoms account for 25% of the mass, but only about 8% of the nuclei would be helium-4 nuclei. Other (trace) nuclei are usually expressed as number ratios to hydrogen.

3.2 Important parameters

The creation of light elements during BBN was dependent on a number of parameters; among those was the neutron-proton ratio (calculable from Standard Model physics) and the baryon-photon ratio.

3.2.1 Neutron-proton ratio

Neutrons can react with positrons or electron neutrinos to create protons and other products in one of the following reactions:

$$n + e^+ \leftrightarrow \text{anti-}\nu_e + p$$
$$n + \nu_e \leftrightarrow p + e^-$$

At times much earlier than 1 sec, the n/p ratio was close to 1:1. As the temperature dropped, the equilibrium shifted in favour of protons due to their slightly lower mass, and the n/p ratio decreased. These reactions continue until expansion of the universe outpaces the reactions, which occurs at about T = 0.7 MeV and is called the freeze out temperature.[5] At freeze out, the neutron-proton ratio was about 1/5. However, free neutrons are unstable with a mean life of 880 sec; some neutrons decayed in the next few minutes before fusing into any nucleus, so the ratio of total neutrons to protons after nucleosynthesis ends is about 1/7. Almost all neutrons that fused instead of decaying ended up combined into helium-4, due to the fact that helium-4 has the highest binding energy per nucleon among light elements. This predicts that about 8% of all atoms should be helium-4, leading to a mass fraction of helium-4 of about 25%, which is in line with observations. Small traces of deuterium and helium-3 remained as there was insufficient time and density for them to react and form helium-4.

3.2.2 Baryon-photon ratio

The baryon-photon ratio, η, is the key parameter determining the abundance of light elements present in the early universe. Baryons can react with light elements in the following reactions:

$$(p,n) + {}^2H \rightarrow ({}^3He, {}^3H)$$
$$({}^3He, {}^3H) + (n,p) \rightarrow {}^4He$$

It is evident that reactions with baryons during BBN would ultimately result in helium-4, and also that the abundance of primordial deuterium is indirectly related to the baryon density or baryon-photon ratio. That is, the larger the baryon-photon ratio the more reactions there will be and the more efficiently deuterium will be eventually transformed into helium-4. This result makes deuterium a very useful tool in measuring the baryon-to-photon ratio.

3.3 Sequence

Big Bang nucleosynthesis began a few seconds after the big bang, when the universe had cooled sufficiently to allow deuterium nuclei to survive disruption by high-energy photons. This time is essentially independent of dark matter content,

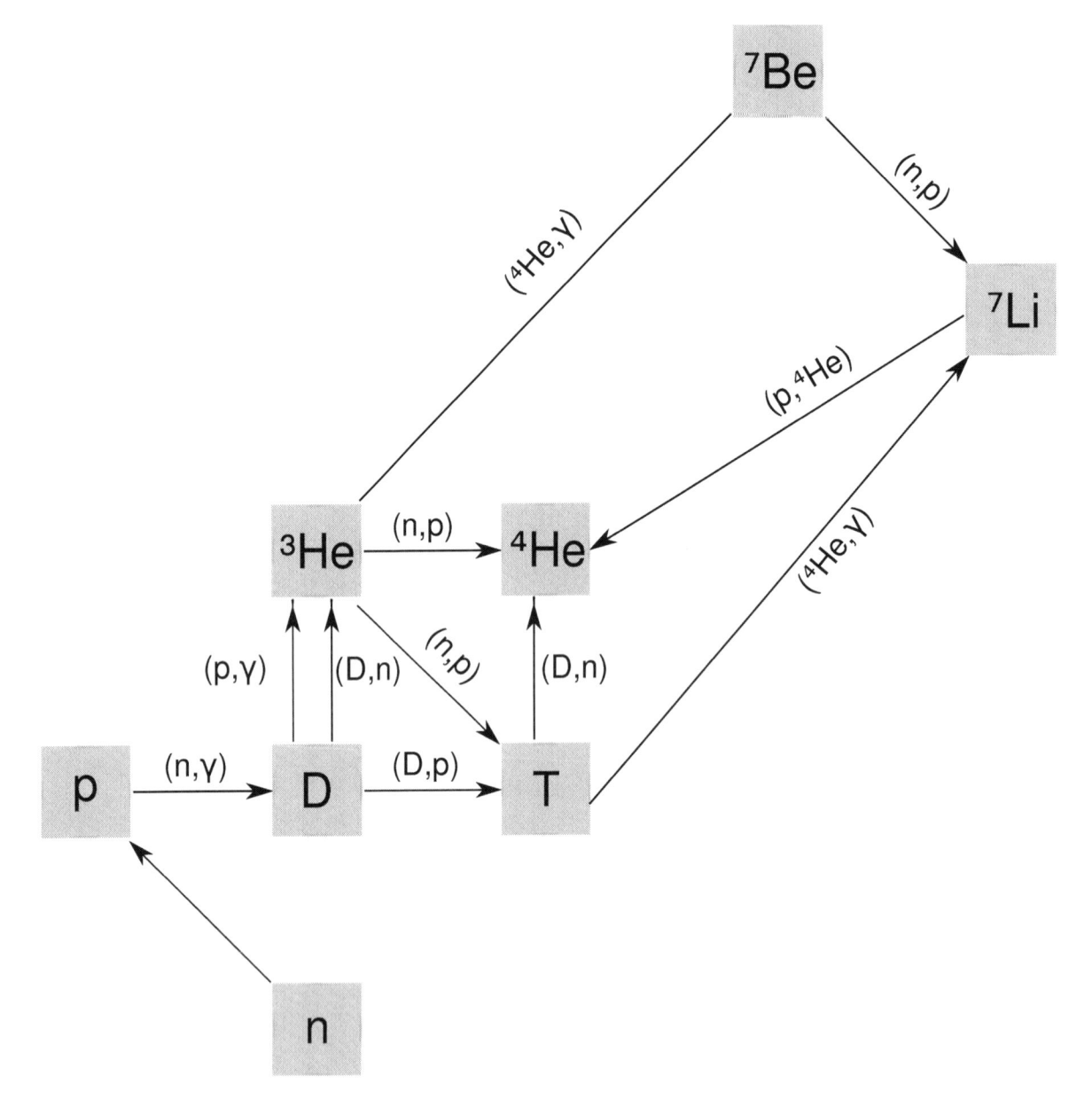

The main nuclear reaction chains for Big Bang nucleosynthesis

since the universe was highly radiation dominated until much later, and this dominant component controls the temperature/time relation. The relative abundances of protons and neutrons follow from simple thermodynamical arguments, combined with the way that the mean temperature of the universe changes over time. If the reactions needed to reach the thermodynamically favoured equilibrium values are too slow compared to the temperature change brought about by the expansion, abundances would have remained at some specific non-equilibrium value. Combining thermodynamics and the changes brought about by cosmic expansion, one can calculate the fraction of protons and neutrons based on the temperature at this point. The answer is that there are about seven protons for every neutron at the beginning of nucleosynthesis. This fraction is in favour of protons, primarily because their lower mass with respect to the neutron favors their production. Free neutrons decay to protons with a half-life of about 10.2 minutes, but this time-scale is longer than the first three minutes of nucleogenesis, during which time a substantial fraction of them were combined with protons into deuterium and then He-4. The sequence of these reaction chains is shown on the image.[6]

One feature of BBN is that the physical laws and constants that govern the behavior of matter at these energies are very well understood, and hence BBN lacks some of the speculative uncertainties that characterize earlier periods in the life of the universe. Another feature is that the process of nucleosynthesis is determined by conditions at the start of this phase of the life of the universe, and proceeds independently of what happened before.

As the universe expands, it cools. Free neutrons and protons are less stable than helium nuclei, and the protons and neutrons have a strong tendency to form helium-4. However, forming helium-4 requires the intermediate step of forming deuterium. Before nucleosynthesis began, the temperature was high enough for many photons to have energy greater than the binding energy of deuterium; therefore any deuterium that was formed was immediately destroyed (a situation known as the **deuterium bottleneck**). Hence, the formation of helium-4 is delayed until the universe became cool enough for deuterium to survive (at about T = 0.1 MeV); after which there was a sudden burst of element formation. However, very shortly thereafter, at twenty minutes after the Big Bang, the universe became too cool for any further nuclear fusion and nucleosynthesis to occur. At this point, the elemental abundances were nearly fixed, and only change was the result of the radioactive decay of some products of BBN (such as tritium).[7]

3.3.1 History of theory

The history of Big Bang nucleosynthesis began with the calculations of Ralph Alpher in the 1940s. Alpher published the seminal Alpher–Bethe–Gamow paper that outlined the theory of light-element production in the early universe.

During the 1970s, there was a major puzzle in that the density of baryons as calculated by Big Bang nucleosynthesis was much less than the observed mass of the universe based on calculations of the expansion rate. This puzzle was resolved in large part by postulating the existence of dark matter.

3.3.2 Heavy elements

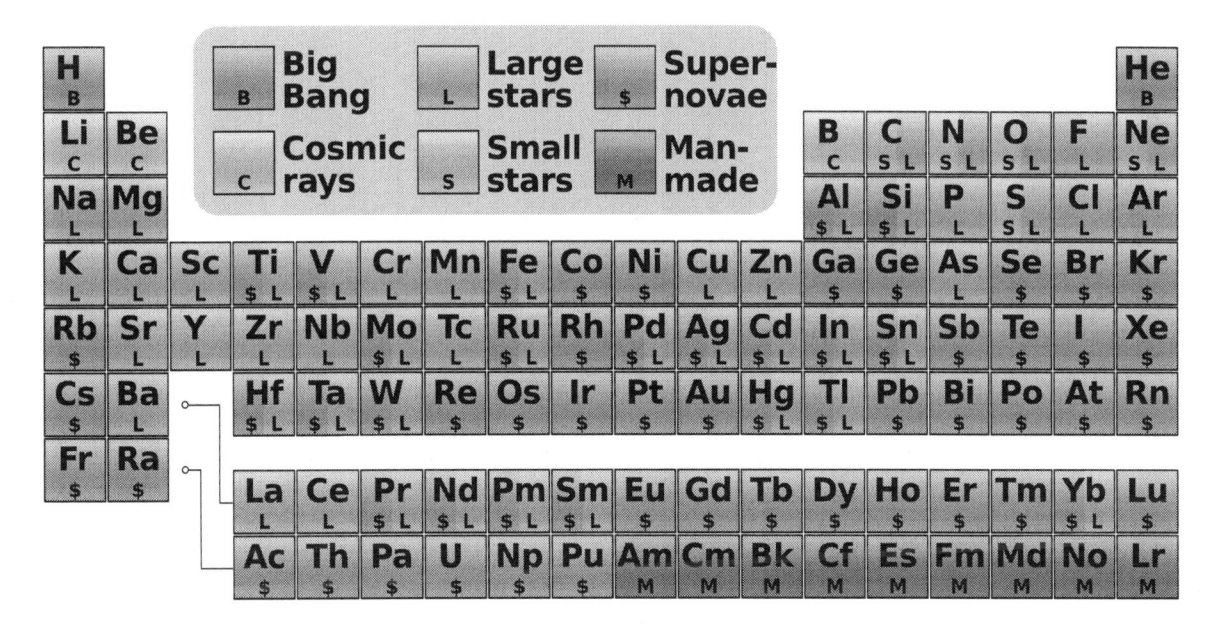

A version of the periodic table indicating the origins – including big bang nucleosynthesis – of the elements. All elements above 103 (lawrencium) are also manmade and are not included.

Big Bang nucleosynthesis produced no elements heavier than beryllium, due to a bottleneck: the absence of a stable nucleus with 8 or 5 nucleons. This deficit of larger atoms also limited the amounts of lithium-7 and beryllium-9 produced during BBN. In stars, the bottleneck is passed by triple collisions of helium-4 nuclei, producing carbon (the triple-alpha process). However, this process is very slow, taking tens of thousands of years to convert a significant amount of helium to carbon in stars, and therefore it made a negligible contribution in the minutes following the Big Bang.

3.3.3 Helium-4

Main article: Helium-4

Big Bang nucleo-synthesis predicts a primordial abundance of about 25% helium-4 by mass, irrespective of the initial conditions of the universe. As long as the universe was hot enough for protons and neutrons to transform into each other easily, their ratio, determined solely by their relative masses, was about 1 neutron to 7 protons (allowing for some decay of neutrons into protons). Once it was cool enough, the neutrons quickly bound with an equal number of protons to form first deuterium, then helium-4. Helium-4 is very stable and is nearly the end of this chain if it runs for only a short time, since helium neither decays nor combines easily to form heavier nuclei (since there are no stable nuclei with mass numbers of 5 or 8, helium does not combine easily with either protons, or with itself). Once temperatures are lowered, out of every 16 nucleons (2 neutrons and 14 protons), 4 of these (25% of the total particles and total mass) combine quickly into one helium-4 nucleus. This produces one helium for every 12 hydrogens, resulting in a universe that is a little over 8% helium by number of atoms, and 25% helium by mass.

One analogy is to think of helium-4 as ash, and the amount of ash that one forms when one completely burns a piece of wood is insensitive to how one burns it. The resort to the BBN theory of the helium-4 abundance is necessary as there is far more helium-4 in the universe than can be explained by stellar nucleosynthesis. In addition, it provides an important test for the Big Bang theory. If the observed helium abundance is much different from 25%, then this would pose a serious challenge to the theory. This would particularly be the case if the early helium-4 abundance was much smaller than 25% because it is hard to destroy helium-4. For a few years during the mid-1990s, observations suggested that this might be the case, causing astrophysicists to talk about a Big Bang nucleosynthetic crisis, but further observations were consistent with the Big Bang theory.[8]

3.3.4 Deuterium

Main article: Deuterium

Deuterium is in some ways the opposite of helium-4 in that while helium-4 is very stable and very difficult to destroy, deuterium is only marginally stable and easy to destroy. The temperatures, time, and densities were sufficient to combine a substantial fraction of the deuterium nuclei to form helium-4 but insufficient to carry the process further using helium-4 in the next fusion step. BBN did not convert all of the deuterium in the universe to helium-4 due to the expansion that cooled the universe and reduced the density and so, cut that conversion short before it could proceed any further. One consequence of this is that unlike helium-4, the amount of deuterium is very sensitive to initial conditions. The denser the initial universe was, the more deuterium would be converted to helium-4 before time ran out, and the less deuterium would remain.

There are no known post-Big Bang processes which can produce significant amounts of deuterium. Hence observations about deuterium abundance suggest that the universe is not infinitely old, which is in accordance with the Big Bang theory.

During the 1970s, there were major efforts to find processes that could produce deuterium, but those revealed ways of producing isotopes other than deuterium. The problem was that while the concentration of deuterium in the universe is consistent with the Big Bang model as a whole, it is too high to be consistent with a model that presumes that most of the universe is composed of protons and neutrons. If one assumes that all of the universe consists of protons and neutrons, the density of the universe is such that much of the currently observed deuterium would have been burned into helium-4. The standard explanation now used for the abundance of deuterium is that the universe does not consist mostly of baryons, but that non-baryonic matter (also known as dark matter) makes up most of the mass of the universe. This explanation is also consistent with calculations that show that a universe made mostly of protons and neutrons would be far more *clumpy* than is observed.

It is very hard to come up with another process that would produce deuterium other than by nuclear fusion. Such a process would require that the temperature be hot enough to produce deuterium, but not hot enough to produce helium-4, and that this process should immediately cool to non-nuclear temperatures after no more than a few minutes. It would also be necessary for the deuterium to be swept away before it reoccurs.

Producing deuterium by fission is also difficult. The problem here again is that deuterium is very unlikely due to nuclear processes, and that collisions between atomic nuclei are likely to result either in the fusion of the nuclei, or in the release of free neutrons or alpha particles. During the 1970s, cosmic ray spallation was proposed as a source of deuterium. That theory failed to account for the abundance of deuterium, but led to explanations of the source of other light elements.

3.4 Measurements and status of theory

The theory of BBN gives a detailed mathematical description of the production of the light "elements" deuterium, helium-3, helium-4, and lithium-7. Specifically, the theory yields precise quantitative predictions for the mixture of these elements, that is, the primordial abundances at the end of the big-bang.

In order to test these predictions, it is necessary to reconstruct the primordial abundances as faithfully as possible, for instance by observing astronomical objects in which very little stellar nucleosynthesis has taken place (such as certain dwarf galaxies) or by observing objects that are very far away, and thus can be seen in a very early stage of their evolution (such as distant quasars).

As noted above, in the standard picture of BBN, all of the light element abundances depend on the amount of ordinary matter (baryons) relative to radiation (photons). Since the universe is presumed to be homogeneous, it has one unique value of the baryon-to-photon ratio. For a long time, this meant that to test BBN theory against observations one had to ask: can *all* of the light element observations be explained with a *single value* of the baryon-to-photon ratio? Or more precisely, allowing for the finite precision of both the predictions and the observations, one asks: is there some *range* of baryon-to-photon values which can account for all of the observations?

More recently, the question has changed: Precision observations of the cosmic microwave background radiation[9][10] with the Wilkinson Microwave Anisotropy Probe (WMAP) and Planck give an independent value for the baryon-to-photon ratio. Using this value, are the BBN predictions for the abundances of light elements in agreement with the observations?

The present measurement of helium-4 indicates good agreement, and yet better agreement for helium-3. But for lithium-7, there is a significant discrepancy between BBN and WMAP/Planck, and the abundance derived from Population II stars. The discrepancy is a factor of 2.4—4.3 below the theoretically predicted value and is considered a problem for the original models,[11] that have resulted in revised calculations of the standard BBN based on new nuclear data, and to various reevaluation proposals for primordial proton-proton nuclear reactions, especially the abundances of $^7Be(n,p)^7Li$ versus $^7Be(d,p)^8Be$.[12]

3.5 Non-standard scenarios

In addition to the standard BBN scenario there are numerous non-standard BBN scenarios. These should not be confused with non-standard cosmology: a non-standard BBN scenario assumes that the Big Bang occurred, but inserts additional physics in order to see how this affects elemental abundances. These pieces of additional physics include relaxing or removing the assumption of homogeneity, or inserting new particles such as massive neutrinos.

There have been, and continue to be, various reasons for researching non-standard BBN. The first, which is largely of historical interest, is to resolve inconsistencies between BBN predictions and observations. This has proved to be of limited usefulness in that the inconsistencies were resolved by better observations, and in most cases trying to change BBN resulted in abundances that were more inconsistent with observations rather than less. The second reason for researching non-standard BBN, and largely the focus of non-standard BBN in the early 21st century, is to use BBN to place limits on unknown or speculative physics. For example, standard BBN assumes that no exotic hypothetical particles were involved in BBN. One can insert a hypothetical particle (such as a massive neutrino) and see what has to happen before BBN predicts abundances which are very different from observations. This has been usefully done to put limits on the mass of a stable tau neutrino.

3.6 See also

- Nucleosynthesis
- Stellar nucleosynthesis
- Ultimate fate of the universe
- Chronology of the universe
- Big Bang

3.7 References

[1] Doglov, A. D. "Big Bang Nucleosynthesis." Nucl.Phys.Proc.Suppl. (2002): 137-43. ArXiv. 17 Jan. 2002. Web. 14 Jan. 2013.

[2] Grupen, Claus. "Big Bang Nucleosynthesis." Astroparticle Physics. Berlin: Springer, 2005. 213-28. Print.

[3] J. Beringer et al. (Particle Data Group), "Big-Bang cosmology" Phys. Rev. D86, 010001 (2012): (21.43)

[4] "Hubble Observations Bring Some Surprises". *The New York Times*. 1992-01-14. Retrieved 2010-04-26.

[5] Gary Steigman (December 2007). "Primordial Nucleosynthesis in the Precision Cosmology Era". *Annual Review of Nuclear and Particle Science*: 463–491. arXiv:0712.1100. Bibcode:2007ARNPS..57..463S. doi:10.1146/annurev.nucl.56.080805.140437.

[6] Bertulani, Carlos A. (2013). *Nuclei in the Cosmos*. World Scientific. ISBN 978-981-4417-66-2.

[7] Weiss, Achim. "Equilibrium and change: The physics behind Big Bang Nucleosynthesis". *Einstein Online*. Archived from the original on 8 February 2007. Retrieved 2007-02-24.

[8] Bludman, S. A. (December 1998). "Baryonic Mass Fraction in Rich Clusters and the Total Mass Density in the Cosmos". *Astrophysical Journal* **508** (2): 535–538. arXiv:astro-ph/9706047. Bibcode:1998ApJ...508..535B. doi:10.1086/306412.

[9] David Toback (2009). "Chapter 12: Cosmic Background Radiation"

[10] David Toback (2009). "Unit 4: The Evolution Of The Universe"

[11] R. H. Cyburt, B. D. Fields & K. A. Olive (2008). "A Bitter Pill: The Primordial Lithium Problem Worsens". arXiv:0808.2818.

[12] Weiss, Achim. "Elements of the past: Big Bang Nucleosynthesis and observation". *Einstein Online*. Archived from the original on 8 February 2007. Retrieved 2007-02-24.
For a recent calculation of BBN predictions, see A. Coc et al. (2004). "Updated Big Bang Nucleosynthesis confronted to WMAP observations and to the Abundance of Light Elements". *Astrophysical Journal* **600** (2): 544. arXiv:astro-ph/0309480. Bibcode:2004ApJ...600..544C. doi:10.1086/380121.
For the observational values, see the following articles:

- Helium-4: K. A. Olive & E. A. Skillman (2004). "A Realistic Determination of the Error on the Primordial Helium Abundance". *Astrophysical Journal* **617** (1): 29. arXiv:astro-ph/0405588. Bibcode:2004ApJ...617...29O. doi:10.1086/425170.
- Helium-3: T. M. Bania, R. T. Rood & D. S. Balser (2002). "The cosmological density of baryons from observations of 3He+ in the Milky Way". *Nature* **415** (6867): 54–7. Bibcode:2002Natur.415...54B. doi:10.1038/415054a. PMID 11780112.
- Deuterium: J. M. O'Meara et al. (2001). "The Deuterium to Hydrogen Abundance Ratio Towards a Fourth QSO: HS0105+1619". *Astrophysical Journal* **552** (2): 718. arXiv:astro-ph/0011179. Bibcode:2001ApJ...552..718O.doi:10.1.
- Lithium-7: C. Charbonnel & F. Primas (2005). "The Lithium Content of the Galactic Halo Stars". *Astronomy & Astrophysics* **442** (3): 961. arXiv:astro-ph/0505247. Bibcode:2005A&A...442..961C. doi:10.1051/0004-6361:20042491. A. Korn et al. (2006). "A probable stellar solution to the cosmological lithium discrepancy". *Nature* **442** (7103): 657–9. arXiv:astro-ph/0608201. Bibcode:2006Natur.442..657K. doi:10.1038/nature05011. PMID 16900193.

3.8 External links

3.8.1 For a general audience

- Weiss, Achim. "Big Bang Nucleosynthesis: Cooking up the first light elements". *Einstein Online*. Archived from the original on 8 February 2007. Retrieved 2007-02-24.

- White, Martin: Overview of BBN

- Wright, Ned: BBN (cosmology tutorial)

- Big Bang nucleosynthesis on arxiv.org

- Burles, Scott; Nollett, Kenneth M.; Turner, Michael S. (1999-03-19). "Big-Bang Nucleosynthesis: Linking Inner Space and Outer Space". arXiv:astro-ph/9903300.

3.8.2 Technical articles

- Burles, Scott, and Kenneth M. Nollett, Michael S. Turner (2001). "What Is The BBN Prediction for the Baryon Density and How Reliable Is It?". *Phys. Rev. D* **63** (6): 063512. arXiv:astro-ph/0008495.Bibcode:2001PhRvD..6. doi:10.1103/PhysRevD.63.063512. **Report-no**: FERMILAB-Pub-00-239-A

- Jedamzik, Karsten, "*Non-Standard Big Bang Nucleosynthesis Scenarios*". Max-Planck-Institut für Astrophysik, Garching.

- Steigman, Gary, Primordial Nucleosynthesis: Successes And Challenges arXiv:astro-ph/0511534; Forensic Cosmology: Probing Baryons and Neutrinos With BBN and the CBR arXiv:hep-ph/0309347; and Big Bang Nucleosynthesis: Probing the First 20 Minutes arXiv:astro-ph/0307244

- R. A. Alpher, H. A. Bethe, G. Gamow, *The Origin of Chemical Elements*, *Physical Review* **73** (1948), 803. The so-called $\alpha\beta\gamma$ paper, in which Alpher and Gamow suggested that the light elements were created by hydrogen ions capturing neutrons in the hot, dense early universe. Bethe's name was added for symmetry

- Gamow, G. (1948)."The Origin of Elements and the Separation of Galaxies".*Physical Review***74**: 505.Bibcode:194G. doi:10.1103/physrev.74.505.2. These two 1948 papers of Gamow laid the foundation for our present understanding of big-bang nucleosynthesis

- Gamow, G. (1948). "The Evolution of the Universe".*Nature***162**: 680.Bibcode:1948Natur.162..680G.doi:10.1038.

- Alpher, R. A. (1948). "A Neutron-Capture Theory of the Formation and Relative Abundance of the Elements". *Physical Review* **74**: 1737. Bibcode:1948PhRv...74.1737A. doi:10.1103/PhysRev.74.1737.

- R. A. Alpher and R. Herman, "On the Relative Abundance of the Elements," *Physical Review* **74** (1948), 1577. This paper contains the first estimate of the present temperature of the universe

- Alpher, R. A.; Herman, R.; Gamow, G. (1948). "Evolution of the Universe".*Nature***162**: 774.Bibcode:1948Natur.. doi:10.1038/162774b0.

- Java Big Bang element abundance calculator

Chapter 4

Abundance of the chemical elements

The **abundance** of a chemical element measures how common is the element relative to all other elements in a given environment. Abundance is measured in one of three ways: by the mass-fraction (the same as weight fraction); by the mole-fraction (fraction of atoms by numerical count, or sometimes fraction of molecules in gases); or by the volume-fraction. Volume-fraction is a common abundance measure in mixed gases such as planetary atmospheres, and is similar in value to molecular mole-fraction for gas mixtures at relatively low densities and pressures, and ideal gas mixtures. Most abundance values in this article are given as mass-fractions.

For example, the abundance of oxygen in pure water can be measured in two ways: the *mass fraction* is about 89%, because that is the fraction of water's mass which is oxygen. However, the *mole-fraction* is 33% because only 1 atom of 3 in water, H_2O, is oxygen.

As another example, looking at the *mass-fraction* abundance of hydrogen and helium in both the Universe as a whole and in the atmospheres of gas-giant planets such as Jupiter, it is 74% for hydrogen and 23-25% for helium; while the *(atomic) mole-fraction* for hydrogen is 92%, and for helium is 8%, in these environments. Changing the given environment to Jupiter's outer atmosphere, where hydrogen is diatomic while helium is not, changes the *molecular* mole-fraction (fraction of total gas molecules), as well as the fraction of atmosphere by volume, of hydrogen to about 86%, and of helium to 13%.[Note 1]

4.1 Abundance of elements in the Universe

See also: Stellar population, Cosmochemistry and Astrochemistry

The elements – that is, ordinary (baryonic) matter made of protons, neutrons, and electrons, are only a small part of the content of the Universe. Cosmological observations suggest that only 4.6% of the universe's energy (including the mass contributed by energy, $E = mc^2 \leftrightarrow m = E/c^2$) comprises the visible baryonic matter that constitutes stars, planets, and living beings. The rest is made up of dark energy (72%) and dark matter (23%).[2] These are forms of matter and energy believed to exist on the basis of scientific theory and observational deductions, but they have not been directly observed and their nature is not well understood.

Most standard (baryonic) matter is found in stars and interstellar clouds, in the form of atoms or ions (plasma), although it can be found in degenerate forms in extreme astrophysical settings, such as the high densities inside white dwarfs and neutron stars.

Hydrogen is the most abundant element in the Universe; helium is second. However, after this, the rank of abundance does not continue to correspond to the atomic number; oxygen has abundance rank 3, but atomic number 8. All others are substantially less common.

The abundance of the lightest elements is well predicted by the standard cosmological model, since they were mostly produced shortly (i.e., within a few hundred seconds) after the Big Bang, in a process known as Big Bang nucleosynthesis.

Heavier elements were mostly produced much later, inside of stars.

Hydrogen and helium are estimated to make up roughly 74% and 24% of all baryonic matter in the universe respectively. Despite comprising only a very small fraction of the universe, the remaining "heavy elements" can greatly influence astronomical phenomena. Only about 2% (by mass) of the Milky Way galaxy's disk is composed of heavy elements.

These other elements are generated by stellar processes.[3][4][5] In astronomy, a "metal" is any element other than hydrogen or helium. This distinction is significant because hydrogen and helium are the only elements that were produced in significant quantities in the Big Bang. Thus, the metallicity of a galaxy or other object is an indication of stellar activity, after the Big Bang.

The following graph (note log scale) shows abundance of elements in our solar system. The table shows the twelve most common elements in our galaxy (estimated spectroscopically), as measured in parts per million, by mass.[1] Nearby galaxies that have evolved along similar lines have a corresponding enrichment of elements heavier than hydrogen and helium. The more distant galaxies are being viewed as they appeared in the past, so their abundances of elements appear closer to the primordial mixture. Since physical laws and processes are uniform throughout the universe, however, it is expected that these galaxies will likewise have evolved similar abundances of elements.

The abundance of elements in the Solar System (see graph) is in keeping with their origin from the Big Bang and nucleosynthesis in a number of progenitor supernova stars. Very abundant hydrogen and helium are products of the Big Bang, while the next three elements are rare since they had little time to form in the Big Bang and are not made in stars (they are, however, produced in small quantities by breakup of heavier elements in interstellar dust, as a result of impact by cosmic rays).

Beginning with carbon, elements have been produced in stars by buildup from alpha particles (helium nuclei), resulting in an alternatingly larger abundance of elements with even atomic numbers (these are also more stable). The effect of odd-numbered chemical elements generally being more rare in the universe was empirically noticed in 1914, and is known as the Oddo-Harkins rule. After hydrogen, these effects cause aluminum to be the most common odd-numbered element in the universe.

Cosmogenesis: In general, such elements up to iron are made in large stars in the process of becoming supernovae. Iron-56 is particularly common, since it is the most stable element that can easily be made from alpha particles (being a product of decay of radioactive nickel-56, ultimately made from 14 helium nuclei). Elements heavier than iron are made in energy-absorbing processes in large stars, and their abundance in the universe (and on Earth) generally decreases with increasing atomic number.

4.1.1 Elemental abundance and nuclear binding energy

Loose correlations have been observed between estimated elemental abundances in the universe and the nuclear binding energy curve. Roughly speaking, the relative stability of various atomic isotopes has exerted a strong influence on the relative abundance of elements formed in the Big Bang, and during the development of the universe thereafter. [7] See the article about nucleosynthesis for the explanation on how certain nuclear fusion processes in stars (such as carbon burning, etc.) create the elements heavier than hydrogen and helium.

A further observed peculiarity is the jagged alternation between relative abundance and scarcity of adjacent atomic numbers in the elemental abundance curve, and a similar pattern of energy levels in the nuclear binding energy curve. This alternation is caused by the higher relative binding energy (corresponding to relative stability) of even atomic numbers compared to odd atomic numbers, and is explained by the Pauli Exclusion Principle.[8] The semi-empirical mass formula (SEMF), also called **Weizsäcker's formula** or the **Bethe-Weizsäcker mass formula**, gives a theoretical explanation of the overall shape of the curve of nuclear binding energy.[9]

4.2 Abundance of elements in the Earth

See also: Earth § Chemical composition

The Earth formed from the same cloud of matter that formed the Sun, but the planets acquired different compositions during the formation and evolution of the solar system. In turn, the natural history of the Earth caused parts of this planet to have differing concentrations of the elements.

The mass of the Earth is approximately 5.98×10^{24} kg. In bulk, by mass, it is composed mostly of iron (32.1%), oxygen (30.1%), silicon (15.1%), magnesium (13.9%), sulfur (2.9%), nickel (1.8%), calcium (1.5%), and aluminium (1.4%); with the remaining 1.2% consisting of trace amounts of other elements.[10]

The bulk composition of the Earth by elemental-mass is roughly similar to the gross composition of the solar system, with the major differences being that Earth is missing a great deal of the volatile elements hydrogen, helium, neon, and nitrogen, as well as carbon which has been lost as volatile hydrocarbons. The remaining elemental composition is roughly typical of the "rocky" inner planets, which formed in the thermal zone where solar heat drove volatile compounds into space. The Earth retains oxygen as the the second-largest component of its mass (and largest atomic-fraction), mainly from this element being retained in silicate minerals which have a very high melting point and low vapor pressure.

4.2.1 Earth's detailed bulk (total) elemental abundance in table form

Click "show" at right, to show more numerical values in a full table. Note that these are ordered by atom-fraction abundance (right-most column), not mass-abundance.

An estimate[11] of the elemental abundances in the total mass of the Earth. Note that numbers are estimates, and they will vary depending on source and method of estimation. Order of magnitude of data can roughly be relied upon. ppb (atoms) is parts per billion, meaning that is the number of atoms of a given element in every billion atoms in the Earth.

4.2.2 Earth's crustal elemental abundance

Main article: Abundance of elements in Earth's crust
The mass-abundance of the nine most abundant elements in the Earth's crust (see main article above) is approximately: oxygen 46%, silicon 28%, aluminum 8.2%, iron 5.6%, calcium 4.2%, sodium 2.5%, magnesium 2.4%, potassium, 2.0%, and titanium 0.61%. Other elements occur at less than 0.15%.

The graph at left illustrates the relative atomic-abundance of the chemical elements in Earth's upper continental crust, which is relatively accessible for measurements and estimation. Many of the elements shown in the graph are classified into (partially overlapping) categories:

1. rock-forming elements (major elements in green field, and minor elements in light green field);

2. rare earth elements (lanthanides, La-Lu, and Y; labeled in blue);

3. major industrial metals (global production >~3×10^7 kg/year; labeled in red);

4. precious metals (labeled in purple);

5. the nine rarest "metals" — the six platinum group elements plus Au, Re, and Te (a metalloid) — in the yellow field.

Note that there are two breaks where the unstable elements technetium (atomic number: 43) and promethium (atomic number: 61) would be. These are both extremely rare, since on Earth they are only produced through the spontaneous fission of very heavy radioactive elements (for example, uranium, thorium, or the trace amounts of plutonium that exist in uranium ores), or by the interaction of certain other elements with cosmic rays. Both of the first two of these elements have been identified spectroscopically in the atmospheres of stars, where they are produced by ongoing nucleosynthetic processes. There are also breaks where the six noble gases would be, since they are not chemically bound in the Earth's crust, and they are only generated by decay chains from radioactive elements and are therefore extremely rare there. The twelve naturally occurring very rare, highly radioactive elements (polonium, astatine, francium, radium, actinium, protactinium, neptunium, plutonium, americium, curium, berkelium, and californium) are not included, since any of these elements that were present at the formation of the Earth have decayed away eons ago, and their quantity today is negligible and is only produced from the radioactive decay of uranium and thorium.

Oxygen and silicon are notably quite common elements in the crust. They have frequently combined with each other to form common silicate minerals.

Crustal rare-earth elemental abundance

"Rare" earth elements is a historical misnomer. The persistence of the term reflects unfamiliarity rather than true rarity. The more abundant rare earth elements are each similar in crustal concentration to commonplace industrial metals such as chromium, nickel, copper, zinc, molybdenum, tin, tungsten, or lead. The two least abundant rare earth elements (thulium and lutetium) are nearly 200 times more common than gold. However, in contrast to the ordinary base and precious metals, rare earth elements have very little tendency to become concentrated in exploitable ore deposits. Consequently, most of the world's supply of rare earth elements comes from only a handful of sources. Furthermore, the rare earth metals are all quite chemically similar to each other, and they are thus quite difficult to separate into quantities of the pure elements.

Differences in abundances of individual rare earth elements in the upper continental crust of the Earth represent the superposition of two effects, one nuclear and one geochemical. First, the rare earth elements with even atomic numbers ($_{58}$Ce, $_{60}$Nd, ...) have greater cosmic and terrestrial abundances than the adjacent rare earth elements with odd atomic numbers ($_{57}$La, $_{59}$Pr, ...). Second, the lighter rare earth elements are more incompatible (because they have larger ionic radii) and therefore more strongly concentrated in the continental crust than the heavier rare earth elements. In most rare earth ore deposits, the first four rare earth elements – lanthanum, cerium, praseodymium, and neodymium – constitute 80% to 99% of the total amount of rare earth metal that can be found in the ore.

4.2.3 Earth's mantle elemental abundance

Main article: Mantle (geology)

The mass-abundance of the eight most abundant elements in the Earth's crust (see main article above) is approximately: oxygen 45%, magnesium 23%, silicon 22%, iron 5.8%, calcium 2.3%, aluminum 2.2%, sodium 0.3%, potassium 0.3%.

The mantle differs in elemental composition from the crust in having a great deal more magnesium and significantly more iron, while having much less aluminum and sodium.

4.2.4 Earth's core elemental abundance

Due to mass segregation, the core of the Earth is believed to be primarily composed of iron (88.8%), with smaller amounts of nickel (5.8%), sulfur (4.5%), and less than 1% trace elements.[10]

4.2.5 Oceanic elemental abundance

For a complete list of the abundance of elements in the ocean, see Abundances of the elements (data page)#Sea water.

4.2.6 Atmospheric elemental abundance

The order of elements by volume-fraction (which is approximately molecular mole-fraction) in the atmosphere is nitrogen (78.1%), oxygen (20.9%),[12] argon (0.96%), followed by (in uncertain order) carbon and hydrogen because water vapor and carbon dioxide, which represent most of these two elements in the air, are variable components. Sulfur, phosphorus, and all other elements are present in significantly lower proportions.

According to the abundance curve graph (above right), argon, a significant if not major component of the atmosphere, does not appear in the crust at all. This is because the atmosphere has a far smaller mass than the crust, so argon remaining in the crust contributes little to mass-fraction there, while at the same time buildup of argon in the atmosphere has become large enough to be significant.

4.2.7 Abundances of elements in urban soils

For a complete list of the abundance of elements in urban soils, see Abundances of the elements (data page)#Urban soils.

Reasons for establishing

In the time of life existence, or at least in the time of the existence of human beings, the abundances of chemical elements within the Earth's crust have not been changed dramatically due to migration and concentration processes except the radioactive elements and their decay products and also noble gases. However, significant changes took place in the distribution of chemical elements. But within the biosphere not only the distribution, but also the abundances of elements have changed during the last centuries.

The rate of a number of geochemical changes taking place during the last decades in the biosphere has become catastrophically high. Such changes are often connected with human activities. To study these changes and to make better informed decisions on diminishing their adverse impact on living organisms, and especially on people, it is necessary to estimate the contemporary abundances of chemical elements in geochemical systems susceptible to the highest anthropogenic impact and having a significant effect on the development and existence of living organisms. One of such systems is the soil of urban landscapes. Settlements occupy less than 10% of the land area, but virtually the entire population of the planet lives within them. The main deposing medium in cities is soil, which ecological and geochemical conditions largely determine the life safety of citizens. So that, one of the priority tasks of the environmental geochemistry is to establish the average contents (abundances) of chemical elements in the soils of settlements.

Methods and results

The geochemical properties of urban soils from more than 300 cities in Europe, Asia, Africa, Australia, and America were evaluated.[13] In each settlement samples were collected uniformly throughout the territory, covering residential, industrial, recreational and other urban areas. The sampling was carried out directly from the soil surface and specifically traversed pits, ditches and wells from the upper soil horizon. The number of samples in each locality ranged from 30 to 1000. The published data and the materials kindly provided by a number of geochemists were also incorporated into the research. Considering the great importance of the defined contents, quantitative and quantitative emission spectral, gravimetric, X-ray fluorescence, and partly neutron activation analyses were carried out in parallel approximately in the samples. In a volume of 3–5% of the total number of samples, sampling and analyses of the inner and external controls were conducted. Calculation of random errors and systematic errors allowed to consider the sampling and analytical laboratory work as good.

For every city the average concentrations of elements in soils were determined. To avoid the errors related to unequal number of samples, each city was then represented by only one "averaged" sample. The statistical processing of this data allowed to calculate the average concentrations, which can be considered as the abundances of chemical elements in urban soils.

This graph illustrates the relative abundance of the chemical elements in urban soils, irregularly decreasing in proportion with the increasing atomic masses. Therefore, the evolution of organisms in this system occurs in the conditions of light elements' prevalence. It corresponds to the conditions of the evolutional development of the living matter on the Earth. The irregularity of element decreasing may be somewhat connected, as stated above, with the technogenic influence. The Oddo-Harkins rule, which holds that elements with an even atomic number are more common than elements with an odd atomic number, is saved in the urban soils but with some technogenic complications. Among the considered abundances the even-atomic elements make 91.48% of the urban soils mass. As it is in the Earth's crust, elements with the 4-divisible atomic masses of leading isotope (oxygen — 16, silicon — 28, calcium — 40, carbon — 12, iron — 56) are sharply prevailing in urban soils.

In spite of significant differences between abundances of several elements in urban soils and those values calculated for the

Earth's crust, the general patterns of element abundances in urban soils repeat those in the Earth's crust in a great measure. The established abundances of chemical elements in urban soils can be considered as their geochemical (ecological and geochemical) characteristic, reflecting the combined impact of technogenic and natural processes occurring during certain time period (the end of the 20th century–beginning of the 21st century). With the development of science and technology the abundances may gradually change. The rate of these changes is still poorly predictable. The abundances of chemical elements may be used during various ecological and geochemical studies.

4.3 Human body elemental abundance

Main article: Chemical makeup of the human body

By mass, human cells consist of 65–90% water (H_2O), and a significant portion of the remainder is composed of carbon-containing organic molecules. Oxygen therefore contributes a majority of a human body's mass, followed by carbon. Almost 99% of the mass of the human body is made up of six elements: oxygen, carbon, hydrogen, nitrogen, calcium, and phosphorus. The next 0.75% is made up of the next five elements: potassium, sulfur, chlorine, sodium, and magnesium. Only 17 elements are known for certain to be necessary to human life, with one additional element (fluorine) thought to be helpful for tooth enamel strength. A few more trace elements may play some role in the health of mammals. Boron and silicon are notably necessary for plants but have uncertain roles in animals. The elements aluminium and silicon, although very common in the earth's crust, are conspicuously rare in the human body.[14]

Periodic table highlighting nutritional elements[15]

Periodic table highlighting dietary elements

4.4 See also

- Abundances of the elements (data page)
- Natural abundance (isotopic abundance)
- Primordial nuclide

4.5 References

4.5.1 Footnotes

[1] Croswell, Ken (February 1996). *Alchemy of the Heavens*. Anchor. ISBN 0-385-47214-5.

[2] WMAP- Content of the Universe

[3] Suess, Hans; Urey, Harold (1956). "Abundances of the Elements".*Reviews of Modern Physics***28**: 53.Bibcode:1956RvMP...28....5. doi:10.1103/RevModPhys.28.53.

[4] Cameron, A.G.W. (1973). "Abundances of the elements in the solar system".*Space Science Reviews***15**: 121.Bibcode:1973SSRv... doi:10.1007/BF00172440.

[5] Anders, E; Ebihara, M (1982). "Solar-system abundances of the elements". *Geochimica et Cosmochimica Acta* **46** (11): 2363. Bibcode:1982GeCoA..46.2363A. doi:10.1016/0016-7037(82)90208-3.

[6] Arnett, David (1996). *Supernovae and Nucleosynthesis* (First ed.). Princeton, New Jersey: Princeton University Press. ISBN 0-691-01147-8. OCLC 33162440.

[7] Bell, Jerry A.; GenChem Editorial/Writing Team (2005). "Chapter 3: Origin of Atoms". *Chemistry: a project of the American Chemical Society*. New York [u.a.]: Freeman. pp. 191–193. ISBN 978-0-7167-3126-9. Correlations between abundance and nuclear binding energy [Subsection title]

[8] Bell, Jerry A.; GenChem Editorial/Writing Team (2005). "Chapter 3: Origin of Atoms". *Chemistry: a project of the American Chemical Society*. New York [u.a.]: Freeman. p. 192. ISBN 978-0-7167-3126-9. The higher abundance of elements with even atomic numbers [Subsection title]

[9] Bailey, David. "Semi-empirical Nuclear Mass Formula". *PHY357: Strings & Binding Energy*. University of Toronto. Retrieved 2011-03-31.

[10] Morgan, J. W.; Anders, E. (1980). "Chemical composition of Earth, Venus, and Mercury". *Proceedings of the National Academy of Sciences* **77** (12): 6973–6977. Bibcode:1980PNAS...77.6973M. doi:10.1073/pnas.77.12.6973. PMC 350422. PMID 16592930.

[11] William F McDonough The composition of the Earth. quake.mit.edu

[12] Zimmer, Carl (3 October 2013). "Earth's Oxygen: A Mystery Easy to Take for Granted". *New York Times*. Retrieved 3 October 2013.

[13] Vladimir Alekseenko; Alexey Alekseenko (2014). "The abundances of chemical elements in urban soils". *Journal of Geochemical Exploration* (Elsevier B.V.) **147**: 245–249. doi:10.1016/j.gexplo.2014.08.003. ISSN 0375-6742.

[14] Table data from Chang, Raymond (2007). *Chemistry, Ninth Edition*. McGraw-Hill. p. 52. ISBN 0-07-110595-6.

[15] Ultratrace minerals. Authors: Nielsen, Forrest H. USDA, ARS Source: Modern nutrition in health and disease / editors, Maurice E. Shils ... et al.. Baltimore : Williams & Wilkins, c1999., p. 283-303. Issue Date: 1999 URI:

4.5.2 Notes

[1] Below Jupiter's outer atmosphere, volume fractions are significantly different from mole fractions due to high temperatures (ionization and disproportionation) and high density where the Ideal Gas Law is inapplicable.

4.5.3 Notations

- http://geopubs.wr.usgs.gov/fact-sheet/fs087-02/
- http://imagine.gsfc.nasa.gov/docs/dict_ei.html

4.6 External links

- List of elements in order of abundance in the Earth's crust (only correct for the twenty most common elements)
- Cosmic abundance of the elements and nucleosynthesis
- webelements.com Lists of elemental abundances for the Universe, Sun, meteorites, Earth, ocean, streamwater

Atoms 4.6%

Dark Matter 23%

Dark Energy 72%

TODAY

Neutrinos 10%

Photons 15%

Atoms 12%

Dark Matter 63%

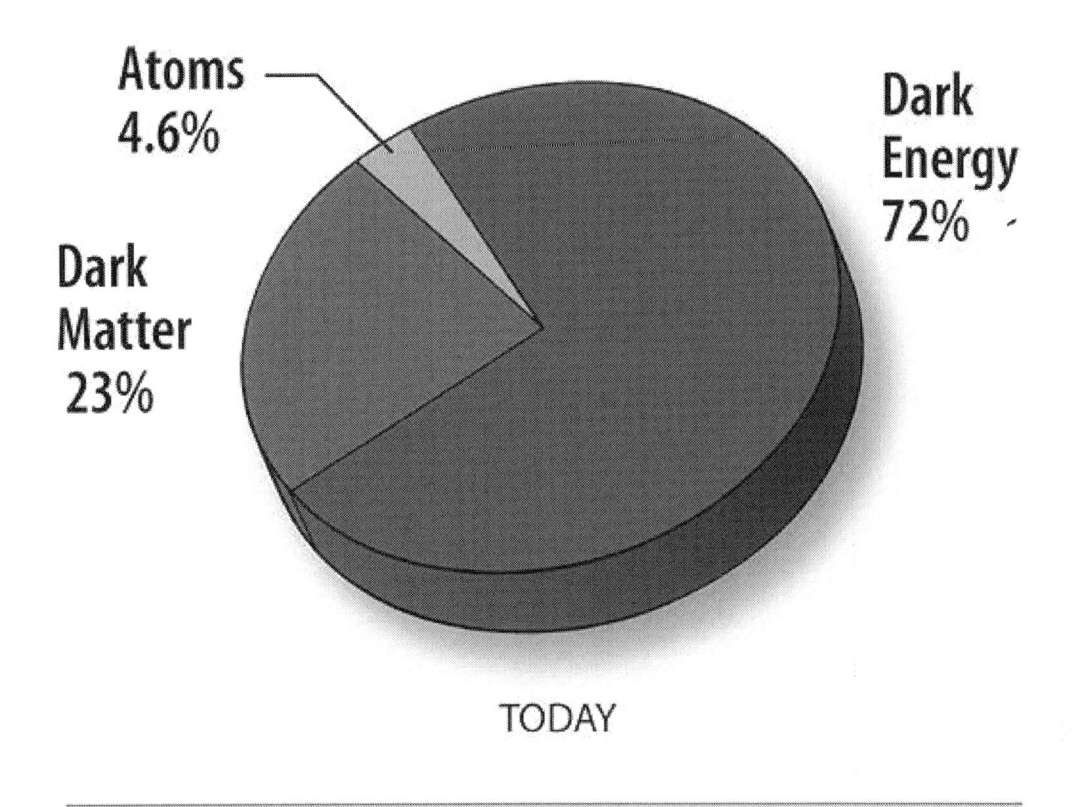

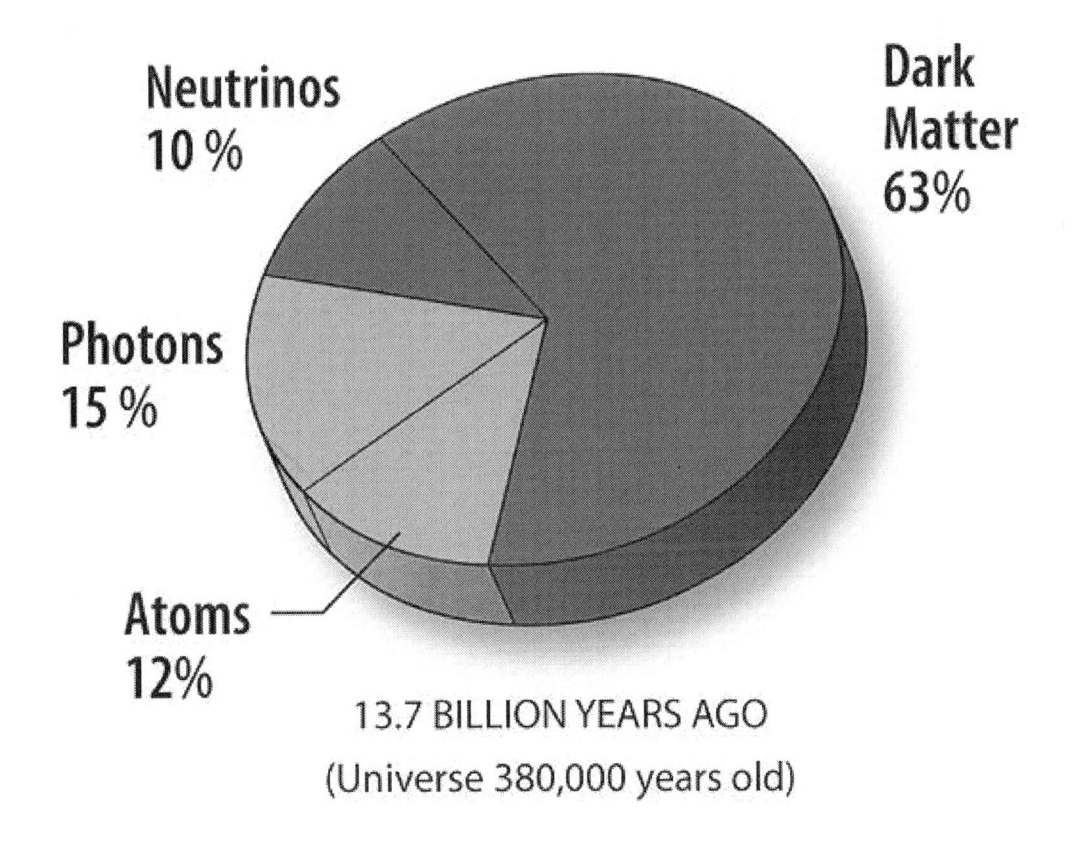

13.7 BILLION YEARS AGO
(Universe 380,000 years old)

Estimated proportions of matter, dark matter and dark energy in the universe. Only the fraction of the mass and energy in the universe labeled "atoms" is composed of chemical elements.

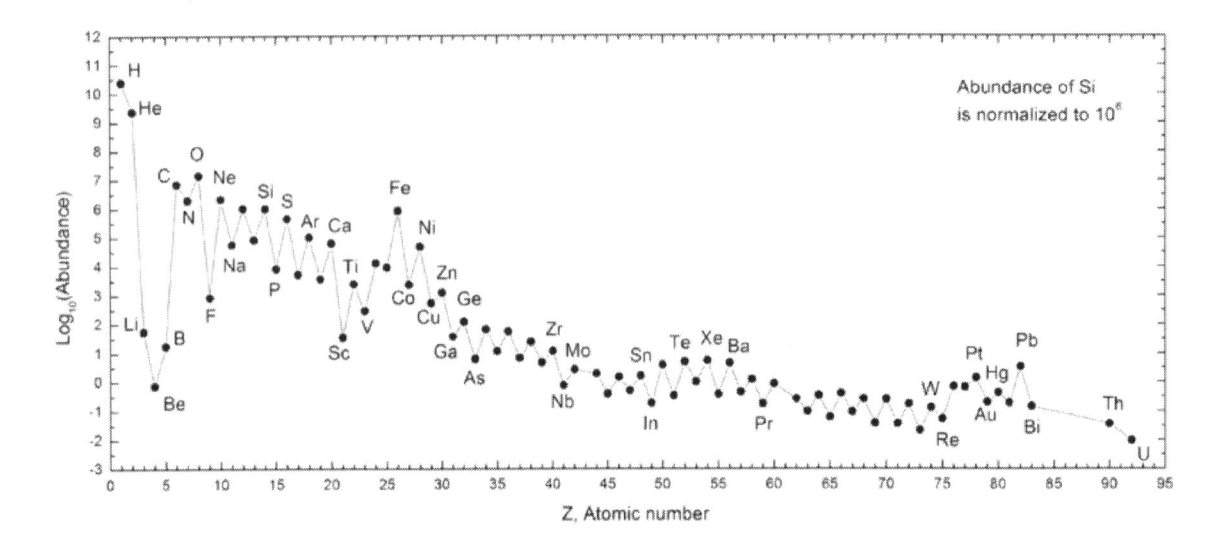

*Estimated abundances of the chemical **elements in the Solar system**. Hydrogen and helium are most common, from the Big Bang. The next three elements (Li, Be, B) are rare because they are poorly synthesized in the Big Bang and also in stars. The two general trends in the remaining stellar-produced elements are: (1) an alternation of abundance in elements as they have even or odd atomic numbers (the Oddo-Harkins rule), and (2) a general decrease in abundance, as elements become heavier. Iron is especially common because it represents the minimum energy nuclide that can be made by fusion of helium in supernovae.*

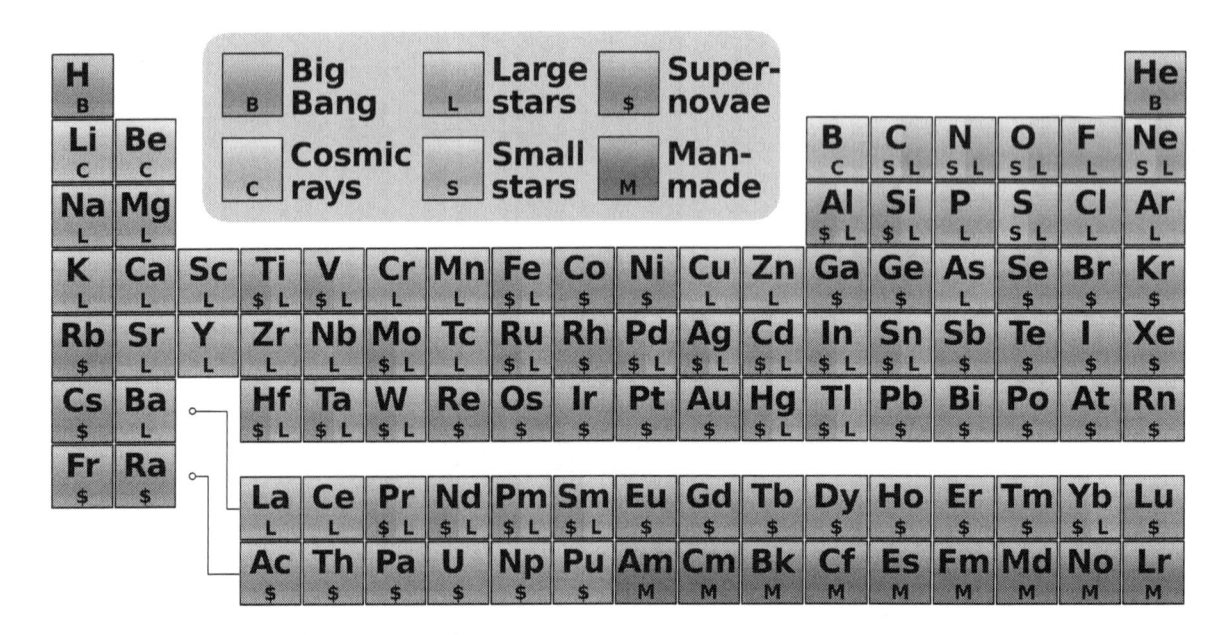

Periodic table showing the cosmogenic origin of each element

4.6. EXTERNAL LINKS

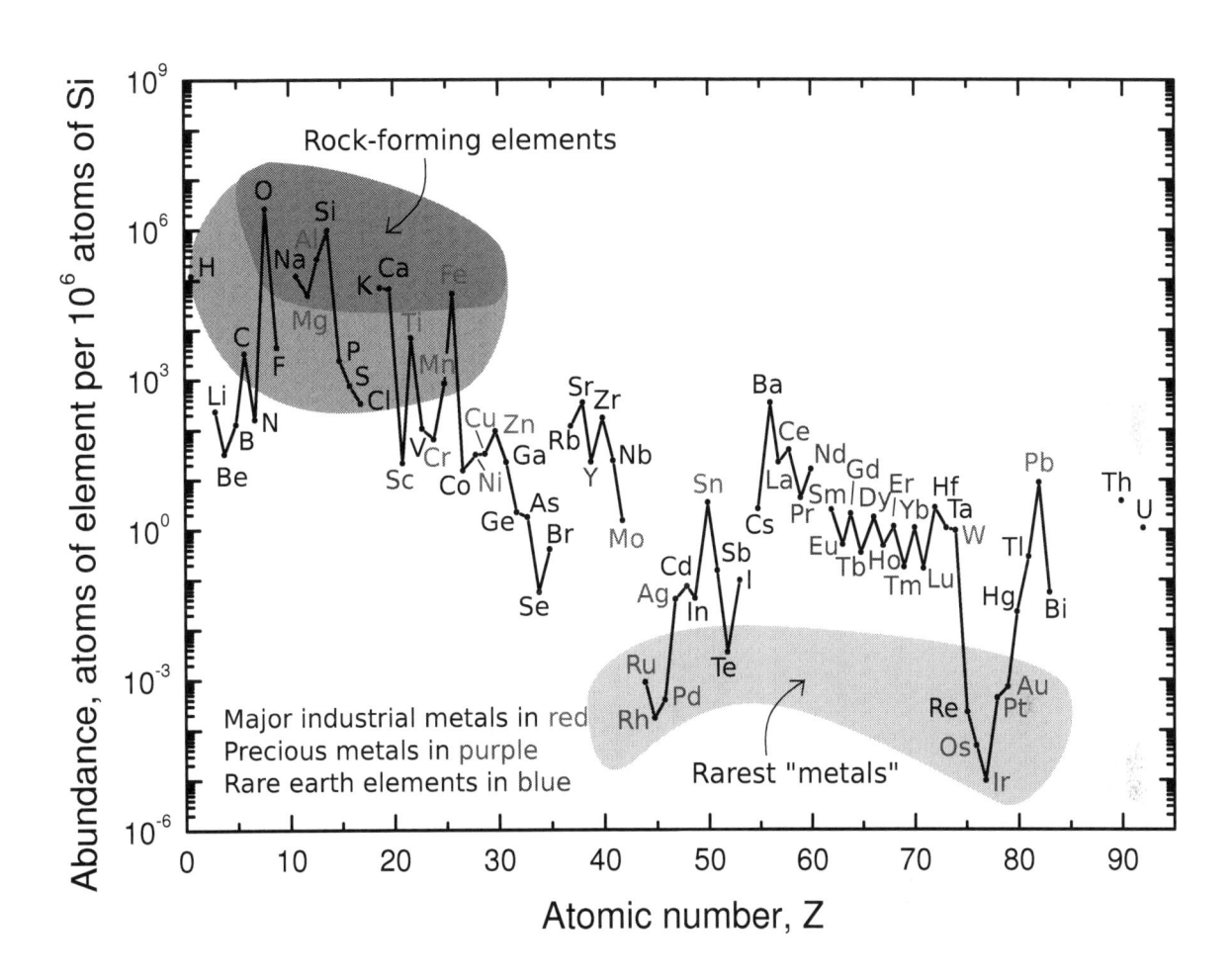

Abundance (atom fraction) of the chemical elements in Earth's upper continental crust as a function of atomic number. The rarest elements in the crust (shown in yellow) are the most dense. They were further rarefied in the crust by being siderophile (iron-loving) elements, in the Goldschmidt classification of elements. Siderophiles were depleted by being relocated into the Earth's core. Their abundance in meteoroid materials is relatively higher. Additionally, tellurium and selenium have been depleted from the crust due to formation of volatile hydrides.

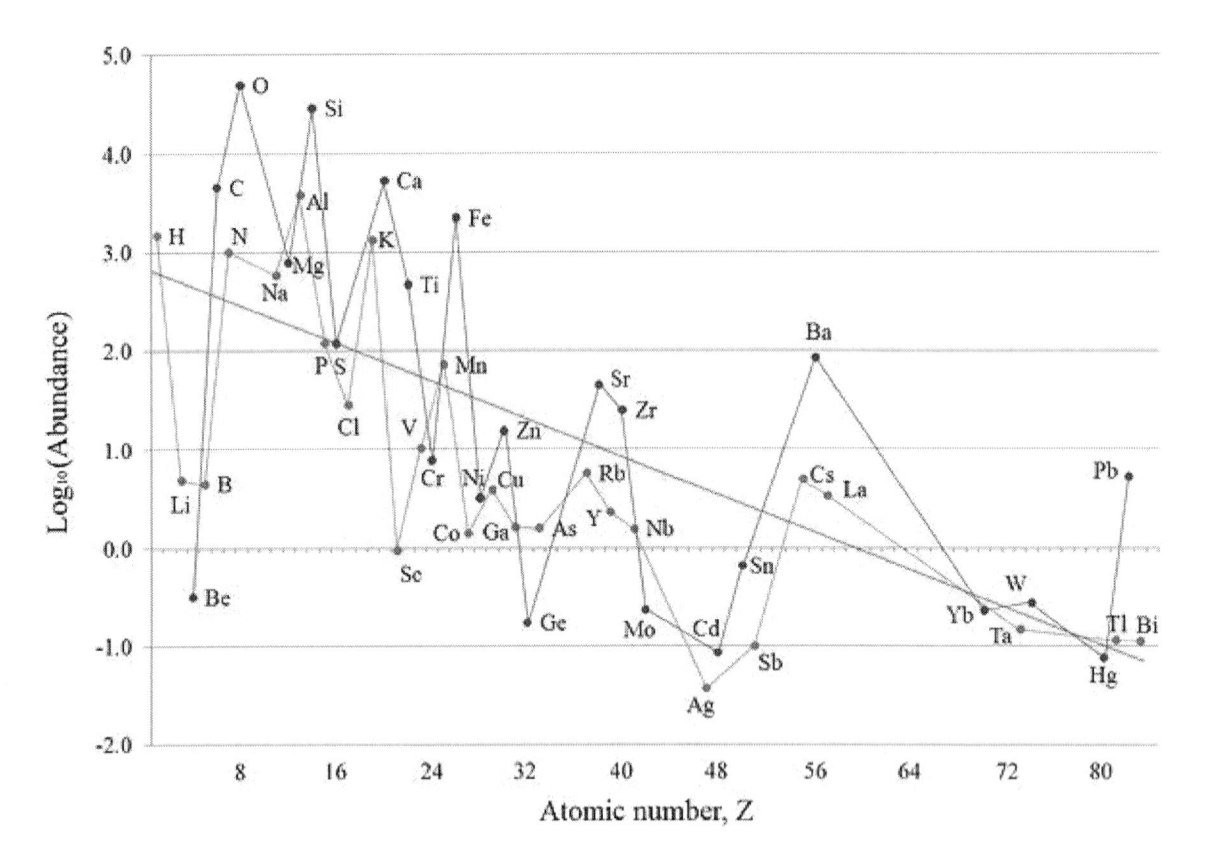

64　　　　　　　　　　　　　　　　　CHAPTER 4. ABUNDANCE OF THE CHEMICAL ELEMENTS

The half-logarithm graph of the abundances of chemical elements in urban soils. *(Alekseenko and Alekseenko, 2014) Chemical elements are distributed extremely irregularly in urban soils, what is also typical for the Earth's crust. Nine elements (O, Si, Ca, C, Al, Fe, H, K, N) make the 97.68% of the considering geochemical system (urban soils). These elements and also Zn, Sr, Zr, Ba, and Pb essentially prevail over the trend line. Part of them could be considered as "inherited" from the concentrations in the Earth's crust; another part is explained as a result of intensive technogenic activity in the cities.*

Chapter 5

Observable universe

The **observable universe** consists of the galaxies and other matter that can, in principle, be observed from Earth at the present time because light and other signals from these objects has had time to reach the Earth since the beginning of the cosmological expansion. Assuming the universe is isotropic, the distance to the edge of the observable universe is roughly the same in every direction. That is, the observable universe is a spherical volume (a ball) centered on the observer. Every location in the Universe has its own observable universe, which may or may not overlap with the one centered on Earth.

The word *observable* used in this sense does not depend on whether modern technology actually permits detection of radiation from an object in this region (or indeed on whether there is any radiation to detect). It simply indicates that it is possible *in principle* for light or other signals from the object to reach an observer on Earth. In practice, we can see light only from as far back as the time of photon decoupling in the recombination epoch. That is when particles were first able to emit photons that were not quickly re-absorbed by other particles. Before then, the Universe was filled with a plasma that was opaque to photons.

The surface of last scattering is the collection of points in space at the exact distance that photons from the time of photon decoupling just reach us today. These are the photons we detect today as cosmic microwave background radiation (CMBR). However, with future technology, it may be possible to observe the still older relic neutrino background, or even more distant events via gravitational waves (which also should move at the speed of light). Sometimes astrophysicists distinguish between the *visible* universe, which includes only signals emitted since recombination—and the *observable* universe, which includes signals since the beginning of the cosmological expansion (the Big Bang in traditional cosmology, the end of the inflationary epoch in modern cosmology). According to calculations, the *comoving distance* (current proper distance) to particles from the CMBR, which represent the radius of the visible universe, is about 14.0 billion parsecs (about 45.7 billion light years), while the comoving distance to the edge of the observable universe is about 14.3 billion parsecs (about 46.6 billion light years),[7] about 2% larger.

The best estimate of the age of the universe as of 2013 is 13.798±0.037 billion years[5] but due to the expansion of space humans are observing objects that were originally much closer but are now considerably farther away (as defined in terms of cosmological proper distance, which is equal to the comoving distance at the present time) than a static 13.8 billion light-years distance.[8] It is estimated that the diameter of the observable universe is about 28 gigaparsecs (93 billion light-years, 8.8×10^{26} metres or 5.5×10^{23} miles),[9] putting the edge of the observable universe at about 46–47 billion light-years away.[10][11]

5.1 The Universe versus the observable universe

Some parts of the Universe are too far away for the light emitted since the Big Bang to have had enough time to reach Earth, so these portions of the Universe lie outside the observable universe. In the future, light from distant galaxies will have had more time to travel, so additional regions will become observable. However, due to Hubble's law regions sufficiently distant from us are expanding away from us faster than the speed of light (special relativity prevents nearby objects in the same local region from moving faster than the speed of light with respect to each other, but there is no such constraint for distant objects when the space between them is expanding; see uses of the proper distance for a discussion) and furthermore the expansion rate appears to be accelerating due to dark energy. Assuming dark energy remains constant (an unchanging cosmological constant), so that the expansion rate of the Universe continues to accelerate, there is a "future visibility limit" beyond which objects will *never* enter our observable universe at any time in the infinite future, because light emitted by objects outside that limit would never reach us. (A subtlety is that, because the Hubble parameter is decreasing with time, there can be cases where a galaxy that is receding from us just a bit faster than light does emit a signal that reaches us eventually[11][12]). This future visibility limit is calculated at a comoving distance of 19 billion parsecs (62 billion light years) assuming the Universe will keep expanding forever, which implies the number of galaxies that we can ever theoretically observe in the infinite future (leaving aside the issue that some may be impossible to observe in practice due to redshift, as discussed in the following paragraph) is only larger than the number currently observable by a factor of 2.36.[7]

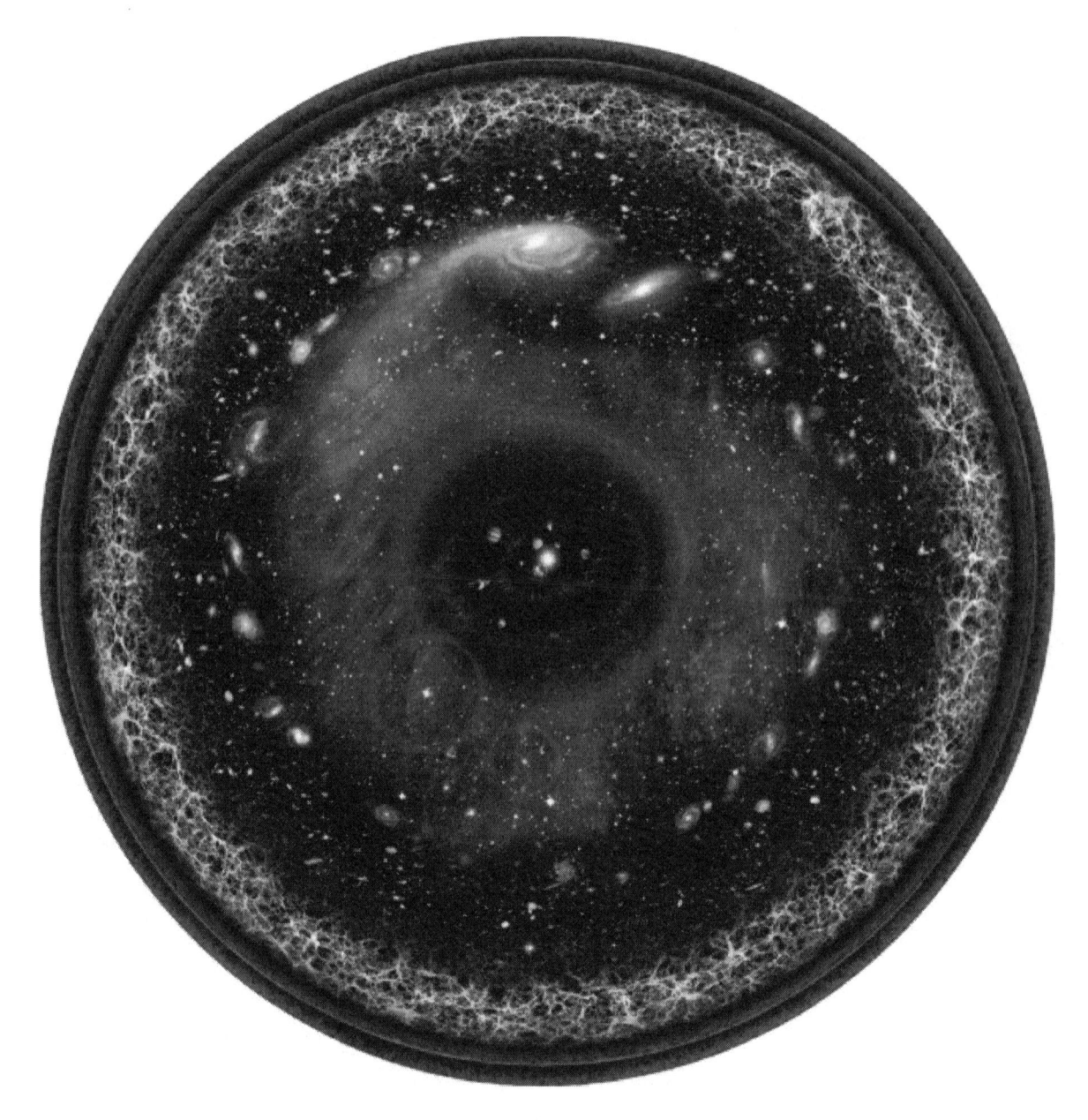

Artist's logarithmic scale conception of the observable universe with the Solar System at the center, inner and outer planets, Kuiper belt, Oort cloud, Alpha Centauri, Perseus Arm, Milky Way galaxy, Andromeda galaxy, nearby galaxies, Cosmic Web, Cosmic microwave radiation and the Big Bang's invisible plasma on the edge.

Though in principle more galaxies will become observable in the future, in practice an increasing number of galaxies will become extremely redshifted due to ongoing expansion, so much so that they will seem to disappear from view and become invisible.[13][14][15] An additional subtlety is that a galaxy at a given comoving distance is defined to lie within the "observable universe" if we can receive signals emitted by the galaxy at any age in its past history (say, a signal sent from the galaxy only 500 million years after the Big Bang), but because of the Universe's expansion, there may be some later age at which a signal sent from the same galaxy can *never* reach us at any point in the infinite future (so for example we might never see what the galaxy looked like 10 billion years after the Big Bang),[16] even though it remains at the same comoving distance (comoving distance is defined to be constant with time—unlike proper distance, which is used to define recession velocity due to the expansion of space), which is less than the comoving radius of the observable universe. This fact can be used to define a type of cosmic event horizon whose distance from us changes over time. For example, the current distance to this horizon is about 16 billion light years, meaning that a signal from an event happening *at present* can eventually reach us in the future if the event is less than 16 billion light years away, but the signal will never reach us if the event is more than 16 billion light years away.[11]

Both popular and professional research articles in cosmology often use the term "universe" to mean "observable universe". This can be justified on the grounds that we can never know anything by direct experimentation about any part of the Universe that is causally disconnected from us, although many credible theories require a total universe much larger than the observable universe. No evidence exists to suggest that the boundary of the observable universe constitutes a boundary on the Universe as a whole, nor do any of the mainstream cosmological models propose that the Universe has any physical boundary in the first place, though some models propose it could be finite but unbounded, like a higher-dimensional analogue of the 2D surface of a sphere that is finite in area but has no edge. It is plausible that the galaxies within our observable universe represent only a minuscule fraction of the galaxies in the Universe. According to the theory of cosmic inflation and its founder, Alan Guth, if it is assumed that inflation began about 10^{-37} seconds after the Big Bang, then with the plausible assumption that the size of the Universe at this time was approximately equal to the speed of light times its age, that would suggest that at present the entire universe's size is at least 3×10^{23} times larger than the size of the observable universe.[17] There are also lower estimates claiming that the entire universe is in excess of 250 times larger than the observable universe.[18]

If the Universe is finite but unbounded, it is also possible that the Universe is *smaller* than the observable universe. In this case, what we take to be very distant galaxies may actually be duplicate images of nearby galaxies, formed by light that has circumnavigated the Universe. It is difficult to test this hypothesis experimentally because different images of a galaxy would show different eras in its history, and consequently might appear quite different. Bielewicz et al.[19] claims to establish a lower bound of 27.9 gigaparsecs (91 billion light-years) on the diameter of the last scattering surface (since this is only a lower bound, the paper leaves open the possibility that the whole universe is much larger, even infinite). This value is based on matching-circle analysis of the WMAP 7 year data. This approach has been disputed.[20]

5.2 Size

The comoving distance from Earth to the edge of the observable universe is about 14 gigaparsecs (46 billion light years or 4.3×10^{26} meters) in any direction. The observable universe is thus a sphere with a diameter of about 29 gigaparsecs[21] (93 Gly or 8.8×10^{26} m).[22] Assuming that space is roughly flat, this size corresponds to a comoving volume of about 1.3×10^4 Gpc^3 (4.1×10^5 Gly^3 or 3.5×10^{80} m^3).

The figures quoted above are distances *now* (in cosmological time), not distances *at the time the light was emitted*. For example, the cosmic microwave background radiation that we see right now was emitted at the time of photon decoupling, estimated to have occurred about 380000 years after the Big Bang,[23][24] which occurred around 13.8 billion years ago. This radiation was emitted by matter that has, in the intervening time, mostly condensed into galaxies, and those galaxies are now calculated to be about 46 billion light-years from us.[7][11] To estimate the distance to that matter at the time the light was emitted, we may first note that according to the Friedmann–Lemaître–Robertson–Walker metric, which is used to model the expanding universe, if at the present time we receive light with a redshift of z, then the scale factor at the time the light was originally emitted is given by[25][26]

$a(t) = \frac{1}{1+z}$.

WMAP nine-year results combined with other measurements give the redshift of photon decoupling as $z = 1091.64 \pm 0.47$.[27]

Hubble Ultra-Deep Field image of a region of the observable universe (equivalent sky area size shown in bottom left corner), near the constellation Fornax. Each spot is a galaxy, consisting of billions of stars. The light from the smallest, most red-shifted galaxies originated nearly 14 billion years ago.

which implies that the scale factor at the time of photon decoupling would be $1/1092.64$. So if the matter that originally emitted the oldest CMBR photons has a *present* distance of 46 billion light years, then at the time of decoupling when the photons were originally emitted, the distance would have been only about 42 *million* light-years.

5.2.1 Misconceptions on its size

Many secondary sources have reported a wide variety of incorrect figures for the size of the visible universe. Some of these figures are listed below, with brief descriptions of possible reasons for misconceptions about them.

13.8 billion light-years The age of the universe is estimated to be 13.8 billion years. While it is commonly understood that nothing can accelerate to velocities equal to or greater than that of light, it is a common misconception that

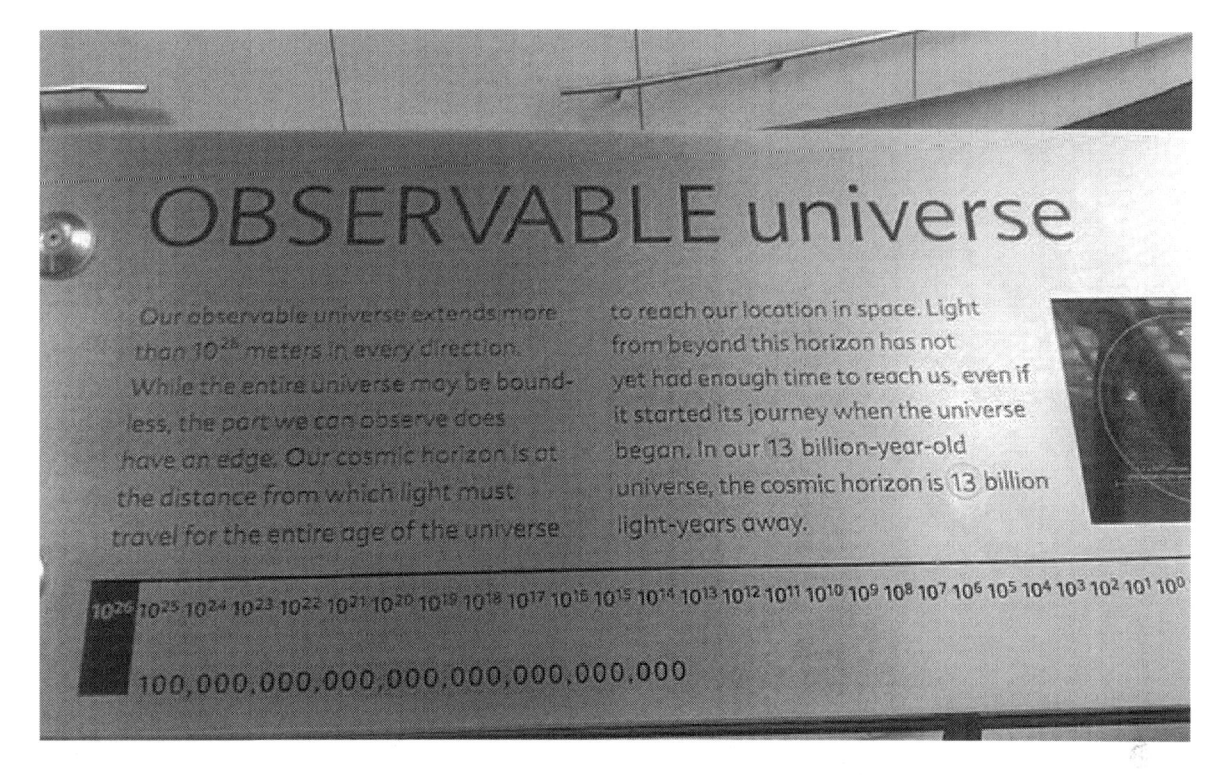

An example of one of the most common misconceptions about the size of the observable universe. Despite the fact that the universe is 13.8 billion years old, the distance to the edge of the observable universe is not 13.8 billion light-years, because the universe is expanding. This plaque appears at the Rose Center for Earth and Space in New York City.

the radius of the observable universe must therefore amount to only 13.8 billion light-years. This reasoning would only make sense if the flat, static Minkowski spacetime conception under special relativity were correct. In the real universe, spacetime is curved in a way that corresponds to the expansion of space, as evidenced by Hubble's law. Distances obtained as the speed of light multiplied by a cosmological time interval have no direct physical significance.[28]

15.8 billion light-years This is obtained in the same way as the 13.8 billion light year figure, but starting from an incorrect age of the universe that the popular press reported in mid-2006.[29][30] For an analysis of this claim and the paper that prompted it, see the following reference at the end of this article.[31]

27.6 billion light-years This is a diameter obtained from the (incorrect) radius of 13.8 billion light-years.

78 billion light-years In 2003, Cornish et al.[32] found this lower bound for the diameter of the *whole* universe (not just the observable part), if we postulate that the universe is finite in size due to its having a nontrivial topology,[33][34] with this lower bound based on the estimated current distance between points that we can see on opposite sides of the cosmic microwave background radiation (CMBR). If the whole universe is smaller than this sphere, then light has had time to circumnavigate it since the big bang, producing multiple images of distant points in the CMBR, which would show up as patterns of repeating circles.[35] Cornish et al. looked for such an effect at scales of up to 24 gigaparsecs (78 Gly or 7.4×10^{26} m) and failed to find it, and suggested that if they could extend their search to all possible orientations, they would then "be able to exclude the possibility that we live in a universe smaller than 24 Gpc in diameter". The authors also estimated that with "lower noise and higher resolution CMB maps (from WMAP's extended mission and from Planck), we will be able to search for smaller circles and extend the limit to ~28 Gpc."[32] This estimate of the maximum lower bound that can be established by future observations corresponds to a radius of 14 gigaparsecs, or around 46 billion light years, about the same as the figure for the radius of the visible universe (whose radius is defined by the CMBR sphere) given in the opening section. A 2012 preprint by most of the same authors as the Cornish et al. paper has extended the current lower bound to a diameter of 98.5% the diameter of the CMBR sphere, or about 26 Gpc.[36]

156 billion light-years This figure was obtained by doubling 78 billion light-years on the assumption that it is a radius.[37] Since 78 billion light-years is already a diameter (the original paper by Cornish et al. says, "By extending the search to all possible orientations, we will be able to exclude the possibility that we live in a universe smaller than 24 Gpc in diameter," and 24 Gpc is 78 billion light years),[32] the doubled figure is incorrect. This figure was very widely reported.[37][38][39] A press release from Montana State University – Bozeman, where Cornish works as an astrophysicist, noted the error when discussing a story that had appeared in *Discover* magazine, saying "*Discover* mistakenly reported that the universe was 156 billion light-years wide, thinking that 78 billion was the radius of the universe instead of its diameter."[40]

180 billion light-years This estimate combines the erroneous 156-billion-light-year figure with evidence that the M33 Galaxy is actually fifteen percent farther away than previous estimates and that, therefore, the Hubble constant is fifteen percent smaller.[41] The 180-billion figure is obtained by adding 15% to 156 billion light years.

5.3 Large-scale structure

Sky surveys and mappings of the various wavelength bands of electromagnetic radiation (in particular 21-cm emission) have yielded much information on the content and character of the universe's structure. The organization of structure appears to follow as a hierarchical model with organization up to the scale of superclusters and filaments. Larger than this, there seems to be no continued structure, a phenomenon that has been referred to as the *End of Greatness*.[42]

5.3.1 Walls, filaments, nodes, and voids

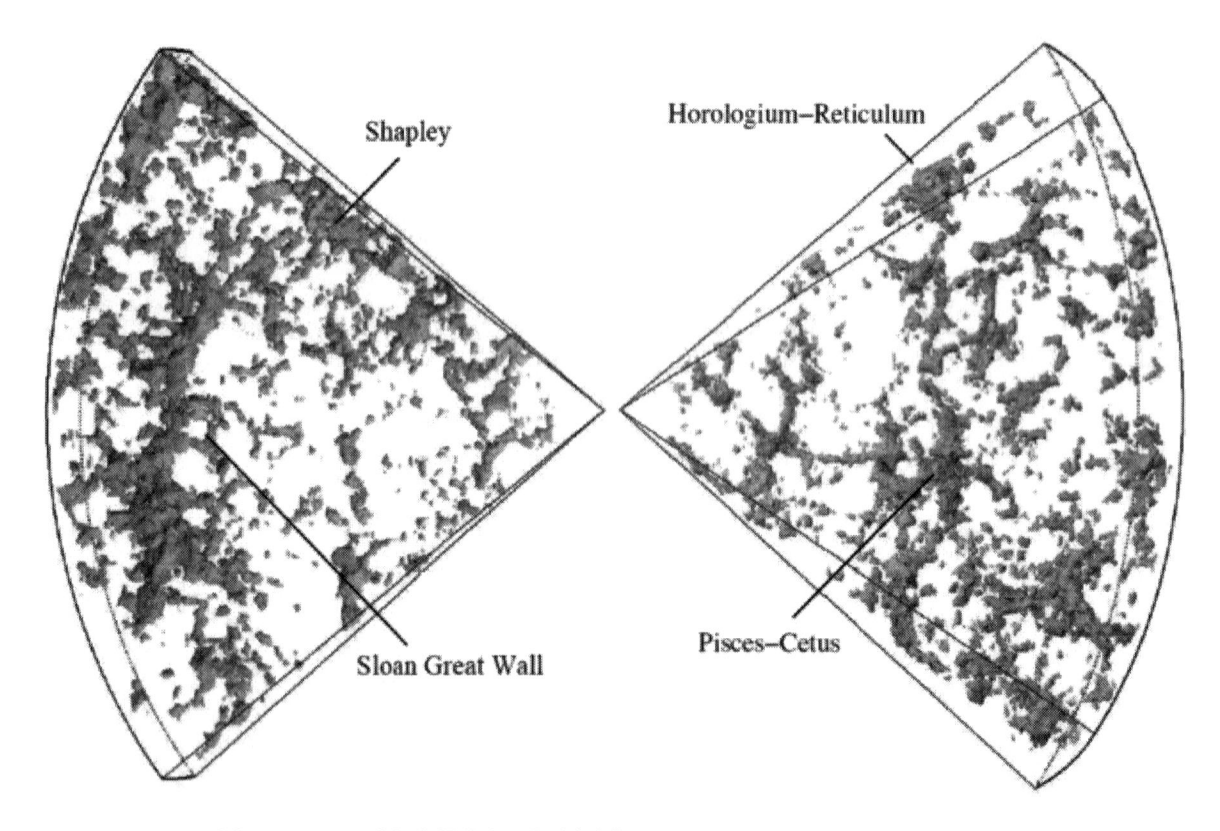

DTFE reconstruction of the inner parts of the 2dF Galaxy Redshift Survey

The organization of structure arguably begins at the stellar level, though most cosmologists rarely address astrophysics on that scale. Stars are organized into galaxies, which in turn form galaxy groups, galaxy clusters, superclusters, sheets, walls and filaments, which are separated by immense voids, creating a vast foam-like structure sometimes called the "cosmic

5.3. LARGE-SCALE STRUCTURE

web". Prior to 1989, it was commonly assumed that virialized galaxy clusters were the largest structures in existence, and that they were distributed more or less uniformly throughout the Universe in every direction. However, since the early 1980s, more and more structures have been discovered. In 1983, Adrian Webster identified the Webster LQG, a large quasar group consisting of 5 quasars. The discovery was the first identification of a large-scale structure, and has expanded the information about the known grouping of matter in the Universe. In 1987, Robert Brent Tully identified the Pisces–Cetus Supercluster Complex, the galaxy filament in which the Milky Way resides. It is about 1 billion light years across. That same year, an unusually large region with no galaxies has been discovered, the Giant Void, which measures 1.3 billion light years across. Based on redshift survey data, in 1989 Margaret Geller and John Huchra discovered the "Great Wall",[43] a sheet of galaxies more than 500 million light-years long and 200 million wide, but only 15 million light-years thick. The existence of this structure escaped notice for so long because it requires locating the position of galaxies in three dimensions, which involves combining location information about the galaxies with distance information from redshifts. Two years later, astronomers Roger G. Clowes and Luis E. Campusano discovered the Clowes–Campusano LQG, a large quasar group measuring two billion light years at its widest point, and was the largest known structure in the Universe at the time of its announcement. In April 2003, another large-scale structure was discovered, the Sloan Great Wall. In August 2007, a possible supervoid was detected in the constellation Eridanus.[44] It coincides with the 'CMB cold spot', a cold region in the microwave sky that is highly improbable under the currently favored cosmological model. This supervoid could cause the cold spot, but to do so it would have to be improbably big, possibly a billion light-years across, almost as big as the Giant Void mentioned above.

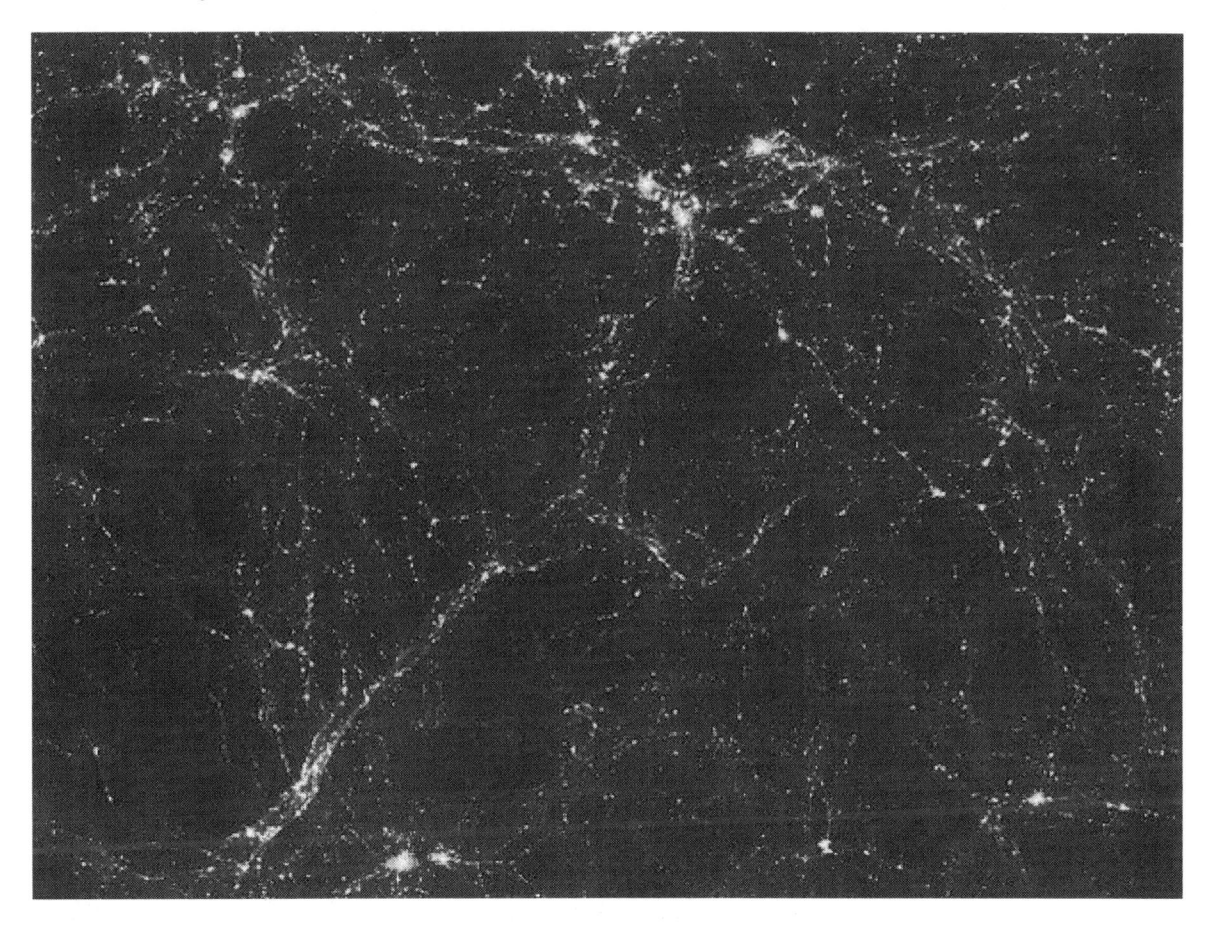

Image (computer simulated) of an area of space more than 50 million light years across, presenting a possible large-scale distribution of light sources in the universe - precise relative contributions of galaxies and quasars are unclear.

Another large-scale structure is the Newfound Blob, a collection of galaxies and enormous gas bubbles that measures about 200 million light years across.

In recent studies the Universe appears as a collection of giant bubble-like voids separated by sheets and filaments of galaxies, with the superclusters appearing as occasional relatively dense nodes. This network is clearly visible in the

2dF Galaxy Redshift Survey. In the figure, a three-dimensional reconstruction of the inner parts of the survey is shown, revealing an impressive view of the cosmic structures in the nearby universe. Several superclusters stand out, such as the Sloan Great Wall.

In 2011, a large quasar group was discovered, U1.11, measuring about 2.5 billion light years across. On January 11, 2013, another large quasar group, the Huge-LQG, was discovered, which was measured to be four billion light-years across, the largest known structure in the Universe that time.[45] In November 2013 astronomers discovered the Hercules–Corona Borealis Great Wall,[46][47] an even bigger structure twice as large as the former. It was defined by mapping of gamma-ray bursts.[46][48]

5.3.2 End of Greatness

The *End of Greatness* is an observational scale discovered at roughly 100 Mpc (roughly 300 million lightyears) where the lumpiness seen in the large-scale structure of the universe is homogenized and isotropized in accordance with the Cosmological Principle.[42] At this scale, no pseudo-random fractalness is apparent.[49] The superclusters and filaments seen in smaller surveys are randomized to the extent that the smooth distribution of the Universe is visually apparent. It was not until the redshift surveys of the 1990s were completed that this scale could accurately be observed.[42]

5.3.3 Observations

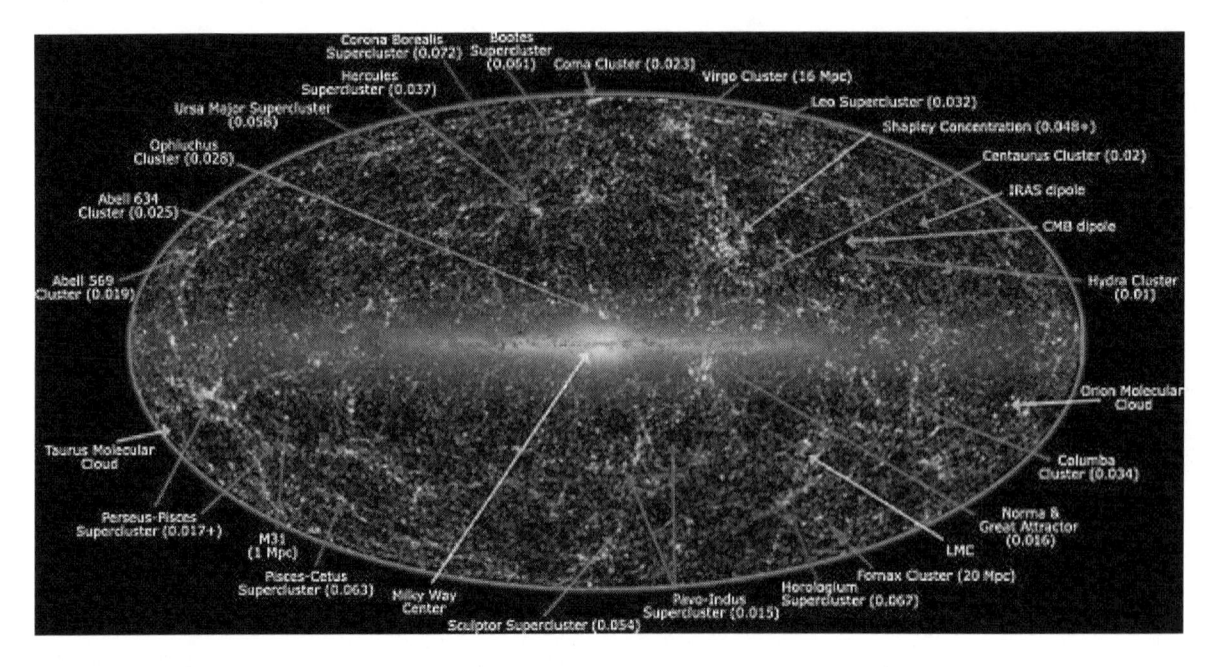

"Panoramic view of the entire near-infrared sky reveals the distribution of galaxies beyond the Milky Way. The image is derived from the 2MASS Extended Source Catalog (XSC)—more than 1.5 million galaxies, and the Point Source Catalog (PSC)--nearly 0.5 billion Milky Way stars. The galaxies are color-coded by 'redshift' obtained from the UGC, CfA, Tully NBGC, LCRS, 2dF, 6dFGS, and SDSS surveys (and from various observations compiled by the NASA Extragalactic Database), or photo-metrically deduced from the K band (2.2 um). Blue are the nearest sources ($z < 0.01$); green are at moderate distances ($0.01 < z < 0.04$) and red are the most distant sources that 2MASS resolves ($0.04 < z < 0.1$). The map is projected with an equal area Aitoff in the Galactic system (Milky Way at center)." [50]

Another indicator of large-scale structure is the 'Lyman-alpha forest'. This is a collection of absorption lines that appear in the spectra of light from quasars, which are interpreted as indicating the existence of huge thin sheets of intergalactic (mostly hydrogen) gas. These sheets appear to be associated with the formation of new galaxies.

Caution is required in describing structures on a cosmic scale because things are often different from how they appear. Gravitational lensing (bending of light by gravitation) can make an image appear to originate in a different direction from its real source. This is caused when foreground objects (such as galaxies) curve surrounding spacetime (as predicted by

5.4. MASS OF ORDINARY MATTER

general relativity), and deflect passing light rays. Rather usefully, strong gravitational lensing can sometimes magnify distant galaxies, making them easier to detect. Weak lensing (gravitational shear) by the intervening universe in general also subtly changes the observed large-scale structure. As of 2004, measurements of this subtle shear showed considerable promise as a test of cosmological models.

The large-scale structure of the Universe also looks different if one only uses redshift to measure distances to galaxies. For example, galaxies behind a galaxy cluster are attracted to it, and so fall towards it, and so are slightly blueshifted (compared to how they would be if there were no cluster) On the near side, things are slightly redshifted. Thus, the environment of the cluster looks a bit squashed if using redshifts to measure distance. An opposite effect works on the galaxies already within a cluster: the galaxies have some random motion around the cluster center, and when these random motions are converted to redshifts, the cluster appears elongated. This creates a *"finger of God"*—the illusion of a long chain of galaxies pointed at the Earth.

5.3.4 Cosmography of our cosmic neighborhood

At the centre of the Hydra-Centaurus Supercluster, a gravitational anomaly called the Great Attractor affects the motion of galaxies over a region hundreds of millions of light-years across. These galaxies are all redshifted, in accordance with Hubble's law. This indicates that they are receding from us and from each other, but the variations in their redshift are sufficient to reveal the existence of a concentration of mass equivalent to tens of thousands of galaxies.

The Great Attractor, discovered in 1986, lies at a distance of between 150 million and 250 million light-years (250 million is the most recent estimate), in the direction of the Hydra and Centaurus constellations. In its vicinity there is a preponderance of large old galaxies, many of which are colliding with their neighbours, or radiating large amounts of radio waves.

In 1987 astronomer R. Brent Tully of the University of Hawaii's Institute of Astronomy identified what he called the Pisces-Cetus Supercluster Complex, a structure one billion light years long and 150 million light years across in which, he claimed, the Local Supercluster was embedded.[51][52]

5.4 Mass of ordinary matter

The mass of the known Universe is often quoted as 10^{50} tonnes or 10^{53} kg.[3] In this context, mass refers to ordinary matter and includes the interstellar medium (ISM) and the intergalactic medium (IGM). However, it excludes dark matter and dark energy. Three calculations substantiate this quoted value for the mass of ordinary matter in the Universe: Estimates based on critical density, extrapolations from number of stars, and estimates based on steady-state. The calculations obviously assume a **finite** universe.

5.4.1 Estimates based on critical density

Critical Density is the energy density where the expansion of the Universe is poised between continued expansion and collapse.[53] Observations of the cosmic microwave background from the Wilkinson Microwave Anisotropy Probe suggest that the spatial curvature of the Universe is very close to zero, which in current cosmological models implies that the value of the density parameter must be very close to a certain critical density value. At this condition, the calculation for ρ_c critical density, is:[54]

$\rho_c = \frac{3H_0^2}{8\pi G}$

where G is the gravitational constant. From The European Space Agency's Planck Telescope results: H_0, is 67.15 kilometers per second per mega parsec. This gives a critical density of 0.85×10^{-26} kg/m^3 (commonly quoted as about 5 hydrogen atoms per cubic meter). This density includes four significant types of energy/mass: ordinary matter (4.8%), neutrinos (0.1%), cold dark matter (26.8%), and dark energy (68.3%).[5] Note that although neutrinos are defined as particles like electrons, they are listed separately because they are difficult to detect and so different from ordinary matter. Thus, the density of ordinary matter is 4.8% of the total critical density calculated or 4.08×10^{-28} kg/m^3. To convert this density to mass we must multiply by volume, a value based on the radius of the "observable universe". Since the

Universe has been expanding for 13.8 billion years, the comoving distance (radius) is now about 46.6 billion light years. Thus, volume ($\frac{4}{3}\pi r^3$) equals 3.58×10^{80} m^3 and mass of ordinary matter equals density (4.08×10^{-28} kg/m^3) times volume (3.58×10^{80} m^3) or 1.46×10^{53} kg.

5.4.2 Extrapolation from number of stars

There is no way to know exactly the number of stars, but from current literature, the range of 10^{22} to 10^{24} is normally quoted.[55][56][57][58] One way to substantiate this range is to estimate the number of galaxies and multiply by the number of stars in an average galaxy. The 2004 Hubble Ultra-Deep Field image contains an estimated 10000 galaxies.[59] The patch of sky in this area, is 3.4 arc minutes on each side. For a relative comparison, it would require over 50 of these images to cover the full moon. If this area is typical for the entire sky, there are over 100 billion galaxies in the Universe.[60] More recently, in 2012, Hubble scientists produced the Hubble Extreme Deep Field image which showed slightly more galaxies for a comparable area.[61] However, in order to compute the number of stars based on these images, we would need additional assumptions: the percent of both large and dwarf galaxies; and, their average number of stars. Thus, a reasonable option is to assume 100 billion average galaxies and 100 billion stars per average galaxy. This results in 10^{22} stars. Next, we need average star mass which can be calculated from the distribution of stars in the Milky Way. Within the Milky Way, if a large number of stars are counted by spectral class, 73% are class M stars which contain only 30% of the Sun's mass. Considering mass and number of stars in each spectral class, the average star is 51.5% of the Sun's mass.[62] The Sun's mass is 2×10^{30} kg. so a reasonable number for the mass of an average star in the Universe is 10^{30} kg. Thus, the mass of all stars equals the number of stars (10^{22}) times an average mass of star (10^{30} kg) or 10^{52} kg. The next calculation adjusts for Interstellar Medium (ISM) and Intergalactic Medium (IGM). ISM is material between stars: gas (mostly hydrogen) and dust. IGM is material between galaxies, mostly hydrogen. Ordinary matter (protons, neutrons and electrons) exists in ISM and IGM as well as in stars. In the reference, "The Cosmic Energy Inventory", the percentage of each part is defined: stars = 5.9%, Interstellar Medium (ISM) = 1.7%, and Intergalactic Medium (IGM) = 92.4%.[63] Thus, to extrapolate the mass of the Universe from the star mass, divide the 10^{52} kg mass calculated for stars by 5.9%. The result is 1.7×10^{53} kg for all the ordinary matter.

5.4.3 Estimates based on steady-state universe

Sir Fred Hoyle calculated the mass of an observable steady-state universe using the formula:[64]

$$\frac{4}{3}\pi\rho\left(\frac{c}{H}\right)^3$$

which can also be stated as [65]

$$\frac{c^3}{2GH}$$

Here H = Hubble constant, ρ = Hoyle's value for the density, G = gravitational constant, and c = speed of light.

This calculation yields approximately 0.92×10^{53} kg; however, this represents **all** energy/matter and is based on the Hubble volume (the volume of a sphere with radius equal to the Hubble length of about 13.8 billion light years). The critical density calculation above was based on the comoving distance radius of 46.6 billion light years. Thus, the Hoyle equation mass/energy result must be adjusted for increased volume. The comoving distance radius gives a volume about 39 times greater (46.7 cubed divided by 13.8 cubed). However, as volume increases, ordinary matter and dark matter would not increase; only dark energy increases with volume. Thus, assuming ordinary matter, neutrinos, and dark matter are 31.7% of the total mass/energy, and dark energy is 68.3%, the amount of total mass/energy for the steady-state calculation would be: mass of ordinary matter and dark matter (31.7% times 0.92×10^{53} kg) plus the mass of dark energy ((68.3% times 0.92×10^{53} kg) times increased volume (39)). This equals: 2.48×10^{54} kg. As noted above for the Critical Density method, ordinary matter is 4.8% of all energy/matter. If the Hoyle result is multiplied by this percent, the result for ordinary matter is 1.20×10^{53} kg.

5.4.4 Comparison of results

In summary, the three independent calculations produced reasonably close results: 1.46×10^{53}, 1.7×10^{53}, and 1.20×10^{53} kg. The average is 1.45×10^{53} kg.

The key assumptions using the Extrapolation from Star Mass method were the number of stars (10^{22}) and the percentage of ordinary matter in stars (5.9%). The key assumptions using the Critical Density method were the comoving distance radius of the Universe (46.6 billion light years) and the percentage of ordinary matter in all matter (4.8%). The key assumptions using the Hoyle steady-state method were the comoving distance radius and the percentage of dark energy in all mass (68.3%). Both the Critical Density and the Hoyle steady-state equations also used the Hubble constant (67.15 (km/s)/Mpc).

5.5 Matter content — number of atoms

Main article: cosmic abundance of elements

Assuming the mass of ordinary matter is about 1.45×10^{53} kg (reference previous section) and assuming all atoms are hydrogen atoms (which in reality make up about 74% of all atoms in our galaxy by mass, see Abundance of the chemical elements), calculating the estimated total number of atoms in the Universe is straightforward. Divide the mass of ordinary matter by the mass of a hydrogen atom (1.45×10^{53} kg divided by 1.67×10^{-27} kg). The result is approximately 10^{80} hydrogen atoms. The chemistry of life may have begun shortly after the Big Bang, 13.8 billion years ago, during a habitable epoch when the Universe was only 10–17 million years old.[66][67][68] According to the panspermia hypothesis, microscopic life—distributed by meteoroids, asteroids and other small Solar System bodies—may exist throughout the Universe.[69] Though life is confirmed only on the Earth, many think that extraterrestrial life is not only plausible, but probable or inevitable.[70][71]

5.6 Most distant objects

The most distant astronomical object yet announced as of January 2011 is a galaxy candidate classified UDFj-39546284. In 2009, a gamma ray burst, GRB 090423, was found to have a redshift of 8.2, which indicates that the collapsing star that caused it exploded when the Universe was only 630 million years old.[72] The burst happened approximately 13 billion years ago,[73] so a distance of about 13 billion light years was widely quoted in the media (or sometimes a more precise figure of 13.035 billion light years),[72] though this would be the "light travel distance" (*see* Distance measures (cosmology)) rather than the "proper distance" used in both Hubble's law and in defining the size of the observable universe (cosmologist Ned Wright argues against the common use of light travel distance in astronomical press releases on this page, and at the bottom of the page offers online calculators that can be used to calculate the current proper distance to a distant object in a flat universe based on either the redshift z or the light travel time). The proper distance for a redshift of 8.2 would be about 9.2 Gpc,[74] or about 30 billion light years. Another record-holder for most distant object is a galaxy observed through and located beyond Abell 2218, also with a light travel distance of approximately 13 billion light years from Earth, with observations from the Hubble telescope indicating a redshift between 6.6 and 7.1, and observations from Keck telescopes indicating a redshift towards the upper end of this range, around 7.[75] The galaxy's light now observable on Earth would have begun to emanate from its source about 750 million years after the Big Bang.[76]

5.7 Horizons

Main article: cosmological horizon

The limit of observability in our universe is set by a set of cosmological horizons which limit, based on various physical constraints, the extent to which we can obtain information about various events in the Universe. The most famous horizon is the particle horizon which sets a limit on the precise distance that can be seen due to the finite age of the Universe. Additional horizons are associated with the possible future extent of observations (larger than the particle horizon owing to the expansion of space), an "optical horizon" at the surface of last scattering, and associated horizons with the surface of last scattering for neutrinos and gravitational waves.

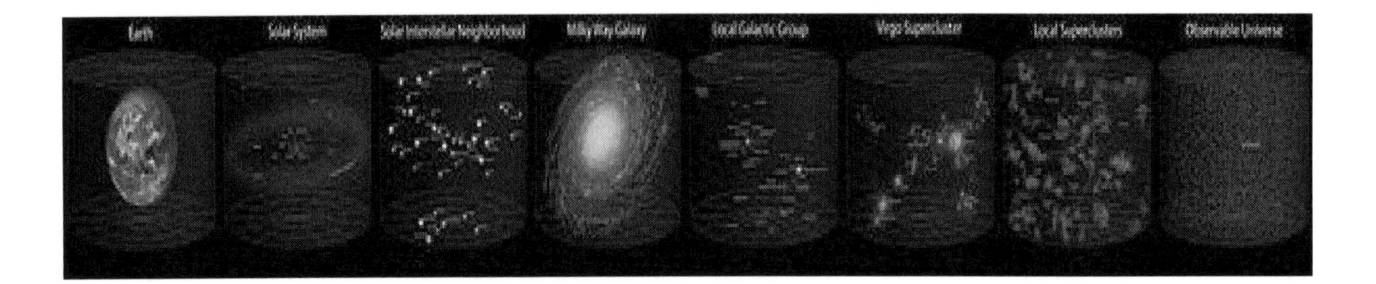

A diagram of our location in the observable universe. (*Click here for an alternative image.*)

5.8 See also

- Big Bang
- Bolshoi Cosmological Simulation
- Causality (physics)
- Chronology of the universe
- Dark flow
- Event horizon of the universe
- Hubble volume
- Illustris project
- Multiverse
- Orders of magnitude (length)
- Timeline of the Big Bang

5.9 References

[1] Itzhak Bars; John Terning (November 2009). *Extra Dimensions in Space and Time*. Springer. pp. 27–. ISBN 978-0-387-77637-8. Retrieved 2011-05-01.

[2] http://www.wolframalpha.com/input/?i=volume+universe

[3] Paul Davies (2006). *The Goldilocks Enigma*. First Mariner Books. p. 43–. ISBN 978-0-618-59226-5. Retrieved 1 July 2013.

5.9. REFERENCES

[4] http://map.gsfc.nasa.gov/universe/uni_matter.html January 13, 2015

[5] Planck collaboration (2013). "Planck 2013 results. XVI. Cosmological parameters". arXiv:1303.5076 [astro-ph.CO].

[6] Fixsen, D. J. (December 2009). "The Temperature of the Cosmic Microwave Background". *The Astrophysical Journal* **707** (2): 916–920. arXiv:0911.1955. Bibcode:2009ApJ...707..916F. doi:10.1088/0004-637X/707/2/916.

[7] Gott III, J. Richard; Mario Jurić; David Schlegel; Fiona Hoyle et al. (2005). "A Map of the Universe" (PDF). *The Astrophysics Journal* **624** (2): 463. arXiv:astro-ph/0310571. Bibcode:2005ApJ...624..463G. doi:10.1086/428890.

[8] Davis, Tamara M.; Charles H. Lineweaver (2004). "Expanding Confusion: common misconceptions of cosmological horizons and the superluminal expansion of the universe". *Publications of the Astronomical Society of Australia* **21** (1): 97. arXiv:astro-ph/0310808. Bibcode:2004PASA...21...97D. doi:10.1071/AS03040.

[9] Itzhak Bars; John Terning (November 2009). *Extra Dimensions in Space and Time*. Springer. pp. 27–. ISBN 978-0-387-77637-8. Retrieved 1 May 2011.

[10] Frequently Asked Questions in Cosmology. Astro.ucla.edu. Retrieved on 2011-05-01.

[11] Lineweaver, Charles; Tamara M. Davis (2005). "Misconceptions about the Big Bang". Scientific American.

[12] Is the universe expanding faster than the speed of light? (see the last two paragraphs)

[13] Krauss, Lawrence M.; Robert J. Scherrer (2007). "The Return of a Static Universe and the End of Cosmology". *General Relativity and Gravitation* **39** (10): 1545–1550. arXiv:0704.0221. Bibcode:2007GReGr..39.1545K. doi:10.1007/s10714-007-0472-9.

[14] Using Tiny Particles To Answer Giant Questions. Science Friday, 3 Apr 2009. According to the transcript, Brian Greene makes the comment "And actually, in the far future, everything we now see, except for our local galaxy and a region of galaxies will have disappeared. The entire universe will disappear before our very eyes, and it's one of my arguments for actually funding cosmology. We've got to do it while we have a chance."

[15] See also Faster than light#Universal expansion and Future of an expanding universe#Galaxies outside the Local Supercluster are no longer detectable.

[16] Loeb, Abraham (2002). "The Long-Term Future of Extragalactic Astronomy". *Physical Review D* **65** (4). arXiv:astro-ph/0107568. Bibcode:2002PhRvD..65d7301L. doi:10.1103/PhysRevD.65.047301.

[17] Alan H. Guth (17 March 1998). *The inflationary universe: the quest for a new theory of cosmic origins*. Basic Books. pp. 186–. ISBN 978-0-201-32840-0. Retrieved 1 May 2011.

[18] Universe Could be 250 Times Bigger Than What is Observable - by Vanessa D'Amico on February 8, 2011 http://www.universetoday.com/83167/universe-could-be-250-times-bigger-than-what-is-observable/

[19] Bielewicz, P.; Banday, A. J.; Gorski, K. M. (2013). "Constraints on the Topology of the Universe". arXiv:1303.4004 [astro-ph.CO].

[20] Mota; Reboucas; Tavakol (2010). "Observable circles-in-the-sky in flat universes". arXiv:1007.3466 [astro-ph.CO].

[21] "WolframAlpha". Retrieved 29 November 2011.

[22] "WolframAlpha". Retrieved 29 November 2011.

[23] "Seven-Year Wilson Microwave Anisotropy Probe (WMAP) Observations: Sky Maps, Systematic Errors, and Basic Results" (PDF). nasa.gov. Retrieved 2010-12-02. (see p. 39 for a table of best estimates for various cosmological parameters)

[24] Abbott, Brian (May 30, 2007). "Microwave (WMAP) All-Sky Survey". Hayden Planetarium. Retrieved 2008-01-13.

[25] Paul Davies (28 August 1992). *The new physics*. Cambridge University Press. pp. 187–. ISBN 978-0-521-43831-5. Retrieved 1 May 2011.

[26] V. F. Mukhanov (2005). *Physical foundations of cosmology*. Cambridge University Press. pp. 58–. ISBN 978-0-521-56398-7. Retrieved 1 May 2011.

[27] Bennett, C. L.; Larson, D.; Weiland, J. L.; Jarosik, N. et al. (1 October 2013). "Nine-year Wilkinson Microwave Anisotropy Probe (WMAP) Observations: Final Maps and Results". *The Astrophysical Journal Supplement Series* **208** (2): 20. arXiv:1212.5225. Bibcode:2013ApJS..208...20B. doi;10.1088/0067-0049/208/2/20.

[28] Ned Wright, "Why the Light Travel Time Distance should not be used in Press Releases".

[29] Universe Might be Bigger and Older than Expected. Space.com (2006-08-07). Retrieved on 2011-05-01.

[30] Big bang pushed back two billion years – space – 04 August 2006 – New Scientist. Space.newscientist.com. Retrieved on 2011-05-01.

[31] Edward L. Wright, "An Older but Larger Universe?"

[32] Cornish; Spergel; Starkman; Eiichiro Komatsu (May 2004) [October 2003 (arXiv)]. "Constraining the Topology of the Universe". *Phys. Rev. Lett.* **92** (20). arXiv:astro-ph/0310233. Bibcode:2004PhRvL..92t1302C. doi:10.1103/PhysRevLett.92.201302. 201302.

[33] Levin, Janna. "In space, do all roads lead to home?". plus.maths.org. Retrieved 2012-08-15.

[34] http://cosmos.phy.tufts.edu/~{}zirbel/ast21/sciam/IsSpaceFinite.pdf

[35] Bob Gardner's "Topology, Cosmology and Shape of Space" Talk, Section 7. Etsu.edu. Retrieved on 2011-05-01.

[36] Vaudrevange; Starkmanl; Cornish; Spergel. "Constraints on the Topology of the Universe: Extension to General Geometries". arXiv:1206.2939. Bibcode:2012PhRvD..86h3526V. doi:10.1103/PhysRevD.86.083526.

[37] SPACE.com – Universe Measured: We're 156 Billion Light-years Wide!

[38] Roy, Robert. (2004-05-24) New study super-sizes the universe – Technology & science – Space – Space.com – msnbc.com. MSNBC. Retrieved on 2011-05-01.

[39] "Astronomers size up the Universe". *BBC News*. 2004-05-28. Retrieved 2010-05-20.

[40] "MSU researcher recognized for discoveries about universe". 2004-12-21. Retrieved 2011-02-08.

[41] Space.com – Universe Might be Bigger and Older than Expected

[42] Robert P Kirshner (2002). *The Extravagant Universe: Exploding Stars, Dark Energy and the Accelerating Cosmos*. Princeton University Press. p. 71. ISBN 0-691-05862-8.

[43] M. J. Geller; J. P. Huchra (1989). "Mapping the universe.". *Science* **246** (4932): 897–903. Bibcode:1989Sci...246..897G. doi:10.1126/science.246.4932.897. PMID 17812575.

[44] Biggest void in space is 1 billion light years across – space – 24 August 2007 – New Scientist. Space.newscientist.com. Retrieved on 2011-05-01.

[45] Wall, Mike (2013-01-11). "Largest structure in universe discovered". Fox News.

[46] Horváth, I; Hakkila, Jon; Bagoly, Z. (2014). "Possible structure in the GRB sky distribution at redshift two". arXiv:1401.0533. Bibcode:2014A&A...561L..12H. doi:10.1051/0004-6361/201323020.

[47] Horvath, I.; Hakkila, J.; Bagoly, Z. (2013). "The largest structure of the Universe, defined by Gamma-Ray Bursts". arXiv:1311.1104 [astro-ph.CO].

[48] Klotz, Irene (2013-11-19). "Universe's Largest Structure is a Cosmic Conundrum". *Discovery*.

[49] LiveScience.com, "The Universe Isn't a Fractal, Study Finds", Natalie Wolchover, 22 August 2012

[50] 1Jarrett, T. H. (2004). "Large Scale Structure in the Local Universe: The 2MASS Galaxy Catalog". *Publications of the Astronomical Society of Australia* **21** (4): 396. arXiv:astro-ph/0405069. Bibcode:2004PASA...21..396J. doi:10.1071/AS04050.

[51] Massive Clusters of Galaxies Defy Concepts of the Universe N.Y. Times Tue. November 10, 1987:

[52] Map of the Pisces-Cetus Supercluster Complex:

[53] Michio Kaku (2005). *Parallel Worlds*. Anchor Books. p. 385. ISBN 978-1-4000-3372-0. Retrieved 1 July 2013.

5.9. REFERENCES

[54] Bernard F. Schutz (2003). *Gravity from the ground up*. Cambridge University Press. pp. 361–. ISBN 978-0-521-45506-0. Retrieved 1 May 2011.

[55] "Astronomers count the stars". BBC News. July 22, 2003. Retrieved 2006-07-18.

[56] "trillions-of-earths-could-be-orbiting-300-sextillion-stars"

[57] van Dokkum, Pieter G.; Charlie Conroy (2010). "A substantial population of low-mass stars in luminous elliptical galaxies". *Nature* **468** (7326): 940–942. arXiv:1009.5992. Bibcode:2010Natur.468..940V. doi:10.1038/nature09578. PMID 21124316.

[58] "How many stars?"

[59] [url= http://hubblesite.org/newscenter/archive/releases/2004/28/text/]| NASA, Hubble News Release STSci - 2004-7

[60] James R Johnson. *Comprehending the Cosmos, a Macro View of the Universe*. p. 36. ISBN 978-1-477-64969-5. Retrieved 1 July 2013.

[61] "Hubble Goes to the eXtreme to Assemble Farthest Ever View of the Universe" (Press release). 25 September 2012. Retrieved 1 July 2013.

[62] James R Johnson. *Comprehending the Cosmos, a Macro View of the Universe*. p. 34. ISBN 978-1-477-64969-5. Retrieved 1 July 2013.

[63] Fukugita, Masataka; Peebles, P. J. E. (2004). "The Cosmic Energy Inventory". *Astrophysical Journal* **616** (2): 643–668. arXiv:astro-ph/0406095. Bibcode:2004ApJ...616..643F. doi:10.1086/425155.

[64] Helge Kragh (1999-02-22). "Chapter 5". *Cosmology and Controversy: The Historical Development of Two Theories of the Universe*. Princeton University Press. p. 212. ISBN 0-691-00546-X.

[65] Valev, Dimitar (2010). "Estimation of the total mass and energy of the universe". arXiv:1004.1035 [physics.gen-ph].

[66] Loeb, Abraham (October 2014). "The Habitable Epoch of the Early Universe". *International Journal of Astrobiology* **13** (04): 337–339. arXiv:1312.0613. Bibcode:2014IJAsB..13..337L. doi:10.1017/S1473550414000196. Retrieved 15 December 2014.

[67] Loeb, Abraham (2 December 2013). "The Habitable Epoch of the Early Universe" (PDF). *Arxiv*. arXiv:1312.0613v3. Retrieved 15 December 2014.

[68] Dreifus, Claudia (2 December 2014). "Much-Discussed Views That Go Way Back - Avi Loeb Ponders the Early Universe, Nature and Life". *New York Times*. Retrieved 3 December 2014.

[69] Rampelotto, P.H. (2010). "Panspermia: A Promising Field Of Research" (PDF). *Astrobiology Science Conference*. Retrieved 3 December 2014.

[70] Race, Margaret S.; Randolph, Richard O. (2002). "The need for operating guidelines and a decision making framework applicable to the discovery of non-intelligent extraterrestrial life". *Advances in Space Research* **30** (6): 1583–1591. Bibcode:2002AdS doi:10.1016/S0273-1177(02)00478-7. ISSN 0273-1177. There is growing scientific confidence that the discovery of extraterrestrial life in some form is nearly inevitable

[71] Cantor, Matt (15 February 2009). "Alien Life 'Inevitable': Astronomer". *newser*. Archived from the original on 3 May 2013. Retrieved 3 May 2013. Scientists now believe there could be as many habitable planets in the cosmos as there are stars, and that makes life's existence elsewhere "inevitable" over billions of years, says one.

[72] New Gamma-Ray Burst Smashes Cosmic Distance Record – NASA Science. Science.nasa.gov. Retrieved on 2011-05-01.

[73] More Observations of GRB 090423, the Most Distant Known Object in the Universe. Universetoday.com (2009-10-28). Retrieved on 2011-05-01.

[74] Meszaros, Attila et al. (2009). "Impact on cosmology of the celestial anisotropy of the short gamma-ray bursts". *Baltic Astronomy* **18**: 293–296. arXiv:1005.1558. Bibcode:2009BaltA..18..293M.

[75] Hubble and Keck team up to find farthest known galaxy in the Universe|Press Releases|ESA/Hubble. Spacetelescope.org (2004-02-15). Retrieved on 2011-05-01.

[76] MSNBC: "Galaxy ranks as most distant object in cosmos"

5.10 Further reading

- Vicent J. Martínez; Jean-Luc Starck; Enn Saar; David L. Donoho et al. (2005). "Morphology Of The Galaxy Distribution From Wavelet Denoising". *The Astrophysical Journal* **634** (2): 744–755. arXiv:astro-ph/0508326. Bibcode:2005ApJ...634..744M. doi:10.1086/497125.

- Mureika, J. R. & Dyer, C. C. (2004). "Review: Multifractal Analysis of Packed Swiss Cheese Cosmologies". *General Relativity and Gravitation* **36** (1): 151–184. arXiv:gr-qc/0505083. Bibcode:2004GReGr..36..151M. doi:10.1023/B:GERG.0000006699.45969.49.

- Gott, III, J. R. et al. (May 2005). "A Map of the Universe". *The Astrophysical Journal* **624** (2): 463–484. arXiv:astro-ph/0310571. Bibcode:2005ApJ...624..463G. doi:10.1086/428890.

- F. Sylos Labini; M. Montuori & L. Pietronero (1998). "Scale-invariance of galaxy clustering". *Physics Reports* **293** (1): 61–226. arXiv:astro-ph/9711073. Bibcode:1998PhR...293...61S. doi:10.1016/S0370-1573(97)00044-6.

5.11 External links

- Calculating the total mass of ordinary matter in the universe, what you always wanted to know
- "Millennium Simulation" of structure forming Max Planck Institute of Astrophysics, Garching, Germany
- Visualisations of large-scale structure: animated spins of groups, clusters, filaments and voids, identified in SDSS data by MSPM (Sydney Institute for Astronomy)
- The Sloan Great Wall: Largest Known Structure? on APOD
- Cosmology FAQ
- Forming Galaxies Captured In The Young Universe By Hubble, VLT & Spitzer
- NASA featured Images and Galleries
- Star Survey reaches 70 sextillion
- Animation of the cosmic light horizon
- Inflation and the Cosmic Microwave Background by Charles Lineweaver
- Logarithmic Maps of the Universe
- List of publications of the 2dF Galaxy Redshift Survey
- List of publications of the 6dF Galaxy Redshift and peculiar velocity survey
- The Universe Within 14 Billion Light Years—NASA Atlas of the Universe (note—this map only gives a rough cosmographical estimate of the expected distribution of superclusters within the observable universe; very little actual mapping has been done beyond a distance of one billion light years):
- Video: "The Known Universe", from the American Museum of Natural History
- NASA/IPAC Extragalactic Database
- Cosmography of the Local Universe at irfu.cea.fr (17:35) (arXiv)
- What is the size of the universe? — Astronoo

Chapter 6

Lambda-CDM model

"Standard cosmological model" redirects here. For other uses, see Standard model (disambiguation).

The **ΛCDM** (**Lambda cold dark matter**) or **Lambda-CDM** model is a parametrization of the Big Bang cosmological model in which the universe contains a cosmological constant, denoted by Lambda (Greek Λ), associated with dark energy, and cold dark matter (abbreviated **CDM**). It is frequently referred to as the **standard model** of Big Bang cosmology, because it is the simplest model that provides a reasonably good account of the following properties of the cosmos:

- the existence and structure of the cosmic microwave background
- the large-scale structure in the distribution of galaxies
- the abundances of hydrogen (including deuterium), helium, and lithium
- the accelerating expansion of the universe observed in the light from distant galaxies and supernovae

The model assumes that general relativity is the correct theory of gravity on cosmological scales. It emerged in the late 1990s as a **concordance cosmology**, after a period of time when disparate observed properties of the universe appeared mutually inconsistent, and there was no consensus on the makeup of the energy density of the universe.

The ΛCDM model can be extended by adding cosmological inflation, quintessence and other elements that are current areas of speculation and research in cosmology.

Some alternative models challenge the assumptions of the ΛCDM model. Examples of these are modified Newtonian dynamics, modified gravity and theories of large-scale variations in the matter density of the universe.[1]

6.1 Overview

Most modern cosmological models are based on the cosmological principle, which states that our observational location in the universe is not unusual or special; on a large-enough scale, the universe looks the same in all directions (isotropy) and from every location (homogeneity).[2]

The model includes an expansion of metric space that is well documented both as the red shift of prominent spectral absorption or emission lines in the light from distant galaxies and as the time dilation in the light decay of supernova luminosity curves. Both effects are attributed to a Doppler shift in electromagnetic radiation as it travels across expanding space. Although this expansion increases the distance between objects that are not under shared gravitational influence, it does not increase the size of the objects (e.g. galaxies) in space. It also allows for distant galaxies to recede from each other at speeds greater than the speed of light; local expansion is less than the speed of light, but expansion summed across great distances can collectively exceed the speed of light.

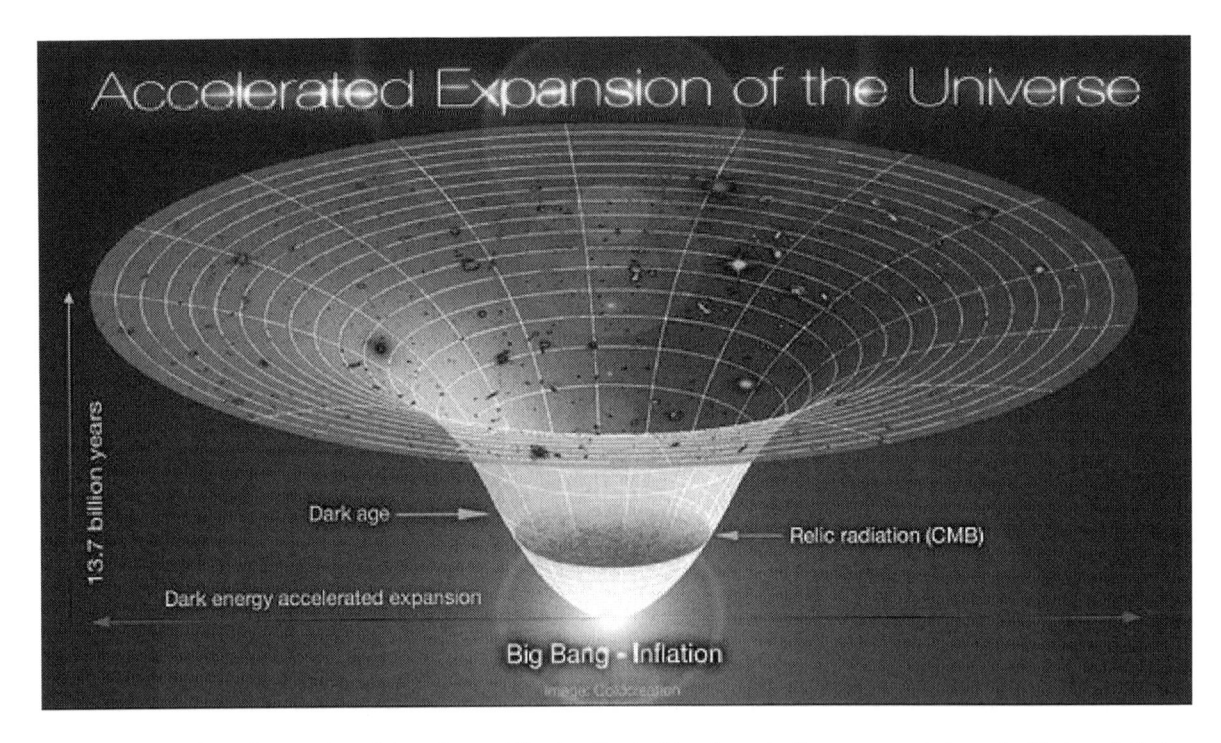

Lambda-CDM, accelerated expansion of the universe. The time-line in this schematic diagram extends from the big bang/inflation era 13.7 Gyr ago to the present cosmological time.

The letter Λ (lambda) stands for the cosmological constant, which is currently associated with a vacuum energy or dark energy in empty space that is used to explain the contemporary accelerating expansion of space against the attractive effects of gravity. A cosmological constant has negative pressure, $p = -\rho c^2$, which contributes to the stress-energy tensor that, according to the general theory of relativity, causes accelerating expansion. The fraction of the total energy density of our (flat or almost flat) universe that is dark energy, Ω_Λ, is currently [2015] estimated to be 69.2 ± 1.2% based on Planck satellite data.[3]

Cold dark matter is a form of matter introduced in order to account for gravitational effects observed in very large-scale structures (the "flat" rotation curves of galaxies; the gravitational lensing of light by galaxy clusters; and enhanced clustering of galaxies) that cannot be accounted for by the quantity of observed matter. Dark matter is described as being cold (i.e. its velocity is far less than the speed of light at the epoch of radiation-matter equality); non-baryonic (i.e. consisting of matter other than protons and neutrons); dissipationless (i.e. cannot cool by radiating photons); and collisionless (i.e. the dark matter particles interact with each other and other particles only through gravity and possibly the weak force). The dark matter component is currently [2013] estimated to constitute about 26.8% of the mass-energy density of the universe.

The remaining 4.9% [2013] comprises all ordinary matter observed as atoms, chemical elements, gas and plasma, the stuff of which visible planets, stars and galaxies are made.

Also, the energy density includes a very small fraction (~ 0.01%) in cosmic microwave background radiation, and not more than 0.5% in relic neutrinos. Although very small today, these were much more important in the distant past, dominating the matter at redshift > 3200.

The model includes a single originating event, the "Big Bang" or initial singularity, which was not an explosion but the abrupt appearance of expanding space-time containing radiation at temperatures of around 10^{15} K. This was immediately (within 10^{-29} seconds) followed by an exponential expansion of space by a scale multiplier of 10^{27} or more, known as cosmic inflation. The early universe remained hot (above 10,000 K) for several hundred thousand years, a state that is detectable as a residual cosmic microwave background, or CMB, a very low energy radiation emanating from all parts of the sky. The "Big Bang" scenario, with cosmic inflation and standard particle physics, is the only current cosmological model consistent with the observed continuing expansion of space, the observed distribution of lighter elements in the universe (hydrogen, helium, and lithium), and the spatial texture of minute irregularities (anisotropies) in the CMB radiation. Cosmic inflation also addresses the "horizon problem" in the CMB; indeed, it seems likely that the universe is larger than the observable particle horizon.

The model uses the FLRW metric, the Friedmann equations and the cosmological equations of state to describe the observable universe from right after the inflationary epoch to present and future.

6.2 Cosmic expansion history

The expansion of the universe is parametrized by a scale factor $a(t)$ which is defined relative to the present day, so $a_0 = 1$; the usual convention in cosmology is that subscript 0 denotes present-day values. In general relativity, a is related to the observed redshift:[4] by

$$a(t_{em}) \equiv (1 + z(t_{em}))^{-1}$$

where t_{em} is the age of the universe at the time the photons were emitted. The time-dependent Hubble parameter, $H(a)$ is defined as:

$$H(a) \equiv \frac{\dot{a}}{a}$$

where $\dot{a}$ is the time-derivative of the scale factor. The first of two Friedmann equations gives the expansion rate in terms of the the matter+radiation density ρ, the curvature, k , and the cosmological constant, Λ:[4]

$$H^2 = \left(\frac{\dot{a}}{a}\right)^2 = \frac{8\pi G}{3}\rho - \frac{kc^2}{a^2} + \frac{\Lambda c^2}{3}$$

where G is the usual gravitational constant. From the Friedmann equations it follows that there is a critical mass-energy density ρ_{crit} giving zero curvature; historically, if dark energy were zero, this would also be the dividing line between eventual recollapse of the universe to a Big Crunch, or unlimited expansion. In the Lambda-CDM model the universe is predicted to expand forever regardless of whether the total density is slightly above or below the critical density, though this may not apply if dark energy is actually time-dependent.

The critical density is given by

$$\rho_{crit} = \frac{3H^2}{8\pi G} = 1.88 \times 10^{-26} h^2 \text{m kg}^{-3}$$

where the reduced Hubble constant, h, is defined as $h \equiv H_0/(100\text{km/s/Mpc})$; it is standard to define the present-day **density parameter** Ω_x for various species as the dimensionless ratio

$$\Omega_x \equiv \frac{\rho_x}{\rho_{crit}} = \frac{8\pi G \rho_x(t = t_0)}{3H_0^2}$$

where the subscript x is one of c for cold dark matter, b for baryons, rad for radiation (photons plus relativistic neutrinos), and DE or Λ for dark energy.

Since the densities of various species scale as different powers of a , e.g. a^{-3} for matter etc, the Friedmann equation can be conveniently rewritten in terms of the various density parameters as

$$H(a) \equiv \frac{\dot{a}}{a} = H_0 \sqrt{\left[(\Omega_c + \Omega_b)a^{-3} + \Omega_{rad}a^{-4} + \Omega_k a^{-2} + \Omega_{DE} a^{-3(1+w)}\right]}$$

where w is the equation of state of dark energy, and assuming negligible neutrino mass (significant neutrino mass requires a more complex equation). The various Ω parameters add up to 1 by construction. In the general case this is integrated by computer to give the expansion history a(t) and also observable distance-redshift relations for any chosen values of the cosmological parameters, which can then be compared with observations such as supernovae and baryon acoustic oscillations.

In the minimal 6-parameter LambdaCDM model, it is assumed that curvature Ω_k is zero and $w = -1$, so this simplifies to

$$H(a) = H_0\sqrt{[\Omega_m a^{-3} + \Omega_{rad} a^{-4} + \Omega_\Lambda]}$$

Observations show that the radiation density is very small today, $\Omega_{rad} \sim 10^{-4}$; if this term is neglected the above has an analytic solution[5]

$$a(t) = (\Omega_m/\Omega_\Lambda)^{1/3} \sinh^{2/3}(t/t_\Lambda)$$

where $t_\Lambda \equiv 2/(3H_0\sqrt{\Omega_\Lambda})$; this is fairly accurate for a > 0.01 or t > 10 Myr. Solving for $a(t) = 1$ gives the present age of the universe t_0 in terms of the other parameters.

It follows that the transition from decelerating to accelerating expansion (the second derivative $\ddot{a}$ crossing zero) occurred when

$$a = (\Omega_m/2\Omega_\Lambda)^{1/3}$$

which evaluates to a ~ 0.6 or z ~ 0.66 for the Planck best-fit parameters.

6.3 Historical development

The discovery of the Cosmic Microwave Background (CMB) in 1965 confirmed a key prediction of the Big Bang cosmology. From that point on, it was generally accepted that the universe started in a hot, dense state and has been expanding over time. The rate of expansion depends on the types of matter and energy present in the universe, and in particular, whether the total density is above or below the so-called critical density. During the 1970s, most attention focused on pure-baryonic models, but there were serious challenges explaining the formation of galaxies, given the small anisotropies in the CMB (upper limits at that time). In the early 1980s, it was realized that this could be resolved if cold dark matter dominated over the baryons, and the theory of cosmic inflation motivated models with critical density. During the 1980s, most research focused on cold dark matter with critical density in matter, around 95% CDM and 5% baryons: these showed success at forming galaxies and clusters of galaxies, but problems remained; notably, the model required a Hubble constant lower than preferred by observations, and observations around 1988-1990 showed more large-scale galaxy clustering than predicted. These difficulties sharpened with the discovery of CMB anisotropy by COBE in 1992, and several modified CDM models, including ΛCDM and mixed cold+hot dark matter, came under active consideration through the mid-1990s. The ΛCDM model then became the leading model following the observations of accelerating expansion in 1998, and was quickly supported by other observations: in 2000, the BOOMERanG microwave background experiment measured the total (matter+energy) density to be close to 100% of critical, whereas in 2001 the 2dFGRS galaxy redshift survey measured the matter density to be near 25%; the large difference between these values supports a positive Λ or dark energy. Much more precise measurements of the microwave background from WMAP in 2003 – 2010 and Planck in 2013 - 2015 have continued to support the model and pin down the parameters, now mostly constrained below 1 percent uncertainty.

There is currently active research into many aspects of the ΛCDM model, both to refine the parameters and possibly detect deviations. In addition, ΛCDM has no explicit physical theory for the origin or physical nature of dark matter or dark energy; the nearly scale-invariant spectrum of the CMB perturbations, and their image across the celestial sphere, are believed to result from very small thermal and acoustic irregularities at the point of recombination. A large majority of astronomers and astrophysicists support the ΛCDM model or close relatives of it, but Milgrom, McGaugh, and Kroupa are leading critics, attacking the dark matter portions of the theory from the perspective of galaxy formation models and supporting the alternative MOND theory, which requires a modification of the Einstein field equations and the Friedmann equations as seen in proposals such as MOG theory or TeVeS theory. Other proposals by theoretical astrophysicists of cosmological alternatives to Einstein's general relativity that attempt to account for dark energy or dark matter include f(R) gravity, scalar–tensor theories such as galileon theories, brane cosmologies, the DGP model, and massive gravity and its extensions such as bimetric gravity.

6.4 Successes

In addition to explaining pre-2000 observations, the model has made a number of successful predictions: notably the existence of the baryon acoustic oscillation feature, discovered in 2005 in the predicted location; and the statistics of weak gravitational lensing, first observed in 2000 by several teams. The polarization of the CMB, discovered in 2002 by DASI [6] is now a dramatic success: in the 2015 Planck data release,[7] there are seven observed peaks in the temperature (TT) power spectrum, six peaks in the temperature-polarization (TE) cross spectrum, and five peaks in the polarization (EE) spectrum, and all agree with the predictions of LambdaCDM.

6.5 Challenges

See also: Hierarchy problem, Physics beyond the Standard Model and Elementary particle § Beyond the Standard Model

Extensive searches for dark matter particles have so far shown no well-agreed detection; the dark energy may be almost impossible to detect in a laboratory, and its value is unnaturally small compared to naive theoretical predictions.

Comparison of the model with observations is very successful on large scales (larger than galaxies, up to the observable horizon), but may have some problems on sub-galaxy scales, possibly predicting too many dwarf galaxies and too much dark matter in the innermost regions of galaxies. These small scales are harder to resolve in computer simulations, so it is not yet clear whether the problem is the simulations, non-standard properties of dark matter, or a more radical error in the model.

6.6 Parameters

The simple ΛCDM model is based on six parameters: physical baryon density; physical dark matter density; the age of the universe; scalar spectral index; curvature fluctuation amplitude; and reionization optical depth.[8] In accordance with Occam's razor, six is the smallest number of parameters needed to give an acceptable fit to current observations; other possible parameters are fixed at "natural" values, e.g. total density = 1.00, dark energy equation of state = −1, neutrino masses are small enough to be negligible. (See below for extended models that allow these to vary.)

The values of these six parameters are mostly not predicted by current theory (though, ideally, they may be related by a future "Theory of Everything"), except that most versions of cosmic inflation predict the scalar spectral index should be slightly smaller than 1, consistent with the estimated value 0.96. The parameter values, and uncertainties, are estimated using large computer searches to locate the region of parameter space providing an acceptable match to cosmological observations. From these six parameters, the other model values, such as the Hubble constant and the dark energy density, can be readily calculated.

Commonly, the set of observations fitted includes the cosmic microwave background anisotropy, the brightness/redshift relation for supernovae, and large-scale galaxy clustering including the baryon acoustic oscillation feature. Other observations, such as the Hubble constant, the abundance of galaxy clusters, weak gravitational lensing and globular cluster ages, are generally consistent with these, providing a check of the model, but are less accurately measured at present.

Parameter values listed below are from the Planck Collaboration Cosmological parameters 68% confidence limits for the base ΛCDM model from Planck CMB power spectra, in combination with lensing reconstruction and external data (BAO+JLA+H_0).[9] See also Planck (spacecraft).

[1] The "physical baryon density" $\Omega_b h^2$ is the "baryon density" Ω_b multiplied by the square of the reduced Hubble constant h,[10] where h is related to the Hubble constant H_0 by the equation H_0=100 h (km/s)/Mpc.[11] Likewise for the difference between "physical dark matter density" and "dark matter density". The baryon density gives the fraction of the critical density made up of baryons. The critical density is the total density of matter/energy needed for the universe to be spatially flat, with measurements indicating that the actual total density Ω_{tot} is very close if not equal to this value, see below.

6.7 Extended models

Possible extensions of the simplest ΛCDM model are to allow quintessence rather than a cosmological constant. In this case, the equation of state of dark energy is allowed to differ from −1. Cosmic inflation predicts tensor fluctuations (gravitational waves). Their amplitude is parameterized by the tensor-to-scalar ratio (denoted r), which is determined by the energy scale of inflation. Other modifications allow for spatial curvature (Ω_{tot} may be different from 1), hot dark matter in the form of neutrinos, or a running spectral index, which are generally viewed as inconsistent with cosmic inflation.

Allowing these parameters will generally *increase* the uncertainties in the parameters quoted above, and may also shift the observed values somewhat.

Some researchers have suggested that there is a running spectral index, but no statistically significant study has revealed one. Theoretical expectations suggest that the tensor-to-scalar ratio r should be between 0 and 0.3, and the latest results are now within those limits.

6.8 See also

- Bolshoi Cosmological Simulation
- List of cosmological computation software
- Dark matter
- Galaxy formation and evolution
- Illustris project
- WIMPs
- The ΛCDM model is also known as the standard model of cosmology, but is not related to the Standard Model of particle physics.

6.9 References

[1] P. Kroupa, B. Famaey, K.S. de Boer, J. Dabringhausen, M. Pawlowski, C.M. Boily, H. Jerjen, D. Forbes, G. Hensler, M. Metz, "Local-Group tests of dark-matter concordance cosmology. Towards a new paradigm for structure formation" A&A 523, 32 (2010).

[2] Andrew Liddle. *An Introduction to Modern Cosmology (2nd ed.)*. London: Wiley, 2003.

[3] Camille M. Carlisle, *Planck Upholds Standard Cosmology*, Sky & Telescope, February 10, 2015

[4] Dodelson, Scott (2008). *Modern cosmology* (4. [print.]. ed.). San Diego, CA [etc.]: Academic Press. ISBN 978-0122191411.

[5] Frieman, Joshua A.; Turner, Michael S.; Huterer, Dragan (September 2008). "Dark Energy and the Accelerating Universe". *Annual Review of Astronomy and Astrophysics* **46** (1): 385–432. arXiv:0803.0982. doi:10.1146/annurev.astro.46.060407.145243.

[6] Kovac, J. M.; Leitch, E. M.; Pryke, C.; Carlstrom, J. E.; Halverson, N. W.; Holzapfel, W. L. (19 December 2002). "Detection of polarization in the cosmic microwave background using DASI". *Nature* **420** (6917): 772–787. arXiv:astro-ph/0209478. doi:10.1038/nature01269.

[7] A bot will complete this citation soon. Click here to jump the queue arXiv:1502.01589.

[8] Spergel, D. N. (2015). "The dark side of the cosmology: dark matter and dark energy". *Science* **347** (6226): 1100–1102. doi:10.1126/science.aaa0980.

[9] Table 4 on p. 31 of Planck Collaboration. "Planck 2015 results. XIII. Cosmological parameters" (PDF). Retrieved 2015-02-18.

[10] Appendix A of the LSST Science Book Version 2.0

[11] p. 7 of Findings of the Joint Dark Energy Mission Figure of Merit Science Working Group

[12] Table 8 on p. 39 of Jarosik, N. et al. (WMAP Collaboration). "Seven-Year Wilkinson Microwave Anisotropy Probe (WMAP) Observations: Sky Maps, Systematic Errors, and Basic Results" (PDF). nasa.gov. Retrieved 2010-12-04. (from NASA's WMAP Documents page)

6.10 Further reading

- Rebolo, R. et al. (2004). "Cosmological parameter estimation using Very Small Array data out to $\ell = 1500$". *Monthly Notices of the Royal Astronomical Society* **353** (3): 747. arXiv:astro-ph/0402466.Bibcode:2004MNRAS.. doi:10.1111/j.1365-2966.2004.08102.x.

- Ostriker, J. P.; Steinhardt, P. J. (1995). "Cosmic Concordance". arXiv:astro-ph/9505066 [astro-ph].

- Ostriker, Jeremiah P.; Mitton, Simon (2013). *Heart of Darkness: Unraveling the mysteries of the invisible universe*. Princeton, NJ: Princeton University Press. ISBN 978-0-691-13430-7.

6.11 External links

- Bolshoi Simulation
- Cosmology tutorial/NedWright
- Millennium Simulation
- WMAP estimated cosmological parameters/Latest Summary

Chapter 7

Mass–energy equivalence

"E=MC2" and "E=mc2" redirect here. For other uses, see E=MC2 (disambiguation).

In physics, **mass–energy equivalence** is the concept that the mass of an object or system is a measure of its energy

The four-metre tall sculpture of Einstein's 1905 formula E=mc² at the 2006 Walk of Ideas, Berlin, Germany.

content. For instance, adding 25 kilowatt-hours (90 megajoules) of *any* form of energy to any object increases its mass by 1 microgram (and, accordingly, its inertia and weight) even though no matter has been added.

A physical system has a property called energy and a corresponding property called mass; the two properties are equivalent in that they are always both present in the same (i.e. constant) proportion to one another. Mass–energy equivalence arose originally from special relativity as a paradox described by Henri Poincaré.[1] The equivalence of energy E and mass m is reliant on the speed of light c and is described by the famous equation:

$$E = mc^2$$

Thus, this mass–energy relation states that the universal proportionality factor between equivalent amounts of energy and mass is equal to the speed of light squared. This also serves to convert units of mass to units of energy, no matter what system of measurement units used.

If a body is stationary, it still has some internal or intrinsic energy, called its rest energy. Rest mass and rest energy are equivalent and remain proportional to one another. When the body is in motion (relative to an observer), its total energy is greater than its rest energy. The rest mass (or rest energy) remains an important quantity in this case because it remains the same regardless of this motion, even for the extreme speeds or gravity considered in special and general relativity; thus it is also called the invariant mass.

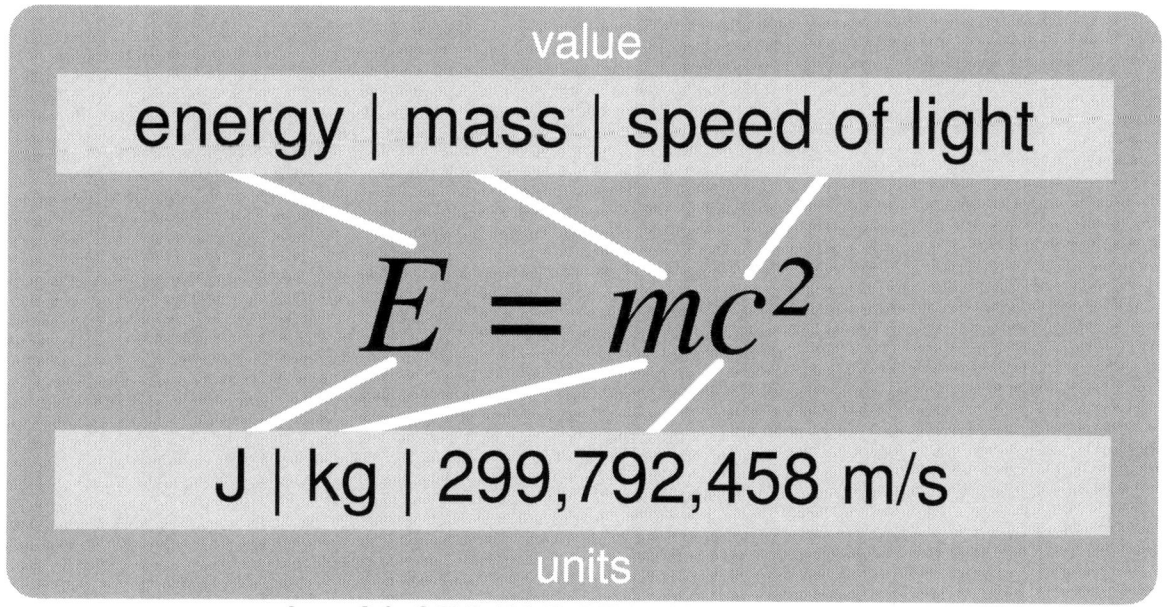

$E = mc^2$ *explicated.*

$c^2 = 89,875,517,873,681,800 \ m^2/s^2$

On the one hand, the equation $E = mc^2$ can be applied to rest mass (m or m_0) and rest energy (E_0) to show their proportionality as $E_0 = m_0 c^2$.[2]

On the other hand, it can also be applied to the total energy (E_{tot} or simply E) and total mass of a moving body. The total mass is also called the relativistic mass m_{rel}. The total energy and total mass are related by $E = m_{rel} c^2$.[3]

Thus, the mass–energy relation $E = mc^2$ can be used to relate the rest energy to the rest mass, or to relate the total energy to the total mass. To instead relate the *total* energy or mass to the *rest* energy or mass, a generalization of the mass–energy relation is required: the energy–momentum relation.

$E = mc^2$ has frequently been invoked as an explanation for the origin of energy in nuclear processes specifically, but such processes can be understood as converting nuclear potential energy in a manner precisely analogous to the way that chemical processes convert electrical potential energy. The more common association of mass–energy equivalence with nuclear processes derives from the fact that the large amounts of energy released in such reactions may exhibit enough mass that the mass loss (which is called the **mass defect**) may be measured, when the released energy (and its mass) have been removed from the system; while the energy released in chemical processes is smaller by roughly six orders of magnitude, and so the resulting **mass defect** is much more difficult to measure. For example, the loss of mass to an atom and a neutron, as a result of the capture of the neutron and the production of a gamma ray, has been used to test mass–energy equivalence to high precision, as the energy of the gamma ray may be compared with the mass defect after capture. In 2005, these were found to agree to 0.0004%, the most precise test of the equivalence of mass and energy to date. This test was performed in the World Year of Physics 2005, a centennial celebration of Albert Einstein's achievements in 1905.[4]

Einstein was not the first to propose a mass–energy relationship (see the History section). However, Einstein was the first scientist to propose the $E = mc^2$ formula and the first to interpret mass–energy equivalence as a fundamental principle that follows from the relativistic symmetries of space and time.

7.1 Nomenclature

The formula was initially written in many different notations, and its interpretation and justification was further developed in several steps.[5][6]

- In *"Does the inertia of a body depend upon its energy content?"* (1905), Einstein used V to mean the speed of light in a vacuum and L to mean the energy lost by a body in the form of radiation.[7] Consequently, the equation $E = mc^2$ was not originally written as a formula but as a sentence in German saying that *if a body gives off the energy L in the form of radiation, its mass diminishes by* L/V^2. A remark placed above it informed that the equation was approximated by neglecting "magnitudes of fourth and higher orders" of a series expansion.[8]

- In May 1907, Einstein explained that the expression for energy ε of a moving mass point assumes the simplest form, when its expression for the state of rest is chosen to be $\varepsilon_0 = \mu V^2$ (where μ is the mass), which is in agreement with the "principle of the equivalence of mass and energy". In addition, Einstein used the formula $\mu = E_0/V^2$, with E_0 being the energy of a system of mass points, in order to describe the energy and mass increase of that system when the velocity of the differently moving mass points is increased.[9]

- In June 1907, Max Planck rewrote Einstein's mass–energy relationship as $M = (E_0 + pV_0)/c^2$, where p is the pressure and V the volume, in order to express the relation between mass, its "latent energy", and thermodynamic energy within the body.[10] Subsequently in October 1907, this was rewritten as $M_0 = E_0/c^2$ and given a quantum interpretation by Johannes Stark, who assumed its validity and correctness (*Gültigkeit*).[11]

- In December 1907, Einstein expressed the equivalence in the form $M = \mu + E_0/c^2$ and concluded: *A mass μ is equivalent, as regards inertia, to a quantity of energy μc^2. [...] It appears far more natural to consider every inertial mass as a store of energy.*[12][13]

- In 1909, Gilbert N. Lewis and Richard C. Tolman used two variations of the formula: $m = E/c^2$ and $m_0 = E_0/c^2$, with E being the energy of a moving body, E_0 its rest energy, m the relativistic mass, and m_0 the invariant mass.[14] The same relations in different notation were used by Hendrik Lorentz in 1913 (published 1914), though he placed the energy on the left-hand side: $\varepsilon = Mc^2$ and $\varepsilon_0 = mc^2$, with ε being the total energy (rest energy plus kinetic energy) of a moving material point, ε_0 its rest energy, M the relativistic mass, and m the invariant (or rest) mass.[15]

- In 1911, Max von Laue gave a more comprehensive proof of $M_0 = E_0/c^2$ from the stress–energy tensor,[16] which was later (1918) generalized by Felix Klein.[17]

- Einstein returned to the topic once again after World War II and this time he wrote $E = mc^2$ in the title of his article[18] intended as an explanation for a general reader by analogy.[19]

7.2 Conservation of mass and energy

Main articles: Conservation of energy and Conservation of mass

Mass and energy can be seen as two names (and two measurement units) for the same underlying, conserved physical quantity.[20] Thus, the laws of conservation of energy and conservation of (total) mass are equivalent and both hold true.[21] Einstein elaborated in a 1946 essay that "the principle of the conservation of mass [...] proved inadequate in the face of the special theory of relativity. It was therefore merged with the energy [conservation] principle—just as, about 60 years before, the principle of the conservation of mechanical energy had been combined with the principle of the conservation of heat [thermal energy]. We might say that the principle of the conservation of energy, having previously swallowed up that of the conservation of heat, now proceeded to swallow that of the conservation of mass—and holds the field alone."[22]

If the conservation of mass law is interpreted as conservation of *rest* mass, it does not hold true in special relativity. The *rest* energy (equivalently, rest mass) of a particle can be converted, not "to energy" (it already *is* energy (mass)), but rather to *other* forms of energy (mass) which require motion, such as kinetic energy, thermal energy, or radiant energy; similarly, kinetic or radiant energy can be converted to other kinds of particles which have rest energy (rest mass). In the transformation process, neither the total amount of mass nor the total amount of energy changes, since both are properties which are connected to each other via a simple constant.[23][24] This view requires that if either energy or (total) mass disappears from a system, it will always be found that both have simply moved off to another place, where they may both be measured as an increase of both energy and mass corresponding to the loss in the first system.

7.2.1 Fast-moving objects and systems of objects

When an object is pushed in the direction of motion, it gains momentum and energy, but when the object is already traveling near the speed of light, it cannot move much faster, no matter how much energy it absorbs. Its momentum and energy continue to increase without bounds, whereas its speed approaches a constant value—the speed of light. This implies that in relativity the momentum of an object cannot be a constant times the velocity, nor can the kinetic energy be a constant times the square of the velocity.

A property called the relativistic mass is defined as the ratio of the momentum of an object to its velocity.[25] Relativistic mass depends on the motion of the object, so that different observers in relative motion see different values for it. If the object is moving slowly, the relativistic mass is nearly equal to the rest mass and both are nearly equal to the usual Newtonian mass. If the object is moving quickly, the relativistic mass is greater than the rest mass by an amount equal to the mass associated with the kinetic energy of the object. As the object approaches the speed of light, the relativistic mass grows infinitely, because the kinetic energy grows infinitely and this energy is associated with mass.

The relativistic mass is always equal to the total energy (rest energy plus kinetic energy) divided by c^2.[3] Because the relativistic mass is exactly proportional to the energy, relativistic mass and relativistic energy are nearly synonyms; the only difference between them is the units. If length and time are measured in natural units, the speed of light is equal to 1, and even this difference disappears. Then mass and energy have the same units and are always equal, so it is redundant to speak about relativistic mass, because it is just another name for the energy. This is why physicists usually reserve the useful short word "mass" to mean rest mass, or invariant mass, and not relativistic mass.

The relativistic mass of a moving object is larger than the relativistic mass of an object that is not moving, because a moving object has extra kinetic energy. The *rest mass* of an object is defined as the mass of an object when it is at rest, so that the rest mass is always the same, independent of the motion of the observer: it is the same in all inertial frames.

For things and systems made up of many parts, like an atomic nucleus, planet, or star, the relativistic mass is the sum of the relativistic masses (or energies) of the parts, because energies are additive in isolated systems. This is not true in systems which are open, however, if energy is subtracted. For example, if a system is *bound* by attractive forces, and the energy gained due to the forces of attraction in excess of the work done is removed from the system, then mass will be lost with this removed energy. For example, the mass of an atomic nucleus is less than the total mass of the protons and neutrons that make it up, but this is only true after this energy from binding has been removed in the form of a gamma ray (which in this system, carries away the mass of the energy of binding). This mass decrease is also equivalent to the energy required to break up the nucleus into individual protons and neutrons (in this case, work and mass would need to be supplied). Similarly, the mass of the solar system is slightly less than the sum of the individual masses of the sun and planets.

For a system of particles going off in different directions, the invariant mass of the system is the analog of the rest mass, and is the same for all observers, even those in relative motion. It is defined as the total energy (divided by c^2) in the center of mass frame (where by definition, the system total momentum is zero). A simple example of an object with moving parts but zero total momentum is a container of gas. In this case, the mass of the container is given by its total energy (including the kinetic energy of the gas molecules), since the system total energy and invariant mass are the same in any reference frame where the momentum is zero, and such a reference frame is also the only frame in which the object can be weighed. In a similar way, the theory of special relativity posits that the thermal energy in all objects (including solids) contributes to their total masses and weights, even though this energy is present as the kinetic and potential energies of the atoms in the object, and it (in a similar way to the gas) is not seen in the rest masses of the atoms that make up the object.

In a similar manner, even photons (light quanta), if trapped in a container space (as a photon gas or thermal radiation), would contribute a mass associated with their energy to the container. Such an extra mass, in theory, could be weighed in the same way as any other type of rest mass. This is true in special relativity theory, even though individually photons have no rest mass. The property that trapped energy *in any form* adds weighable mass to systems that have no net momentum is one of the characteristic and notable consequences of relativity. It has no counterpart in classical Newtonian physics, in which radiation, light, heat, and kinetic energy never exhibit weighable mass under any circumstances.

Just as the relativistic mass of an isolated system is conserved through time, so also is its invariant mass. It is this property which allows the conservation of all types of mass in systems, and also conservation of all types of mass in reactions where matter is destroyed (annihilated), leaving behind the energy that was associated with it (which is now in non-material form, rather than material form). Matter may appear and disappear in various reactions, but mass and energy are both unchanged in this process.

7.3 Applicability of the strict mass–energy equivalence formula, $E = mc^2$

As is noted above, two different definitions of mass have been used in special relativity, and also two different definitions of energy. The simple equation $E = mc^2$ is not generally applicable to all these types of mass and energy, except in the special case that the total additive momentum is zero for the system under consideration. In such a case, which is always guaranteed when observing the system from either its center of mass frame or its center of momentum frame, $E = mc^2$ is always true for any type of mass and energy that are chosen. Thus, for example, in the center of mass frame, the total energy of an object or system is equal to its rest mass times c^2, a useful equality. This is the relationship used for the container of gas in the previous example. It is *not* true in other reference frames where the center of mass is in motion. In these systems or for such an object, its total energy will depend on both its rest (or invariant) mass, and also its (total) momentum.[26]

In inertial reference frames other than the rest frame or center of mass frame, the equation $E = mc^2$ remains true if the energy is the relativistic energy *and* the mass is the relativistic mass. It is also correct if the energy is the rest or invariant energy (also the minimum energy), *and* the mass is the rest mass, or the invariant mass. However, connection of the **total or relativistic energy** (E_r) with the **rest or invariant mass** (m_0) requires consideration of the system total momentum, in systems and reference frames where the total momentum has a non-zero value. The formula then required to connect the two different kinds of mass and energy, is the extended version of Einstein's equation, called the relativistic energy–momentum relation:[27]

$$E_r^2 - |\vec{p}|^2 c^2 = m_0^2 c^4$$
$$E_r^2 - (pc)^2 = (m_0 c^2)^2$$

or

$$E_r = \sqrt{(m_0 c^2)^2 + (pc)^2}$$

Here the $(pc)^2$ term represents the square of the Euclidean norm (total vector length) of the various momentum vectors in the system, which reduces to the square of the simple momentum magnitude, if only a single particle is considered. This equation reduces to $E = mc^2$ when the momentum term is zero. For photons where $m_0 = 0$, the equation reduces to $Er = pc$.

7.4 Meanings of the strict mass–energy equivalence formula, $E = mc^2$

Mass–energy equivalence states that any object has a certain energy, even when it is stationary. In Newtonian mechanics, a motionless body has no kinetic energy, and it may or may not have other amounts of internal stored energy, like chemical energy or thermal energy, in addition to any potential energy it may have from its position in a field of force. In Newtonian mechanics, all of these energies are much smaller than the mass of the object times the speed of light squared.

In relativity, all of the energy that moves along with an object (that is, all the energy which is present in the object's rest frame) contributes to the total mass of the body, which measures how much it resists acceleration. Each potential and kinetic energy makes a proportional contribution to the mass. As noted above, even if a box of ideal mirrors "contains" light, then the individually massless photons still contribute to the total mass of the box, by the amount of their energy divided by c^2.[28]

In relativity, removing energy is removing mass, and for an observer in the center of mass frame, the formula $m = E/c^2$ indicates how much mass is lost when energy is removed. In a nuclear reaction, the mass of the atoms that come out is less than the mass of the atoms that go in, and the difference in mass shows up as heat and light which has the same

7.4. MEANINGS OF THE STRICT MASS–ENERGY EQUIVALENCE FORMULA, $E = MC^2$

The mass–energy equivalence formula was displayed on Taipei 101 during the event of the World Year of Physics 2005.

relativistic mass as the difference (and also the same invariant mass in the center of mass frame of the system). In this case, the E in the formula is the energy released and removed, and the mass m is how much the mass decreases. In the same way, when any sort of energy is added to an isolated system, the increase in the mass is equal to the added energy divided by c^2. For example, when water is heated it gains about 1.11×10^{-17} kg of mass for every joule of heat added to the water.

An object moves with different speed in different frames, depending on the motion of the observer, so the kinetic energy in both Newtonian mechanics and relativity is *frame dependent*. This means that the amount of relativistic energy, and therefore the amount of relativistic mass, that an object is measured to have depends on the observer. The *rest mass* is defined as the mass that an object has when it is not moving (or when an inertial frame is chosen such that it is not moving). The term also applies to the invariant mass of systems when the system as a whole is not "moving" (has no net momentum). The rest and invariant masses are the smallest possible value of the mass of the object or system. They also are conserved quantities, so long as the system is isolated. Because of the way they are calculated, the effects of moving observers are subtracted, so these quantities do not change with the motion of the observer.

The rest mass is almost never additive: the rest mass of an object is not the sum of the rest masses of its parts. The rest mass of an object is the total energy of all the parts, including kinetic energy, as measured by an observer that sees the center of the mass of the object to be standing still. The rest mass adds up only if the parts are standing still and do not attract or repel, so that they do not have any extra kinetic or potential energy. The other possibility is that they have a positive kinetic energy and a negative potential energy that exactly cancels.

7.4.1 Binding energy and the "mass defect"

Main article: binding energy

Whenever any type of energy is removed from a system, the mass associated with the energy is also removed, and the system therefore loses mass. This mass defect in the system may be simply calculated as $\Delta m = \Delta E/c^2$, and this was the form of the equation historically first presented by Einstein in 1905. However, use of this formula in such circumstances has led to the false idea that mass has been "converted" to energy. This may be particularly the case when the energy (and mass) removed from the system is associated with the *binding energy* of the system. In such cases, the binding energy is observed as a "mass defect" or deficit in the new system.

The fact that the released energy is not easily weighed in many such cases, may cause its mass to be neglected as though it no longer existed. This circumstance has encouraged the false idea of conversion of *mass* to energy, rather than the correct idea that the binding energy of such systems is relatively large, and exhibits a measurable mass, which is removed when the binding energy is removed. This energy is often released in the form of light and heat, which is too quickly and widely dispersed to be easily weighed, though it does carry mass.

The difference between the rest mass of a bound system and of the unbound parts is the binding energy of the system, if this energy has been removed after binding. For example, a water molecule weighs a little less than two free hydrogen atoms and an oxygen atom; the minuscule mass difference is the energy that is needed to split the molecule into three individual atoms (divided by c^2), and which was given off as heat when the molecule formed (this heat had mass). Likewise, a stick of dynamite in theory weighs a little bit more than the fragments after the explosion, but this is true only so long as the fragments are cooled and the heat removed. In this case the mass difference is the energy/heat that is released when the dynamite explodes, and when this heat escapes, the mass associated with it escapes, only to be deposited in the surroundings which absorb the heat (so that total mass is conserved).

Such a change in mass may only happen when the system is open, and the energy and mass escapes. Thus, if a stick of dynamite is blown up in a hermetically sealed chamber, the mass of the chamber and fragments, the heat, sound, and light would still be equal to the original mass of the chamber and dynamite. If sitting on a scale, the weight and mass would not change. This would in theory also happen even with a nuclear bomb, if it could be kept in an ideal box of infinite strength, which did not rupture or pass radiation.[24] Thus, a 21.5 kiloton (9×10^{13} joule) nuclear bomb produces about one gram of heat and electromagnetic radiation, but the mass of this energy would not be detectable in an exploded bomb in an ideal box sitting on a scale; instead, the contents of the box would be heated to millions of degrees without changing total mass and weight. If then, however, a transparent window (passing only electromagnetic radiation) were opened in such an ideal box after the explosion, and a beam of X-rays and other lower-energy light allowed to escape the box, it would eventually be found to weigh one gram less than it had before the explosion. This weight loss and mass loss would happen as the box was cooled by this process, to room temperature. However, any surrounding mass which had absorbed the X-rays (and other "heat") would *gain* this gram of mass from the resulting heating, so the mass "loss" would represent merely its relocation. Thus, no mass (or, in the case of a nuclear bomb, no matter) would be "converted" to energy in such a process. Mass and energy, as always, would both be separately conserved.

7.4.2 Massless particles

Massless particles have zero rest mass. Their relativistic mass is simply their relativistic energy, divided by c^2, or $m_{rel} = E/c^2$.[29][30] The energy for photons is $E = hf$, where h is Planck's constant and f is the photon frequency. This frequency and thus the relativistic energy are frame-dependent.

If an observer runs away from a photon in the direction it travels from a source, having it catch up with the observer, then when the photon catches up it will be seen as having less energy than it had at the source. The faster the observer is traveling with regard to the source when the photon catches up, the less energy the photon will have. As an observer approaches the speed of light with regard to the source, the photon looks redder and redder, by relativistic Doppler effect (the Doppler shift is the relativistic formula), and the energy of a very long-wavelength photon approaches zero. This is why a photon is *massless*; this means that the rest mass of a photon is zero.

7.4.3 Massless particles contribute rest mass and invariant mass to systems

Two photons moving in different directions cannot both be made to have arbitrarily small total energy by changing frames, or by moving toward or away from them. The reason is that in a two-photon system, the energy of one photon is decreased by chasing after it, but the energy of the other will increase with the same shift in observer motion. Two photons not moving in the same direction will exhibit an inertial frame where the combined energy is smallest, but not zero. This is called the center of mass frame or the center of momentum frame; these terms are almost synonyms (the center of mass frame is the special case of a center of momentum frame where the center of mass is put at the origin). The most that chasing a pair of photons can accomplish to decrease their energy is to put the observer in a frame where the photons have equal energy and are moving directly away from each other. In this frame, the observer is now moving in the same direction and speed as the center of mass of the two photons. The total momentum of the photons is now zero, since their momenta are equal and opposite. In this frame the two photons, as a system, have a mass equal to their total energy divided by c^2. This mass is called the invariant mass of the pair of photons together. It is the smallest mass and energy the system may be seen to have, by any observer. It is only the invariant mass of a two-photon system that can be used to make a single particle with the same rest mass.

If the photons are formed by the collision of a particle and an antiparticle, the invariant mass is the same as the total energy of the particle and antiparticle (their rest energy plus the kinetic energy), in the center of mass frame, where they will automatically be moving in equal and opposite directions (since they have equal momentum in this frame). If the photons are formed by the disintegration of a *single* particle with a well-defined rest mass, like the neutral pion, the invariant mass of the photons is equal to rest mass of the pion. In this case, the center of mass frame for the pion is just the frame where the pion is at rest, and the center of mass does not change after it disintegrates into two photons. After the two photons are formed, their center of mass is still moving the same way the pion did, and their total energy in this frame adds up to the mass energy of the pion. Thus, by calculating the invariant mass of pairs of photons in a particle detector, pairs can be identified that were probably produced by pion disintegration.

A similar calculation illustrates that the invariant mass of systems is conserved, even when massive particles (particles with rest mass) within the system are converted to massless particles (such as photons). In such cases, the photons contribute invariant mass to the system, even though they individually have no invariant mass or rest mass. Thus, an electron and positron (each of which has rest mass) may undergo annihilation with each other to produce two photons, each of which is massless (has no rest mass). However, in such circumstances, no system mass is lost. Instead, the system of both photons moving away from each other has an invariant mass, which acts like a rest mass for any system in which the photons are trapped, or that can be weighed. Thus, not only the quantity of relativistic mass, but also the quantity of invariant mass does not change in transformations between "matter" (electrons and positrons) and energy (photons).

7.4.4 Relation to gravity

In physics, there are two distinct concepts of mass: the gravitational mass and the inertial mass. The gravitational mass is the quantity that determines the strength of the gravitational field generated by an object, as well as the gravitational force acting on the object when it is immersed in a gravitational field produced by other bodies. The inertial mass, on the other hand, quantifies how much an object accelerates if a given force is applied to it. The mass–energy equivalence in special relativity refers to the inertial mass. However, already in the context of Newton gravity, the Weak Equivalence Principle is postulated: the gravitational and the inertial mass of every object are the same. Thus, the mass–energy equivalence, combined with the Weak Equivalence Principle, results in the prediction that all forms of energy contribute to the gravitational field generated by an object. This observation is one of the pillars of the general theory of relativity.

The above prediction, that all forms of energy interact gravitationally, has been subject to experimental tests. The first observation testing this prediction was made in 1919.[31] During a solar eclipse, Arthur Eddington observed that the light from stars passing close to the Sun was bent. The effect is due to the gravitational attraction of light by the Sun. The observation confirmed that the energy carried by light indeed is equivalent to a gravitational mass. Another seminal experiment, the Pound–Rebka experiment, was performed in 1960.[32] In this test a beam of light was emitted from the top of a tower and detected at the bottom. The frequency of the light detected was higher than the light emitted. This result confirms that the energy of photons increases when they fall in the gravitational field of the Earth. The energy, and therefore the gravitational mass, of photons is proportional to their frequency as stated by the Planck's relation.

7.5 Application to nuclear physics

Task Force One, *the world's first nuclear-powered task force.* Enterprise, Long Beach *and* Bainbridge *in formation in the Mediterranean, 18 June 1964.* Enterprise *crew members are spelling out Einstein's Mass–Energy Equivalence formula* $E = mc^2$ *on the flight deck.*

Max Planck pointed out that the mass–energy equivalence formula implied that bound systems would have a mass less than the sum of their constituents, once the binding energy had been allowed to escape. However, Planck was thinking about chemical reactions, where the binding energy is too small to measure. Einstein suggested that radioactive materials such as radium would provide a test of the theory, but even though a large amount of energy is released per atom in radium, due to the half-life of the substance (1602 years), only a small fraction of radium atoms decay over an experimentally measurable period of time.

Once the nucleus was discovered, experimenters realized that the very high binding energies of the atomic nuclei should allow calculation of their binding energies, simply from mass differences. But it was not until the discovery of the neutron in 1932, and the measurement of the neutron mass, that this calculation could actually be performed (see nuclear binding energy for example calculation). A little while later, the first transmutation reactions (such as[33] the Cockcroft–Walton experiment: $^7Li + p \rightarrow 2\ ^4He$) verified Einstein's formula to an accuracy of ±0.5%. In 2005, Rainville et al. published a direct test of the energy-equivalence of mass lost in the binding energy of a neutron to atoms of particular isotopes of silicon and sulfur, by comparing the mass lost to the energy of the emitted gamma ray associated with the neutron capture. The binding mass-loss agreed with the gamma ray energy to a precision of ±0.00004%, the most accurate test of $E = mc^2$ to date.[4]

The mass–energy equivalence formula was used in the understanding of nuclear fission reactions, and implies the great amount of energy that can be released by a nuclear fission chain reaction, used in both nuclear weapons and nuclear power. By measuring the mass of different atomic nuclei and subtracting from that number the total mass of the protons and neutrons as they would weigh separately, one gets the exact binding energy available in an atomic nucleus. This is used to calculate the energy released in any nuclear reaction, as the difference in the total mass of the nuclei that enter and exit the reaction.

7.6 Practical examples

Einstein used the CGS system of units (centimeters, grams, seconds, dynes, and ergs), but the formula is independent of the system of units. In natural units, the numerical value of the speed of light is set to equal 1, and the formula expresses an equality of numerical values: $E = m$. In the SI system (expressing the ratio E/m in joules per kilogram using the value of c in meters per second):

$E / m = c^2 = (299{,}792{,}458 \text{ m/s})^2 = 89{,}875{,}517{,}873{,}681{,}764$ J/kg ($\approx 9.0 \times 10^{16}$ joules per kilogram).

So the energy equivalent of one gram (1/1000 of a kilogram) of mass is equivalent to:

- 89.9 terajoules
- 25.0 million kilowatt-hours ($\approx$ 25 GW·h)
- 21.5 billion kilocalories ($\approx$ 21 Tcal) [34]
- 85.2 billion BTUs[34]

or to the energy released by combustion of the following:

- 21.5 kilotons of TNT-equivalent energy ($\approx$ 21 kt) [34]
- 568,000 US gallons of automotive gasoline

Any time energy is generated, the process can be evaluated from an $E = mc^2$ perspective. For instance, the "Gadget"-style bomb used in the Trinity test and the bombing of Nagasaki had an explosive yield equivalent to 21 kt of TNT. About 1 kg of the approximately 6.15 kg of plutonium in each of these bombs fissioned into lighter elements totaling almost exactly one gram less, after cooling. The electromagnetic radiation and kinetic energy (thermal and blast energy) released in this explosion carried the missing one gram of mass.[35] This occurs because nuclear binding energy is released whenever elements with more than 62 nucleons fission.

Another example is hydroelectric generation. The electrical energy produced by Grand Coulee Dam's turbines every 3.7 hours represents one gram of mass. This mass passes to the electrical devices (such as lights in cities) which are powered by the generators, where it appears as a gram of heat and light.[36] Turbine designers look at their equations in terms of pressure, torque, and RPM. However, Einstein's equations show that all energy has mass, and thus the electrical energy produced by a dam's generators, and the heat and light which result from it, all retain their mass, which is equivalent to the energy. The potential energy—and equivalent mass—represented by the waters of the Columbia River as it descends to the Pacific Ocean would be converted to heat due to viscous friction and the turbulence of white water rapids and waterfalls were it not for the dam and its generators. This heat would remain as mass on site at the water, were it not for the equipment which converted some of this potential and kinetic energy into electrical energy, which can be moved from place to place (taking mass with it).

Whenever energy is added to a system, the system gains mass:

- A spring's mass increases whenever it is put into compression or tension. Its added mass arises from the added potential energy stored within it, which is bound in the stretched chemical (electron) bonds linking the atoms within the spring.

- Raising the temperature of an object (increasing its heat energy) increases its mass. For example, consider the world's primary mass standard for the kilogram, made of platinum/iridium. If its temperature is allowed to change by 1 °C, its mass will change by 1.5 picograms (1 pg = 1×10^{-12} g).[37]

- A spinning ball will weigh more than a ball that is not spinning. Its increase of mass is exactly the equivalent of the mass of energy of rotation, which is itself the sum of the kinetic energies of all the moving parts of the ball. For example, the Earth itself is more massive due to its daily rotation, than it would be with no rotation. This rotational energy (2.14×10^{29} J) represents 2.38 billion metric tons of added mass.[38]

Note that no net mass or energy is really created or lost in any of these examples and scenarios. Mass/energy simply moves from one place to another. These are some examples of the *transfer* of energy and mass in accordance with the *principle of mass–energy conservation*.

7.7 Efficiency

Although mass cannot be converted to energy,[24] in some reactions matter particles (which contain a form of rest energy) can be destroyed and converted to other types of energy which are more usable and obvious as forms of energy, such as light and energy of motion (heat, etc.). However, the total amount of energy and mass does not change in such a transformation. Even when particles are not destroyed, a certain fraction of the ill-defined "matter" in ordinary objects can be destroyed, and its associated energy liberated and made available as the more dramatic energies of light and heat, even though no identifiable real particles are destroyed, and even though (again) the total energy is unchanged (as also the total mass). Such conversions between types of energy (resting to active energy) happen in nuclear weapons, in which the protons and neutrons in atomic nuclei lose a small fraction of their average mass, but this mass loss is not due to the destruction of any protons or neutrons (or even, in general, lighter particles like electrons). Also the mass is not destroyed, but simply removed from the system. in the form of heat and light from the reaction.

In nuclear reactions, typically only a small fraction of the total mass–energy of the bomb is converted into the mass–energy of heat, light, radiation and motion, which are "active" forms which can be used. When an atom fissions, it loses only about 0.1% of its mass (which escapes from the system and does not disappear), and additionally, in a bomb or reactor not all the atoms can fission. In a modern fission-based atomic bomb, the efficiency is only about 40%, so only 40% of the fissionable atoms actually fission, and only about 0.03% of the fissile core mass appears as energy in the end. In nuclear fusion, more of the mass is released as usable energy, roughly 0.3%. But in a fusion bomb, the bomb mass is partly casing and non-reacting components, so that in practicality, again (coincidentally) no more than about 0.03% of the total mass of the entire weapon is released as usable energy (which, again, retains the "missing" mass). See nuclear weapon yield for practical details of this ratio in modern nuclear weapons.

In theory, it should be possible to destroy matter and convert all of the rest-energy associated with matter into heat and light (which would of course have the same mass), but none of the theoretically known methods are practical. One way to convert all the energy within matter into usable energy is to annihilate matter with antimatter. But antimatter is rare in our universe, and must be made first. Due to inefficient mechanisms of production, making antimatter always requires far more usable energy than would be released when it was annihilated.

Since most of the mass of ordinary objects resides in protons and neutrons, in order to convert all of the energy of ordinary matter into a more useful type of energy, the protons and neutrons must be converted to lighter particles, or else particles with no rest-mass at all. In the Standard Model of particle physics, the number of protons plus neutrons is nearly exactly conserved. Still, Gerard 't Hooft showed that there is a process which will convert protons and neutrons to antielectrons and neutrinos.[39] This is the weak SU(2) instanton proposed by Belavin Polyakov Schwarz and Tyupkin.[40] This process, can in principle destroy matter and convert all the energy of matter into neutrinos and usable energy, but it is normally extraordinarily slow. Later it became clear that this process will happen at a fast rate at very high temperatures,[41] since then instanton-like configurations will be copiously produced from thermal fluctuations. The temperature required is so high that it would only have been reached shortly after the big bang.

Many extensions of the standard model contain magnetic monopoles, and in some models of grand unification, these monopoles catalyze proton decay, a process known as the Callan–Rubakov effect.[42] This process would be an efficient mass–energy conversion at ordinary temperatures, but it requires making monopoles and anti-monopoles first. The energy required to produce monopoles is believed to be enormous, but magnetic charge is conserved, so that the lightest monopole is stable. All these properties are deduced in theoretical models—magnetic monopoles have never been observed, nor have they been produced in any experiment so far.

A third known method of total matter–energy "conversion" (which again in practice only means conversion of one type of energy into a different type of energy), is using gravity, specifically black holes. Stephen Hawking theorized[43] that black holes radiate thermally with no regard to how they are formed. So it is theoretically possible to throw matter into a black hole and use the emitted heat to generate power. According to the theory of Hawking radiation, however, the black hole used will radiate at a higher rate the smaller it is, producing usable powers at only small black hole masses, where usable may for example be something greater than the local background radiation. It is also worth noting that the ambient irradiated power would change with the mass of the black hole, increasing as the mass of the black hole decreases, or decreasing as the mass increases, at a rate where power is proportional to the inverse square of the mass. In a "practical" scenario, mass and energy could be dumped into the black hole to regulate this growth, or keep its size, and thus power output, near constant. This could result from the fact that mass and energy are lost from the hole with its thermal radiation.

7.8 Background

7.8.1 Mass–velocity relationship

In developing special relativity, Einstein found that the kinetic energy of a moving body is

$$E_k = m_0(\gamma - 1)c^2 = \frac{m_0 c^2}{\sqrt{1 - \frac{v^2}{c^2}}} - m_0 c^2,$$

with v the velocity, m_0 the rest mass, and γ the Lorentz factor.

He included the second term on the right to make sure that for small velocities the energy would be the same as in classical mechanics, thus satisfying the correspondence principle:

$$E_k = \frac{1}{2} m_0 v^2 + \cdots$$

Without this second term, there would be an additional contribution in the energy when the particle is not moving.

Einstein found that the total momentum of a moving particle is:

$$P = \frac{m_0 v}{\sqrt{1 - \frac{v^2}{c^2}}}.$$

and it is this quantity which is conserved in collisions. The ratio of the momentum to the velocity is the relativistic mass, m.

$$m = \frac{m_0}{\sqrt{1 - \frac{v^2}{c^2}}}$$

And the relativistic mass and the relativistic kinetic energy are related by the formula:

$$E_k = mc^2 - m_0 c^2.$$

Einstein wanted to omit the unnatural second term on the right-hand side, whose only purpose is to make the energy at rest zero, and to declare that the particle has a total energy which obeys:

$$E = mc^2$$

which is a sum of the rest energy $m_0 c^2$ and the kinetic energy. This total energy is mathematically more elegant, and fits better with the momentum in relativity. But to come to this conclusion, Einstein needed to think carefully about collisions. This expression for the energy implied that matter at rest has a huge amount of energy, and it is not clear whether this energy is physically real, or just a mathematical artifact with no physical meaning.

In a collision process where all the rest-masses are the same at the beginning as at the end, either expression for the energy is conserved. The two expressions only differ by a constant which is the same at the beginning and at the end of the collision. Still, by analyzing the situation where particles are thrown off a heavy central particle, it is easy to see that the inertia of the central particle is reduced by the total energy emitted. This allowed Einstein to conclude that the inertia of a heavy particle is increased or diminished according to the energy it absorbs or emits.

7.8.2 Relativistic mass

Main article: Mass in special relativity

After Einstein first made his proposal, it became clear that the word mass can have two different meanings. Some denote the *relativistic mass* with an explicit index:

$$m_{\text{rel}} = \frac{m_0}{\sqrt{1 - \frac{v^2}{c^2}}}.$$

This mass is the ratio of momentum to velocity, and it is also the relativistic energy divided by c^2 (it is not Lorentz-invariant, in contrast to m_0). The equation $E = m_{\text{rel}} c^2$ holds for moving objects. When the velocity is small, the relativistic mass and the rest mass are almost exactly the same.

- $E = mc^2$ either means $E = m_0 c^2$ for an object at rest, or $E = m_{\text{rel}} c^2$ when the object is moving.

Also Einstein (following Hendrik Lorentz and Max Abraham) used velocity- and direction-dependent mass concepts (longitudinal and transverse mass) in his 1905 electrodynamics paper and in another paper in 1906.[44][45] However, in his first paper on $E = mc^2$ (1905), he treated m as what would now be called the *rest mass*.[7] Some claim that (in later years) he did not like the idea of "relativistic mass".[2] When modern physicists say "mass", they are usually talking about rest mass, since if they meant "relativistic mass", they would just say "energy".

Considerable debate has ensued over the use of the concept "relativistic mass" and the connection of "mass" in relativity to "mass" in Newtonian dynamics. For example, one view is that only rest mass is a viable concept and is a property of the particle; while relativistic mass is a conglomeration of particle properties and properties of spacetime. A perspective that avoids this debate, due to Kjell Vøyenli, is that the Newtonian concept of mass as a particle property and the relativistic concept of mass have to be viewed as embedded in their own theories and as having no precise connection.[46][47]

7.8.3 Low speed expansion

We can rewrite the expression $E = \gamma m_0 c^2$ as a Taylor series:

$$E = m_0 c^2 \left[1 + \frac{1}{2}\left(\frac{v}{c}\right)^2 + \frac{3}{8}\left(\frac{v}{c}\right)^4 + \frac{5}{16}\left(\frac{v}{c}\right)^6 + \ldots \right].$$

For speeds much smaller than the speed of light, higher-order terms in this expression get smaller and smaller because v/c is small. For low speeds we can ignore all but the first two terms:

$$E \approx m_0 c^2 + \frac{1}{2} m_0 v^2.$$

The total energy is a sum of the rest energy and the Newtonian kinetic energy.

The classical energy equation ignores both the $m_0 c^2$ part, and the high-speed corrections. This is appropriate, because all the high-order corrections are small. Since only *changes* in energy affect the behavior of objects, whether we include the $m_0 c^2$ part makes no difference, since it is constant. For the same reason, it is possible to subtract the rest energy from the total energy in relativity. By considering the emission of energy in different frames, Einstein could show that the rest energy has a real physical meaning.

The higher-order terms are extra correction to Newtonian mechanics which become important at higher speeds. The Newtonian equation is only a low-speed approximation, but an extraordinarily good one. All of the calculations used in putting astronauts on the moon, for example, could have been done using Newton's equations without any of the higher-order corrections. The total mass energy equivalence should also include the rotational and vibrational kinetic energies as well as the linear kinetic energy at low speeds.

7.9 History

While Einstein was the first to have correctly deduced the mass–energy equivalence formula, he was not the first to have related energy with mass. But nearly all previous authors thought that the energy which contributes to mass comes only from electromagnetic fields.[48][49][50][51]

7.9.1 Newton: matter and light

In 1717 Isaac Newton speculated that light particles and matter particles were inter-convertible in "Query 30" of the *Opticks*, where he asks:

> Are not the gross bodies and light convertible into one another, and may not bodies receive much of their activity from the particles of light which enter their composition?

7.9.2 Swedenborg: matter composed of "pure and total motion"

In 1734 the Swedish scientist and theologian Emanuel Swedenborg in his *Principia* theorized that all matter is ultimately composed of dimensionless points of "pure and total motion." He described this motion as being without force, direction or speed, but having the potential for force, direction and speed everywhere within it.[52][53]

7.9.3 Electromagnetic mass

Main article: Electromagnetic mass

There were many attempts in the 19th and the beginning of the 20th century—like those of J. J. Thomson (1881), Oliver Heaviside (1888), and George Frederick Charles Searle (1897), Wilhelm Wien (1900), Max Abraham (1902), Hendrik Antoon Lorentz (1904) — to understand how the mass of a charged object depends on the electrostatic field.[48][49] This concept was called electromagnetic mass, and was considered as being dependent on velocity and direction as well. Lorentz (1904) gave the following expressions for longitudinal and transverse electromagnetic mass:

$$m_L = \frac{m_0}{\left(\sqrt{1-\frac{v^2}{c^2}}\right)^3}, \quad m_T = \frac{m_0}{\sqrt{1-\frac{v^2}{c^2}}}$$

where

$$m_0 = \frac{4}{3}\frac{E_{em}}{c^2}$$

7.9.4 Radiation pressure and inertia

Main article: Electromagnetic mass § Inertia of energy and radiation paradoxes

Another way of deriving some sort of electromagnetic mass was based on the concept of radiation pressure. In 1900, Henri Poincaré associated electromagnetic radiation energy with a "fictitious fluid" having momentum and mass

$$m_{em} = E_{em}/c^2.$$

By that, Poincaré tried to save the center of mass theorem in Lorentz's theory, though his treatment led to radiation paradoxes.[51]

Friedrich Hasenöhrl showed in 1904, that electromagnetic cavity radiation contributes the "apparent mass"

$$m_0 = \frac{4}{3}\frac{E_{em}}{c^2}$$

to the cavity's mass. He argued that this implies mass dependence on temperature as well.[54]

7.9.5 Einstein: mass–energy equivalence

Albert Einstein did not formulate exactly the formula $E = mc^2$ in his 1905 *Annus Mirabilis* paper "Does the Inertia of an object Depend Upon Its Energy Content?";[7] rather, the paper states that if a body gives off the energy L in the form of radiation, its mass diminishes by L/c^2. (Here, "radiation" means electromagnetic radiation, or light, and mass means the

7.9. HISTORY

ordinary Newtonian mass of a slow-moving object.) This formulation relates only a change Δm in mass to a change L in energy without requiring the absolute relationship.

Objects with zero mass presumably have zero energy, so the extension that all mass is proportional to energy is obvious from this result. In 1905, even the hypothesis that changes in energy are accompanied by changes in mass was untested. Not until the discovery of the first type of antimatter (the positron in 1932) was it found that all of the mass of pairs of resting particles could be converted to radiation.

The first derivation by Einstein (1905)

Already in his relativity paper "On the electrodynamics of moving bodies", Einstein derived the correct expression for the kinetic energy of particles:

$$E_k = mc^2 \left(\frac{1}{\sqrt{1 - \frac{v^2}{c^2}}} - 1 \right)$$

Now the question remained open as to which formulation applies to bodies at rest. This was tackled by Einstein in his paper "Does the inertia of a body depend upon its energy content?". Einstein used a body emitting two light pulses in opposite directions, having energies of E_0 before and E_1 after the emission as seen in its rest frame. As seen from a moving frame, this becomes H_0 and H_1. Einstein obtained:

$$(H_0 - E_0) - (H_1 - E_1) = E \left(\frac{1}{\sqrt{1 - \frac{v^2}{c^2}}} - 1 \right)$$

then he argued that $H - E$ can only differ from the kinetic energy K by an additive constant, which gives

$$K_0 - K_1 = E \left(\frac{1}{\sqrt{1 - \frac{v^2}{c^2}}} - 1 \right)$$

Neglecting effects higher than third order in v/c after a Taylor series expansion of the right side of this gives:

$$K_0 - K_1 = \frac{E}{c^2} \frac{v^2}{2}.$$

Einstein concluded that the emission reduces the body's mass by E/c^2, and that the mass of a body is a measure of its energy content.

The correctness of Einstein's 1905 derivation of $E = mc^2$ was criticized by Max Planck (1907), who argued that it is only valid to first approximation. Another criticism was formulated by Herbert Ives (1952) and Max Jammer (1961), asserting that Einstein's derivation is based on begging the question.[5][55] On the other hand, John Stachel and Roberto Torretti (1982) argued that Ives' criticism was wrong, and that Einstein's derivation was correct.[56] Hans Ohanian (2008) agreed with Stachel/Torretti's criticism of Ives, though he argued that Einstein's derivation was wrong for other reasons.[57] For a recent review, see Hecht (2011).[6]

Alternative version

An alternative version of Einstein's thought experiment was proposed by Fritz Rohrlich (1990), who based his reasoning on the Doppler effect.[58] Like Einstein, he considered a body at rest with mass M. If the body is examined in a frame moving with nonrelativistic velocity v, it is no longer at rest and in the moving frame it has momentum $P = Mv$. Then he supposed the body emits two pulses of light to the left and to the right, each carrying an equal amount of energy $E/2$. In its rest frame, the object remains at rest after the emission since the two beams are equal in strength and carry opposite momentum.

But if the same process is considered in a frame moving with velocity v to the left, the pulse moving to the left will be redshifted while the pulse moving to the right will be blue shifted. The blue light carries more momentum than the red light, so that the momentum of the light in the moving frame is not balanced: the light is carrying some net momentum to the right.

The object has not changed its velocity before or after the emission. Yet in this frame it has lost some right-momentum to the light. The only way it could have lost momentum is by losing mass. This also solves Poincaré's radiation paradox, discussed above.

The velocity is small, so the right-moving light is blueshifted by an amount equal to the nonrelativistic Doppler shift factor $1 - v/c$. The momentum of the light is its energy divided by c, and it is increased by a factor of v/c. So the right-moving light is carrying an extra momentum ΔP given by:

$$\Delta P = \frac{v}{c}\frac{E}{2c}.$$

The left-moving light carries a little less momentum, by the same amount ΔP. So the total right-momentum in the light is twice ΔP. This is the right-momentum that the object lost.

$$2\Delta P = v\frac{E}{c^2}.$$

The momentum of the object in the moving frame after the emission is reduced to this amount:

$$P' = Mv - 2\Delta P = \left(M - \frac{E}{c^2}\right)v.$$

So the change in the object's mass is equal to the total energy lost divided by c^2. Since any emission of energy can be carried out by a two step process, where first the energy is emitted as light and then the light is converted to some other form of energy, any emission of energy is accompanied by a loss of mass. Similarly, by considering absorption, a gain in energy is accompanied by a gain in mass.

Relativistic center-of-mass theorem (1906)

Like Poincaré, Einstein concluded in 1906 that the inertia of electromagnetic energy is a necessary condition for the center-of-mass theorem to hold. On this occasion, Einstein referred to Poincaré's 1900 paper and wrote:[59]

> Although the merely formal considerations, which we will need for the proof, are already mostly contained in a work by H. Poincaré[2], for the sake of clarity I will not rely on that work.[60]

In Einstein's more physical, as opposed to formal or mathematical, point of view, there was no need for fictitious masses. He could avoid the *perpetuum mobile* problem, because on the basis of the mass–energy equivalence he could show that the transport of inertia which accompanies the emission and absorption of radiation solves the problem. Poincaré's rejection of the principle of action–reaction can be avoided through Einstein's $E = mc^2$, because mass conservation appears as a special case of the energy conservation law.

7.9.6 Others

During the nineteenth century there were several speculative attempts to show that mass and energy were proportional in various ether theories.[61] In 1873 Nikolay Umov pointed out a relation between mass and energy for ether in the form of $E = kmc^2$, where $0.5 \leq k \leq 1$.[62] The writings of Samuel Tolver Preston,[63][64] and a 1903 paper by Olinto De Pretto,[65][66] presented a mass–energy relation. De Pretto's paper received recent press coverage when Umberto Bartocci discovered that there were only three degrees of separation linking De Pretto to Einstein, leading Bartocci to conclude that Einstein was probably aware of De Pretto's work.[67]

Preston and De Pretto, following Le Sage, imagined that the universe was filled with an ether of tiny particles which are always moving at speed c. Each of these particles have a kinetic energy of mc^2 up to a small numerical factor. The nonrelativistic kinetic energy formula did not always include the traditional factor of 1/2, since Leibniz introduced kinetic energy without it, and the 1/2 is largely conventional in prerelativistic physics.[68] By assuming that every particle has a mass which is the sum of the masses of the ether particles, the authors would conclude that all matter contains an amount of kinetic energy either given by $E = mc^2$ or $2E = mc^2$ depending on the convention. A particle ether was usually considered unacceptably speculative science at the time,[69] and since these authors did not formulate relativity, their reasoning is completely different from that of Einstein, who used relativity to change frames.

Independently, Gustave Le Bon in 1905 speculated that atoms could release large amounts of latent energy, reasoning from an all-encompassing qualitative philosophy of physics.[70][71]

7.9.7 Radioactivity and nuclear energy

It was quickly noted after the discovery of radioactivity in 1897, that the total energy due to radioactive processes is about one *million times* greater than that involved in any known molecular change. However, it raised the question where this energy is coming from. After eliminating the idea of absorption and emission of some sort of Lesagian ether particles, the existence of a huge amount of latent energy, stored within matter, was proposed by Ernest Rutherford and Frederick Soddy in 1903. Rutherford also suggested that this internal energy is stored within normal matter as well. He went on to speculate in 1904:[72][73]

> If it were ever found possible to control at will the rate of disintegration of the radio-elements, an enormous amount of energy could be obtained from a small quantity of matter.

Einstein's equation is in no way an explanation of the large energies released in radioactive decay (this comes from the powerful nuclear forces involved; forces that were still unknown in 1905). In any case, the enormous energy released from radioactive decay (which had been measured by Rutherford) was much more easily measured than the (still small) change in the gross mass of materials, as a result. Einstein's equation, by theory, can give these energies by measuring mass differences before and after reactions, but in practice, these mass differences in 1905 were still too small to be measured in bulk. Prior to this, the ease of measuring radioactive decay energies with a calorimeter was thought possibly likely to allow measurement of changes in mass difference, as a check on Einstein's equation itself. Einstein mentions in his 1905 paper that mass–energy equivalence might perhaps be tested with radioactive decay, which releases enough energy (the quantitative amount known roughly by 1905) to possibly be "weighed," when missing from the system (having been given off as heat). However, radioactivity seemed to proceed at its own unalterable (and quite slow, for radioactives known then) pace, and even when simple nuclear reactions became possible using proton bombardment, the idea that these great amounts of usable energy could be liberated at will with any practicality, proved difficult to substantiate. Rutherford was reported in 1933 to have declared that this energy could not be exploited efficiently: "Anyone who expects a source of power from the transformation of the atom is talking moonshine."[74]

This situation changed dramatically in 1932 with the discovery of the neutron and its mass, allowing mass differences for single nuclides and their reactions to be calculated directly, and compared with the sum of masses for the particles that made up their composition. In 1933, the energy released from the reaction of lithium-7 plus protons giving rise to 2 alpha particles (as noted above by Rutherford), allowed Einstein's equation to be tested to an error of ±0.5%. However, scientists still did not see such reactions as a source of power.

After the very public demonstration of huge energies released from nuclear fission after the atomic bombings of Hiroshima and Nagasaki in 1945, the equation $E = mc^2$ became directly linked in the public eye with the power and peril of nuclear

The popular connection between Einstein, E = mc², and the atomic bomb was prominently indicated on the cover of Time *magazine in July 1946 by the writing of the equation on the mushroom cloud.*

weapons. The equation was featured as early as page 2 of the Smyth Report, the official 1945 release by the US government on the development of the atomic bomb, and by 1946 the equation was linked closely enough with Einstein's work that the cover of *Time* magazine prominently featured a picture of Einstein next to an image of a mushroom cloud emblazoned with the equation.[75] Einstein himself had only a minor role in the Manhattan Project: he had cosigned a letter to the U.S. President in 1939 urging funding for research into atomic energy, warning that an atomic bomb was theoretically possible. The letter persuaded Roosevelt to devote a significant portion of the wartime budget to atomic research. Without a security clearance, Einstein's only scientific contribution was an analysis of an isotope separation method in theoretical terms. It was inconsequential, on account of Einstein not being given sufficient information (for security reasons) to fully work on the problem.[76]

While $E = mc^2$ is useful for understanding the amount of energy potentially released in a fission reaction, it was not strictly necessary to develop the weapon, once the fission process was known, and its energy measured at 200 MeV (which was directly possible, using a quantitative Geiger counter, at that time). As the physicist and Manhattan Project participant Robert Serber put it: "Somehow the popular notion took hold long ago that Einstein's theory of relativity, in particular his famous equation $E = mc^2$, plays some essential role in the theory of fission. Albert Einstein had a part in alerting the United States government to the possibility of building an atomic bomb, but his theory of relativity is not required in discussing fission. The theory of fission is what physicists call a non-relativistic theory, meaning that relativistic effects are too small to affect the dynamics of the fission process significantly."[77] However the association between $E = mc^2$ and nuclear energy has since stuck, and because of this association, and its simple expression of the ideas of Albert Einstein himself, it has become "the world's most famous equation".[78]

While Serber's view of the strict lack of need to use mass–energy equivalence in designing the atomic bomb is correct, it does not take into account the pivotal role which this relationship played in making the fundamental leap to the initial hypothesis that large atoms were energetically *allowed* to split into approximately equal parts (before this energy was in fact measured). In late 1938, while on the winter walk on which they solved the meaning of Hahn's experimental results and introduced the idea that would be called atomic fission, Lise Meitner and Otto Robert Frisch made direct use of Einstein's equation to help them understand the quantitative energetics of the reaction which overcame the "surface tension-like" forces holding the nucleus together, and allowed the fission fragments to separate to a configuration from which their charges could force them into an energetic "fission". To do this, they made use of "packing fraction", or nuclear binding energy values for elements, which Meitner had memorized. These, together with use of $E = mc^2$ allowed them to realize on the spot that the basic fission process was energetically possible:

> ...We walked up and down in the snow, I on skis and she on foot. ...and gradually the idea took shape... explained by Bohr's idea that the nucleus is like a liquid drop; such a drop might elongate and divide itself... We knew there were strong forces that would resist, ..just as surface tension. But nuclei differed from ordinary drops. At this point we both sat down on a tree trunk and started to calculate on scraps of paper. ...the Uranium nucleus might indeed be a very wobbly, unstable drop, ready to divide itself... But, ...when the two drops separated they would be driven apart by electrical repulsion, about 200 MeV in all. Fortunately Lise Meitner remembered how to compute the masses of nuclei... and worked out that the two nuclei formed... would be lighter by about one-fifth the mass of a proton. Now whenever mass disappears energy is created, according to Einstein's formula E = mc^2, and... the mass was just equivalent to 200 MeV; it all fitted![79][80]

7.10 See also

- Energy density

- Index of energy articles

- Index of wave articles

- Outline of energy

7.11 References

[1] http://fr.wikisource.org/wiki/La_Théorie_de_Lorentz_et_le_principe_de_réaction

[2] See e.g. Lev B.Okun, *The concept of Mass*, Physics Today **42** (6), June 1969, p. 31–36, http://www.physicstoday.org/vol-42/iss-6/vol42no6p31_36.pdf

[3] Paul Allen Tipler, Ralph A. Llewellyn (January 2003), *Modern Physics*, W. H. Freeman and Company, pp. 87–88, ISBN 0-7167-4345-0

[4] Rainville, S. et al. World Year of Physics: A direct test of E = mc2.*Nature*438, 1096–1097 (22 December 2005) ldoi:10.1038/43810; Published online 21 December 2005.

[5] Jammer, Max (1997) [1961], *Concepts of Mass in Classical and Modern Physics*, New York: Dover, ISBN 0-486-29998-8

[6] Hecht, Eugene (2011), "How Einstein confirmed E0=mc2",*American Journal of Physics***79**(6): 591–600,Bibcode:2011AmJPh.., doi:10.1119/1.3549223

[7] Einstein, A. (1905), "Ist die Trägheit eines Körpers von seinem Energieinhalt abhängig?", *Annalen der Physik* **18** (13): 639–643, Bibcode:1905AnP...323..639E, doi:10.1002/andp.19053231314. See also the English translation.

[8] See the sentence on the last page (p. 641) of the original German edition, above the equation $K_0 - K_1 = L/V^2\ v^2/2$. See also the sentence above the last equation in the English translation, $K_0 - K_1 = (1/2)(L/c^2)v^2$, and the comment on the symbols used in *About this edition* that follows the translation.

[9] Einstein, Albert (1907), "Über die vom Relativitätsprinzip geforderte Trägheit der Energie" (PDF), *Annalen der Physik* **328** (7): 371–384, Bibcode:1907AnP...328..371E, doi:10.1002/andp.19073280713

[10] Planck, Max (1907), "Zur Dynamik bewegter Systeme", *Sitzungsberichte der Königlich-Preussischen Akademie der Wissenschaften, Berlin*, Erster Halbband (29): 542–570

English Wikisource translation: On the Dynamics of Moving Systems

[11] Stark, J. (1907), "Elementarquantum der Energie, Modell der negativen und der positiven Elekrizität", *Physikalische Zeitschrift* **24** (8): 881

[12] Einstein, Albert (1908), "Über das Relativitätsprinzip und die aus demselben gezogenen Folgerungen" (PDF), *Jahrbuch der Radioaktivität und Elektronik* **4**: 411–462, Bibcode:1908JRE.....4..411E

[13] Schwartz, H. M. (1977), "Einstein's comprehensive 1907 essay on relativity, part II", *American Journal of Physics* **45** (9): 811–817, Bibcode:1977AmJPh..45..811S, doi:10.1119/1.11053

[14] Lewis, Gilbert N. & Tolman, Richard C. (1909), "The Principle of Relativity, and Non-Newtonian Mechanics", *Proceedings of the American Academy of Arts and Sciences* **44** (25): 709–726, doi:10.2307/20022495

[15] Lorentz, Hendrik Antoon (1914), *Das Relativitätsprinzip. Drei Vorlesungen gehalten in Teylers Stiftung zu Haarlem (1913)*, Leipzig and Berlin: B.G. Teubner

[16] Laue, Max von (1911), "Zur Dynamik der Relativitätstheorie",*Annalen der Physik***340**(8): 524–542,Bibcode:1911AnP...340..52, doi:10.1002/andp.19113400808

English Wikisource translation: On the Dynamics of the Theory of Relativity

[17] Klein, Felix (1918), "Über die Integralform der Erhaltungssätze und die Theorie der räumlich-geschlossenen Welt", *Göttinger Nachrichten*: 394–423

[18] A.Einstein $E = mc^2$: *the most urgent problem of our time* Science illustrated, vol. 1 no. 1, April issue, pp. 16–17, 1946 (item 417 in the "Bibliography"

[19] M.C.Shields *Bibliography of the Writings of Albert Einstein to May 1951* in Albert Einstein: Philosopher-Scientist by Paul Arthur Schilpp (Editor) Albert Einstein Philosopher – Scientist

[20] "Einstein was unequivocally against the traditional idea of conservation of mass. He had concluded that mass and energy were essentially one and the same; 'inert[ial] mass is simply latent energy.'[ref...]. He made his position known publicly time and again[ref...]...", Eugene Hecht, "Einstein on mass and energy" Am. J. Phys., Vol. 77, No. 9, September 2009, online.

7.11. REFERENCES

[21] "There followed also the principle of the equivalence of mass and energy, with the laws of conservation of mass and energy becoming one and the same.", Albert Einstein, "Considerations Concerning the Fundaments of Theoretical Physics", Science, Washington, DC, vol. 91, no. 2369, May 24th, 1940 scanned image online

[22] page 14 (preview online) of Albert Einstein, *The Theory of Relativity (And Other Essays)*, Citadel Press, 1950.

[23] In F. Fernflores. The Equivalence of Mass and Energy. Stanford Encyclopedia of Philosophy.

[24] E. F. Taylor and J. A. Wheeler, *Spacetime Physics*, W.H. Freeman and Co., NY. 1992. ISBN 0-7167-2327-1, see pp. 248–9 for discussion of mass remaining constant after detonation of nuclear bombs, until heat is allowed to escape.

[25] Note that the relativistic mass, in contrast to the rest mass m_0, *is not a relativistic invariant, and that the velocity* $v = dx^{(4)}/dt$ *is not a Minkowski four-vector, in contrast to the quantity* $\tilde{v} = dx^{(4)}/d\tau$, *where* $d\tau = dt \cdot \sqrt{1 - (v^2/c^2)}$ *is the differential of the proper time. However, the energy–momentum four-vector* $p^{(4)} = m_0 \cdot dx^{(4)}/d\tau$ *is a genuine Minkowski four-vector, and the intrinsic origin of the square root in the definition of the relativistic mass is the distinction between* $d\tau$ *and* dt.

[26] Relativity DeMystified, D. McMahon, Mc Graw Hill (USA), 2006, ISBN 0-07-145545-0

[27] Dynamics and Relativity, J.R. Forshaw, A.G. Smith, Wiley, 2009, ISBN 978-0-470-01460-8

[28] Hans, H. S.; Puri, S. P. (2003), *Mechanics* (2 ed.), Tata McGraw-Hill, p. 433, ISBN 0-07-047360-9, Chapter 12 page 433

[29] Mould, Richard A. (2002), *Basic relativity* (2 ed.), Springer, p. 126, ISBN 0-387-95210-1, Chapter 5 page 126

[30] Chow, Tail L. (2006), *Introduction to electromagnetic theory: a modern perspective*, Jones & Bartlett Learning, p. 392, ISBN 0-7637-3827-1, Chapter 10 page 392

[31] Dyson, F.W.; Eddington, A.S. & Davidson, C.R. (1920), "A Determination of the Deflection of Light by the Sun's Gravitational Field, from Observations Made at the Solar eclipse of May 29, 1919", *Phil. Trans. Roy. Soc. A* **220** (571–581): 291–333, Bibcode:1920RSPTA.220..291D, doi:10.1098/rsta.1920.0009

[32] Pound, R. V.; Rebka Jr. G. A. (April 1, 1960), "Apparent weight of photons", *Physical Review Letters* **4** (7): 337–341, Bibcode:1960PhRvL...4..337P, doi:10.1103/PhysRevLett.4.337

[33] Cockcroft–Walton experiment

[34] Conversions used: 1956 International (Steam) Table (IT) values where one calorie ≡ 4.1868 J and one BTU ≡ 1055.05585262 J. Weapons designers' conversion value of one gram TNT ≡ 1000 calories used.

[35] The 6.2 kg core comprised 0.8% gallium by weight. Also, about 20% of the Gadget's yield was due to fast fissioning in its natural uranium tamper. This resulted in 4.1 moles of Pu fissioning with 180 MeV per atom actually contributing prompt kinetic energy to the explosion. Note too that the term *"Gadget"-style* is used here instead of "Fat Man" because this general design of bomb was very rapidly upgraded to a more efficient one requiring only 5 kg of the Pu/gallium alloy.

[36] Assuming the dam is generating at its peak capacity of 6,809 MW.

[37] Assuming a 90/10 alloy of Pt/Ir by weight, a C_p of 25.9 for Pt and 25.1 for Ir, a Pt-dominated average C_p of 25.8, 5.134 moles of metal, and 132 J·K^{-1} for the prototype. A variation of ±1.5 picograms is of course, much smaller than the actual uncertainty in the mass of the international prototype, which is ±2 micrograms.

[38] InfraNet Lab (2008-12-07). Harnessing the Energy from the Earth's Rotation. Article on Earth rotation energy. Divided by c^2. InfraNet Lab, 7 December 2008. Retrieved from http://infranetlab.org/blog/harnessing-energy-earth%E2%80%99s-rotation

[39] G. 't Hooft, "Computation of the quantum effects due to a four-dimensional pseudoparticle", Physical Review D14:3432–3450 (1976).

[40] A. Belavin, A. M. Polyakov, A. Schwarz, Yu. Tyupkin, "Pseudoparticle Solutions to Yang Mills Equations", Physics Letters 59B:85 (1975).

[41] F. Klinkhammer, N. Manton, "A Saddle Point Solution in the Weinberg Salam Theory", Physical Review D 30:2212.

[42] Rubakov V. A. "Monopole Catalysis of Proton Decay", Reports on Progress in Physics 51:189–241 (1988).

[43] S.W. Hawking "Black Holes Explosions?" *Nature* 248:30 (1974).

[44] Einstein, A. (1905),"Zur Elektrodynamik bewegter Körper"(PDF),*Annalen der Physik* **17**(10): 891–921,Bibcode:1905AnP...32, doi:10.1002/andp.19053221004. English translation.

[45] Einstein, A. (1906), "Über eine Methode zur Bestimmung des Verhältnisses der transversalen und longitudinalen Masse des Elektrons" (PDF), *Annalen der Physik* **21** (13): 583–586, Bibcode:1906AnP...326..583E, doi:10.1002/andp.19063261310

[46] Max Jammer (1999), *Concepts of mass in contemporary physics and philosophy*, Princeton University Press, p. 51, ISBN 0-691-01017-X

[47] Eriksen, Erik; Vøyenli, Kjell (1976), "The classical and relativistic concepts of mass", *Foundations of Physics* (Springer) **6**: 115–124, Bibcode:1976FoPh....6..115E, doi:10.1007/BF00708670

[48] Jannsen, M., Mecklenburg, M. (2007), V. F. Hendricks et al., eds., "From classical to relativistic mechanics: Electromagnetic models of the electron.", *Interactions: Mathematics, Physics and Philosophy* (Dordrecht: Springer): 65–134

[49] Whittaker, E.T. (1951–1953), 2. Edition: *A History of the theories of aether and electricity, vol. 1: The classical theories / vol. 2: The modern theories 1900–1926*, London: Nelson

[50] Miller, Arthur I. (1981), *Albert Einstein's special theory of relativity. Emergence (1905) and early interpretation (1905–1911)*, Reading: Addison–Wesley, ISBN 0-201-04679-2

[51] Darrigol, O. (2005), "The Genesis of the theory of relativity" (PDF), *Séminaire Poincaré* **1**: 1–22, doi:10.1007/3-7643-7436-5_1

[52] Swedenborg, Emanuel (1734), "De Simplici Mundi vel Puncto naturali", *Principia Rerum Naturalia* (in Latin), Leipzig, p. 32

[53] Swedenborg, Emanuel (1845), *The Principia; or The First Principles of Natural Things*, Translated by Augustus Clissold, London: W. Newbery, pp. 55–57

[54] Philip Ball (Aug 23, 2011). "Did Einstein discover E = mc^2?". Physics World.

[55] Ives, Herbert E. (1952), "Derivation of the mass–energy relation", *Journal of the Optical Society of America* **42** (8): 540–543, doi:10.1364/JOSA.42.000540

[56] Stachel, John; Torretti, Roberto (1982), "Einstein's first derivation of mass–energy equivalence", *American Journal of Physics* **50** (8): 760–763, Bibcode:1982AmJPh..50..760S, doi:10.1119/1.12764

[57] Ohanian, Hans (2008), "Did Einstein prove E=mc2?", *Studies in History and Philosophy of Science Part B* **40** (2): 167–173, arXiv:0805.1400, doi:10.1016/j.shpsb.2009.03.002

[58] Rohrlich, Fritz (1990), "An elementary derivation of E=mc2",*American Journal of Physics***58**(4): 348–349,Bibcode:1990AmJP, doi:10.1119/1.16168

[59] Einstein, A. (1906), "Das Prinzip von der Erhaltung der Schwerpunktsbewegung und die Trägheit der Energie" (PDF), *Annalen der Physik* **20** (8): 627–633, Bibcode:1906AnP...325..627E, doi:10.1002/andp.19063250814

[60] Einstein 1906: Trotzdem die einfachen formalen Betrachtungen, die zum Nachweis dieser Behauptung durchgeführt werden müssen, in der Hauptsache bereits in einer Arbeit von H. Poincaré enthalten sind[2], werde ich mich doch der Übersichtlichkeit halber nicht auf jene Arbeit stützen.

[61] Helge Kragh, "Fin-de-Siècle Physics: A World Picture in Flux" in *Quantum Generations: A History of Physics in the Twentieth Century* (Princeton, NJ: Princeton University Press, 1999).

[62] *Умов Н. А.* Избранные сочинения. М. — Л., 1950. (Russian)

[63] Preston, S. T., Physics of the Ether, E. & F. N. Spon, London, (1875).

[64] Bjerknes: S. Tolver Preston's Explosive Idea $E = mc^2$.

[65] MathPages: Who Invented Relativity?

[66] De Pretto, O. *Reale Instituto Veneto Di Scienze, Lettere Ed Arti*, LXIII, II, 439–500, reprinted in Bartocci.

[67] Umberto Bartocci, *Albert Einstein e Olinto De Pretto—La vera storia della formula più famosa del mondo*, editore Andromeda, Bologna, 1999.

[68] Prentiss, J.J. (August 2005), "Why is the energy of motion proportional to the square of the velocity?", *American Journal of Physics* **73** (8): 705, Bibcode:2005AmJPh..73..701P, doi:10.1119/1.1927550

[69] John Worrall, review of the book *Conceptions of Ether. Studies in the History of Ether Theories* by Cantor and Hodges, The British Journal of the Philosophy of Science vol 36, no 1, March 1985, p. 84. The article contrasts a particle ether with a wave-carrying ether, the latter *was* acceptable.

[70] Le Bon: The Evolution of Forces.

[71] Bizouard: Poincaré $E = mc^2$ l'équation de Poincaré, Einstein et Planck.

[72] Rutherford, Ernest (1904), *Radioactivity*, Cambridge: University Press, pp. 336–338

[73] Heisenberg, Werner (1958), *Physics And Philosophy: The Revolution In Modern Science*, New York: Harper & Brothers, pp. 118–119

[74] "We might in these processes obtain very much more energy than the proton supplied, but on the average we could not expect to obtain energy in this way. It was a very poor and inefficient way of producing energy, and anyone who looked for a source of power in the transformation of the atoms was talking moonshine. But the subject was scientifically interesting because it gave insight into the atoms." *The Times* archives, September 12, 1933, "The British association—breaking down the atom"

[75] Cover. *Time* magazine, July 1, 1946.

[76] Isaacson, *Einstein: His Life and Universe*.

[77] Robert Serber, *The Los Alamos Primer: The First Lectures on How to Build an Atomic Bomb* (University of California Press, 1992), page 7. Note that the quotation is taken from Serber's 1992 version, and is not in the original 1943 Los Alamos Primer of the same name.

[78] David Bodanis, $E = mc^2$: *A Biography of the World's Most Famous Equation* (New York: Walker, 2000).

[79] A quote from Frisch about the discovery day. Accessed April 4, 2009.

[80] Sime, Ruth (1996), *Lise Meitner: A Life in Physics*, California Studies in the History of Science **13**, Berkeley: University of California Press, pp. 236–237, ISBN 0-520-20860-9

- Lasky, Ronald C. (April 23, 2007), "What is the significance of $E = mc^2$? And what does it mean?", *Scientific American* (Scientific American)

7.12 External links

- A shortcut to $E=mc^2$ – An easy to understand, high-school level derivation of the $E=mc^2$ formula.

- Einstein on the Inertia of Energy – MathPages

- Mass and Energy – Conversations About Science with Theoretical Physicist Matt Strassler

- Ask an Astrophysicist | Energy–Matter Conversion, NASA, 1997

- The Equivalence of Mass and Energy – Entry in the Stanford Encyclopedia of Philosophy

- Gail Wilson (May 2014) Scientists discover how to turn light into matter after 80-year quest Imperial College

- Living Reviews in Relativity – An open access, peer-referred, solely online physics journal publishing invited reviews covering all areas of relativity research.

- Merrifield, Michael; Copeland, Ed; Bowley, Roger. "$E=mc^2$ – Mass–Energy Equivalence". *Sixty Symbols*. Brady Haran for the University of Nottingham.

- Einstein on mass and energy, Eugene Hecht, Am. J. Phys. 77, 799 (2009). For example, "Early on, Einstein embraced the idea of a speed-dependent mass but changed his mind in 1906 and thereafter carefully avoided that notion entirely. He shunned, and explicitly rejected, what later came to be known as 'relativistic mass'. ... He consistently related the rest energy of a system to its invariant inertial mass."

Chapter 8

Gravitational lens

A **gravitational lens** refers to a distribution of matter (such as a cluster of galaxies) between a distant source and an observer, that is capable of bending the light from the source, as it travels towards the observer. This effect is known as gravitational lensing and is one of the predictions of Albert Einstein's general theory of relativity.[1]

Although Orest Chwolson (1924) or Frantisek Klin (1936) are sometimes credited as being the first ones to discuss the effect in print, the effect is more commonly associated with Einstein, who published a more famous article on the subject in 1936.

Fritz Zwicky posited in 1937 that the effect could allow galaxy clusters to act as gravitational lenses. It was not until 1979 that this effect was confirmed by observation of the so-called "Twin QSO" SBS 0957+561.

8.1 Description

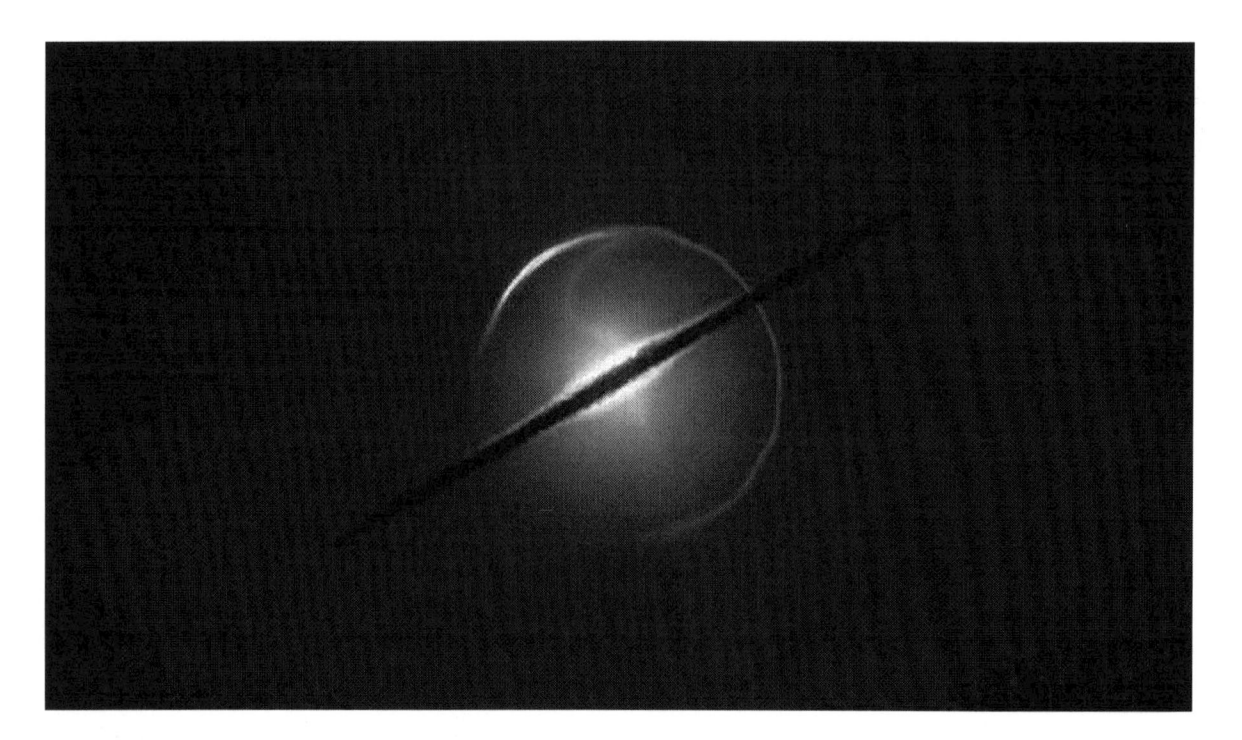

Gravitational lensing - intervening galaxy modifies appearance of a galaxy far behind it (video; artist's concept).

8.1. DESCRIPTION

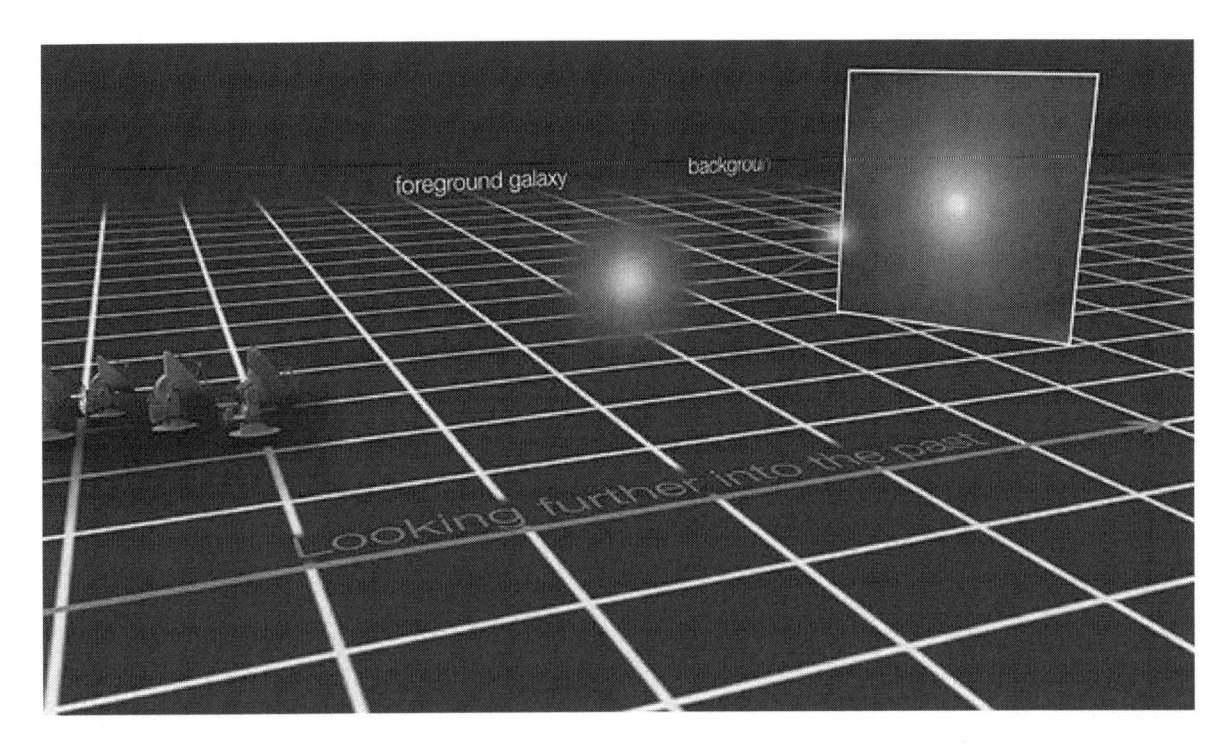

This schematic image shows how light from a distant galaxy is distorted by the gravitational effects of a foreground galaxy, which acts like a lens and makes the distant source appear distorted, but magnified, forming characteristic rings of light, known as Einstein rings.

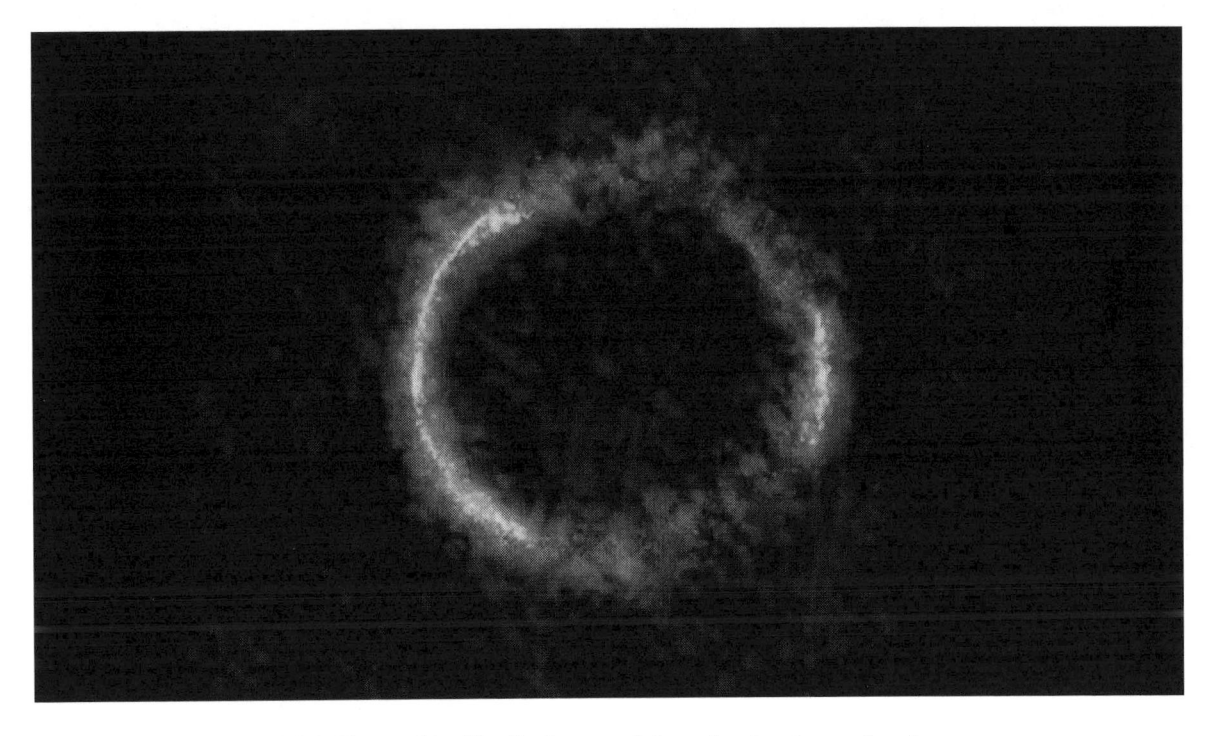

An analysis of the distortion of SDP.81 caused by this effect has revealed star-forming clumps of matter.

Unlike an optical lens, maximum 'bending' occurs closest to, and minimum 'bending' furthest from, the center of a gravitational lens. Consequently, a gravitational lens has no single focal point, but a focal line instead. If the (light) source, the massive lensing object, and the observer lie in a straight line, the original light source will appear as a ring around the massive lensing object. If there is any misalignment the observer will see an arc segment instead. This phenomenon was first mentioned in 1924 by the St. Petersburg physicist Orest Chwolson,[2] and quantified by Albert Einstein in 1936. It is usually referred to in the literature as an **Einstein ring**, since Chwolson did not concern himself with the flux or radius of the ring image. More commonly, where the lensing mass is complex (such as a galaxy group or cluster) and does not cause a spherical distortion of space–time, the source will resemble partial arcs scattered around the lens. The observer may then see multiple distorted images of the same source; the number and shape of these depending upon the relative positions of the source, lens, and observer, and the shape of the gravitational well of the lensing object.[3]

There are three classes of gravitational lensing:[4]

1. Strong lensing: where there are easily visible distortions such as the formation of Einstein rings, arcs, and multiple images.

2. Weak lensing: where the distortions of background sources are much smaller and can only be detected by analyzing large numbers of sources to find coherent distortions of only a few percent. The lensing shows up statistically as a preferred stretching of the background objects perpendicular to the direction to the center of the lens. By measuring the shapes and orientations of large numbers of distant galaxies, their orientations can be averaged to measure the shear of the lensing field in any region. This, in turn, can be used to reconstruct the mass distribution in the area: in particular, the background distribution of dark matter can be reconstructed. Since galaxies are intrinsically elliptical and the weak gravitational lensing signal is small, a very large number of galaxies must be used in these surveys. These weak lensing surveys must carefully avoid a number of important sources of systematic error: the intrinsic shape of galaxies, the tendency of a camera's point spread function to distort the shape of a galaxy and the tendency of atmospheric seeing to distort images must be understood and carefully accounted for. The results of these surveys are important for cosmological parameter estimation, to better understand and improve upon the Lambda-CDM model, and to provide a consistency check on other cosmological observations. They may also provide an important future constraint on dark energy.

3. Microlensing: where no distortion in shape can be seen but the amount of light received from a background object changes in time. The lensing object may be stars in the Milky Way in one typical case, with the background source being stars in a remote galaxy, or, in another case, an even more distant quasar. The effect is small, such that (in the case of strong lensing) even a galaxy with a mass more than 100 billion times that of the Sun will produce multiple images separated by only a few arcseconds. Galaxy clusters can produce separations of several arcminutes. In both cases the galaxies and sources are quite distant, many hundreds of megaparsecs away from our Galaxy.

Gravitational lenses act equally on all kinds of electromagnetic radiation, not just visible light. Weak lensing effects are being studied for the cosmic microwave background as well as galaxy surveys. Strong lenses have been observed in radio and x-ray regimes as well. If a strong lens produces multiple images, there will be a relative time delay between two paths: that is, in one image the lensed object will be observed before the other image.

8.2 History

Spacetime around a massive object (such as a galaxy cluster or a black hole) is curved, and as a result light rays from a background source (such as a galaxy) propagating through spacetime are bent. The lensing effect can magnify and distort the image of the background source.

According to general relativity, mass "warps" space–time to create gravitational fields and therefore bend light as a result. This theory was confirmed in 1919 during a solar eclipse, when Arthur Eddington and Frank Watson Dyson observed the light from stars passing close to the Sun was slightly bent, so that stars appeared slightly out of position.[5]

Einstein realized that it was also possible for astronomical objects to bend light, and that under the correct conditions, one would observe multiple images of a single source, called a **gravitational lens** or sometimes a **gravitational mirage**.

However, as he only considered gravitational lensing by single stars, he concluded that the phenomenon would most likely remain unobserved for the foreseeable future. In 1937, Fritz Zwicky first considered the case where a galaxy could act as a source, something that according to his calculations should be well within the reach of observations.

It was not until 1979 that the first gravitational lens would be discovered. It became known as the "Twin QSO" since it initially looked like two identical quasistellar objects; it is officially named **SBS 0957+561**. This gravitational lens was discovered by Dennis Walsh, Bob Carswell, and Ray Weymann using the Kitt Peak National Observatory 2.1 meter telescope.[6]

Bending light around a massive object from a distant source. The orange arrows show the apparent position of the background source. The white arrows show the path of the light from the true position of the source.

In the 1980s, astronomers realized that the combination of CCD imagers and computers would allow the brightness of millions of stars to be measured each night. In a dense field, such as the galactic center or the Magellanic clouds, many microlensing events per year could potentially be found. This led to efforts such as Optical Gravitational Lensing Experiment, or OGLE, that have characterized hundreds of such events.

8.3 Explanation in terms of space–time curvature

See also: Kepler problem in general relativity

In general relativity, light follows the curvature of spacetime, hence when light passes around a massive object, it is bent. This means that the light from an object on the other side will be bent towards an observer's eye, just like an ordinary lens. Since light always moves at a constant speed, lensing changes the direction of the velocity of the light, but not the magnitude.

Light rays are the boundary between the future, the spacelike, and the past regions. The gravitational attraction can be viewed as the motion of undisturbed objects in a background curved *geometry* or alternatively as the response of objects to a *force* in a flat geometry. The angle of deflection is:

$$\theta = \frac{4GM}{rc^2}$$

toward the mass M at a distance r from the affected radiation, where G is the universal constant of gravitation and c is the

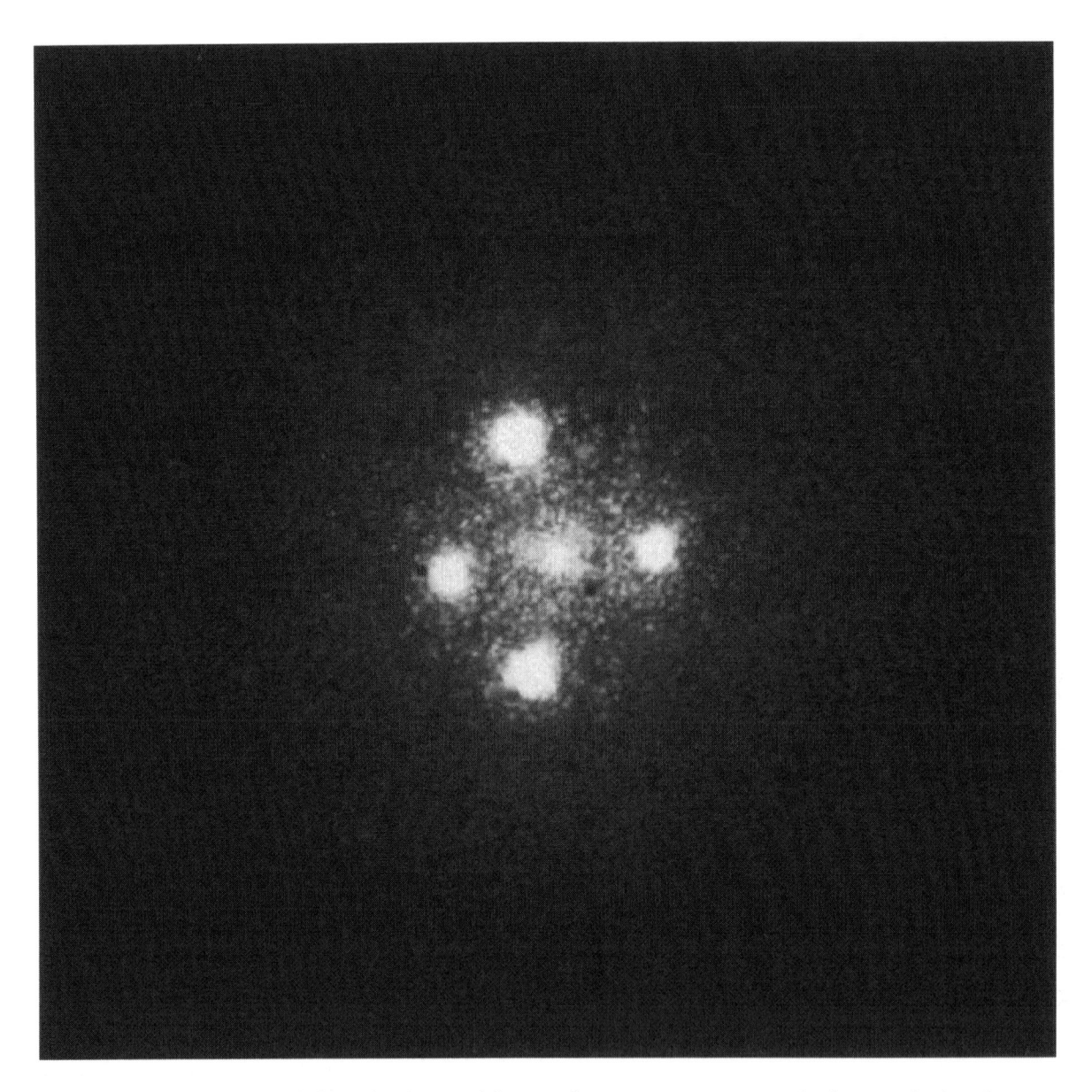

In the formation known as Einstein's Cross, four images of the same distant quasar appear around a foreground galaxy due to strong gravitational lensing.

speed of light in a vacuum.

Since the Schwarzschild radius r_s is defined as $r_s = 2Gm/c^2$, this can also be expressed in simple form as

$$\theta = 2\frac{r_s}{r}$$

8.4 Search for gravitational lenses

Most of the gravitational lenses in the past have been discovered accidentally. A search for gravitational lenses in the northern hemisphere (Cosmic Lens All Sky Survey, CLASS), done in radio frequencies using the Very Large Array (VLA) in New Mexico, led to the discovery of 22 new lensing systems, a major milestone. This has opened a whole new avenue for research ranging from finding very distant objects to finding values for cosmological parameters so we can

8.4. SEARCH FOR GRAVITATIONAL LENSES

Simulated gravitational lensing (black hole going past a background galaxy).

understand the universe better.

A similar search in the southern hemisphere would be a very good step towards complementing the northern hemisphere search as well as obtaining other objectives for study. If such a search is done using well-calibrated and well-parameterized instrument and data, a result similar to the northern survey can be expected. The use of the Australia Telescope 20 GHz (AT20G) Survey data collected using the Australia Telescope Compact Array (ATCA) stands to be such a collection of data. As the data were collected using the same instrument maintaining a very stringent quality of data we should expect to obtain good results from the search. The AT20G survey is a blind survey at 20 GHz frequency in the radio domain of the electromagnetic spectrum. Due to the high frequency used, the chances finding gravitational lenses increases as the relative number of compact core objects (e.g. Quasars) are higher (Sadler et al. 2006). This is important as the lensing is easier to detect and identify in simple objects compared to objects with complexity in them. This search involves the use of interferometric methods to identify candidates and follow them up at higher resolution to identify them. Full detail of the project is currently under works for publication.

In a 2009 article on Science Daily a team of scientists led by a cosmologist from the U.S. Department of Energy's Lawrence Berkeley National Laboratory has made major progress in extending the use of gravitational lensing to the study of much older and smaller structures than was previously possible by stating that weak gravitational lensing improves measurements of distant galaxies.[7]

Astronomers from the Max Planck Institute for Astronomy in Heidelberg, Germany, the results of which are accepted for publication on Oct 21, 2013 in the Astrophysical Journal Letters (arXiv.org), discovered what at the time was the most distant gravitational lens galaxy termed as **J1000+0221** using NASA's Hubble Space Telescope.[8][9] While it remains

This image from the NASA/ESA Hubble Space Telescope shows the galaxy cluster MACS J1206.

the most distant quad-image lensing galaxy known, an even more distant two-image lensing galaxy was subsequently discovered by an international team of astronomers using a combination of Hubble Space Telescope and Keck telescope imaging and spectroscopy. The discovery and analysis of the IRC 0218 lens was published in the Astrophysical Journal Letters on June 23, 2014.[10]

A research published Sep 30, 2013 in the online edition of Physical Review Letters, led by McGill University in Montreal, Québec, Canada, has discovered the B-modes, that are formed due to gravitational lensing effect, using National Science Foundation's South Pole Telescope and with help from the Herschel space observatory. This discovery would open the possibilities of testing the theories of how our universe originated.[11][12]

8.4. SEARCH FOR GRAVITATIONAL LENSES

Abell 2744 HFF
HST WFC3 ACS

Abell 2744 galaxy cluster - extremely distant galaxies revealed by gravitational lensing (16 October 2014).[13][14]

8.5 Solar gravitational lens

Albert Einstein predicted in 1936 that rays of light from the same direction that skirt the edges of the Sun would converge to a focal point approximately 542 AU from the Sun.[15] Thus, the Sun could act as a gravitational lens for magnifying distant objects in a way that provides some flexibility in aiming unlike the coincidence-based lens usage of more distant objects, such as intermediate galaxies. A probe's location could shift around as needed to select different targets relative to the Sun (acting as a lens). This distance is far beyond the progress and equipment capabilities of space probes such as Voyager 1, and beyond the known planets and dwarf planets, though over thousands of years 90377 Sedna will move further away on its highly elliptical orbit. The high gain for potentially detecting signals through this lens, such as microwaves at the 21-cm hydrogen line, led to the suggestion by Frank Drake in the early days of SETI that a probe could be sent to this distance. A multipurpose probe SETISAIL and later FOCAL was proposed to the ESA in 1993, but is expected to be a difficult task. If a probe does pass 542 AU, the gain and image-forming capabilities of the lens will continue to improve at further distances as the rays that come to a focus at these distances pass further away from the distortions of the Sun's corona.[16]

8.6 Measuring weak lensing

Kaiser et al. (1995),[17] Luppino & Kaiser (1997)[18] and Hoekstra et al. (1998) prescribed a method to invert the effects of the Point Spread Function (PSF) smearing and shearing, recovering a shear estimator uncontaminated by the systematic distortion of the PSF. This method (KSB+) is the most widely used method in current weak lensing shear measurements.

Galaxies have random rotations and inclinations. As a result, the shear effects in weak lensing need to be determined by statistically preferred orientations. The primary source of error in lensing measurement is due to the convolution of the PSF with the lensed image. The KSB method measures the ellipticity of a galaxy image. The shear is proportional to the ellipticity. The objects in lensed images are parameterized according to their weighted quadrupole moments. For a perfect ellipse, the weighted quadrupole moments are related to the weighted ellipticity. KSB calculate how a weighted ellipticity measure is related to the shear and use the same formalism to remove the effects of the PSF.

KSB's primary advantages are its mathematical ease and relatively simple implementation. However, KSB is based on a key assumption that the PSF is circular with an anisotropic distortion. It's fine for current cosmic shear surveys, but the next generation of surveys (e.g. LSST) may need much better accuracy than KSB can provide. Because during that time, the statistical errors from the data are negligible, the systematic errors will dominate.

8.7 Gallery

- "Smiley" image of galaxy cluster (SDSS J1038+4849) & gravitational lensing (an Einstein ring) (HST).[1]

- Abell 1689 - actual gravitational lensing effects (Hubble Space Telescope).

- Dark matter distribution - weak gravitational lensing (Hubble Space Telescope).

- Gravitational lens discovered at redshift z = 1.53.[2]

1. ^ Loff, Sarah; Dunbar, Brian (February 10, 2015). "Hubble Sees A Smiling Lens". *NASA*. Retrieved February 10, 2015.

2. ^ "Most distant gravitational lens helps weigh galaxies". *ESA/Hubble Press Release*. Retrieved 18 October 2013.

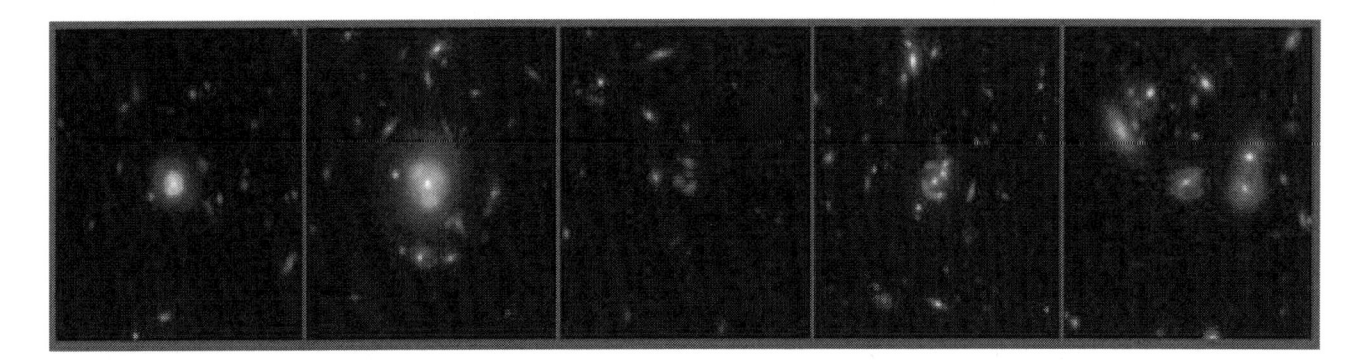

Gravitationally-lensed distant star-forming galaxy.[19]

8.8 See also

- Einstein cross
- Einstein ring
- SN Refsdal

8.9 Historical papers and references

- Chwolson, O (1924). "Über eine mögliche Form fiktiver Doppelsterne". *Astronomische Nachrichten* **221** (20): 329. Bibcode:1924AN....221..329C. doi:10.1002/asna.19242212003.

- Einstein, Albert (1936). "Lens-like Action of a Star by the Deviation of Light in the Gravitational Field". *Science* **84** (2188): 506–7. Bibcode:1936Sci....84..506E. doi:10.1126/science.84.2188.506. JSTOR 1663250. PMID 17769014.

- Renn, Jürgen; Tilman Sauer; John Stachel (1997). "The Origin of Gravitational Lensing: A Postscript to Einstein's 1936 Science paper". *Science* **275** (5297): 184–6. Bibcode:1997Sci...275..184R. doi:10.1126/science.275.5297.184. PMID 8985006.

8.10 References

Notes

[1] Overbye, Dennis (March 5, 2015). "Astronomers Observe Supernova and Find They're Watching Reruns". *New York Times*. Retrieved March 5, 2015.

[2] Gravity Lens – Part 2 (Great Moments in Science, ABS Science)

[3] Dieter Brill, "Black Hole Horizons and How They Begin", Astronomical Review (2012); Online Article, cited Sept.2012.

[4] Melia, Fulvio (2007). *The Galactic Supermassive Black Hole*. Princeton University Press. pp. 255–256. ISBN 0-691-13129-5.

[5] Dyson, F. W.; Eddington, A. S.; Davidson, C. (1 January 1920). "A Determination of the Deflection of Light by the Sun's Gravitational Field, from Observations Made at the Total Eclipse of May 29, 1919". *Philosophical Transactions of the Royal Society A: Mathematical, Physical and Engineering Sciences* **220**(571-581): 291–333.Bibcode:1920RSPTA.220..291D.doi:10.1098/rsta.1920..

[6] Walsh, D.; Carswell, R. F.; Weymann, R. J. (31 May 1979). "0957 + 561 A, B: twin quasistellar objects or gravitational lens?". *Nature* **279** (5712): 381–384. Bibcode:1979Natur.279..381W. doi:10.1038/279381a0. PMID 16068158.

[7] Cosmology: Weak gravitational lensing improves measurements of distant galaxies

[8] Sci-News.com (21 Oct 2013). "Most Distant Gravitational Lens Discovered". Sci-News.com. Retrieved 22 October 2013.

[9] van der Wel, A. et al. (2013). "Discovery of a Quadruple Lens in CANDELS with a Record Lens Redshift". *ApJ Letters*. arXiv:1309.2826. Bibcode:2013ApJ...777L..17V. doi:10.1088/2041-8205/777/1/L17.

[10] Wong, K. et al. (2014). "Discovery of a Strong Lensing Galaxy Embedded in a Cluster at z = 1.62". *ApJ Letters* **789**: L31. arXiv:1405.3661. Bibcode:2014ApJ...789L..31W. doi:10.1088/2041-8205/789/2/L31.

[11] NASA/Jet Propulsion Laboratory (October 22, 2013). "Long-sought pattern of ancient light detected". *ScienceDaily*. Retrieved October 23, 2013.

[12] Hanson, D. et al. (Sep 30, 2013). "Detection of B-Mode Polarization in the Cosmic Microwave Background with Data from the South Pole Telescope". *Physical Review Letters*. 14 **111**. arXiv:1307.5830. Bibcode:2013PhRvL.111n1301H. doi:10.1103/PhysRevLett.111.141301.

[13] Clavin, Whitney; Jenkins, Ann; Villard, Ray (7 January 2014). "NASA's Hubble and Spitzer Team up to Probe Faraway Galaxies". *NASA*. Retrieved 8 January 2014.

[14] Chou, Felecia; Weaver, Donna (16 October 2014). "RELEASE 14-283 - NASA's Hubble Finds Extremely Distant Galaxy through Cosmic Magnifying Glass". *NASA*. Retrieved 17 October 2014.

[15] "Lens-Like Action of a Star by the Deviation of Light in the Gravitational Field". *Science* **84** (2188): 506–507. 1936. Bibcode:1936Sci....84..506E. doi:10.1126/science.84.2188.506. PMID 17769014.

[16] Claudio Maccone (2009). *Deep Space Flight and Communications: Exploiting the Sun as a Gravitational Lens*. Springer.

[17] Kaiser, Nick; Squires, Gordon; Broadhurst, Tom (August 1995). "A Method for Weak Lensing Observations". *The Astrophysical Journal* **449**: 460. arXiv:astro-ph/9411005. Bibcode:1995ApJ...449..460K. doi:10.1086/176071.

[18] Luppino, G. A.; Kaiser, Nick (20 January 1997). "Detection of Weak Lensing by a Cluster of Galaxies at = 0.83". *The Astrophysical Journal* **475** (1): 20–28. arXiv:astro-ph/9601194. Bibcode:1997ApJ...475...20L. doi:10.1086/303508.

[19] "ALMA Rewrites History of Universe's Stellar Baby Boom". *ESO*. Retrieved 2 April 2013.

Bibliography

- "*Accidental Astrophysicists*". Science News, June 13, 2008.

- "*XFGLenses*". A Computer Program to visualize Gravitational Lenses, Francisco Frutos-Alfaro

- "*G-LenS*". A Point Mass Gravitational Lens Simulation, Mark Boughen.

- Newbury, Pete, "*Gravitational Lensing*". Institute of Applied Mathematics, The University of British Columbia.

- Cohen, N., "Gravity's Lens: Views of the New Cosmology", Wiley and Sons, 1988.

- "*Q0957+561 Gravitational Lens*". Harvard.edu.

- "*Gravitational lensing*". Gsfc.nasa.gov.

- Bridges, Andrew, "*Most distant known object in universe discovered*". Associated Press. February 15, 2004. (Farthest galaxy found by gravitational lensing, using Abell 2218 and Hubble Space Telescope.)

- Analyzing Corporations ... and the Cosmos An unusual career path in gravitational lensing.

- "*HST images of strong gravitational lenses*". Harvard-Smithsonian Center for Astrophysics.

- "*A planetary microlensing event*" and "*A Jovian-mass Planet in Microlensing Event OGLE-2005-BLG-071*", the first extra-solar planet detections using microlensing.

- Gravitational lensing on arxiv.org

- NRAO CLASS home page
- AT20G survey
- *A diffraction limit on the gravitational lens effect* (Bontz, R. J. and Haugan, M. P. "Astrophysics and Space Science" vol. 78, no. 1, p. 199-210. August 1981)

Further reading

- Blandford & Narayan; Narayan, R (1992). "Cosmological applications of gravitational lensing". *ARA&A* **30** (1): 311–358. Bibcode:1992ARA&A..30..311B. doi:10.1146/annurev.aa.30.090192.001523.
- Matthias Bartelmann and Peter Schneider (2000-08-17). "Weak Gravitational Lensing" (PDF).
- Khavinson, Dmitry; Neumann, Genevra (June–July 2008). "From Fundamental Theorem of Algebra to Astrophysics: A "Harmonious" Path" (PDF). *Notices* (AMS) **55** (6): 666–675..
- Petters, Arlie O.; Levine, Harold; Wambsganss, Joachim (2001). *Singularity Theory and Gravitational Lensing*. Progress in Mathematical Physics **21**. Birkhäuser.
- *Tools for the evaluation of the possibilities of using parallax measurements of gravitationally lensed sources* (Stein Vidar Hagfors Haugan. June 2008)

8.11 External links

- Video: Evalyn Gates – Einstein's Telescope: The Search for Dark Matter and Dark Energy in the Universe, presentation in Portland, Oregon, on April 19, 2009, from the author's recent book tour.
- Audio: Fraser Cain and Dr. Pamela Gay – Astronomy Cast: Gravitational Lensing, May 2007

8.12 Featured in science-fiction works

- Existence, by David Brin, 2012

Chapter 9

Physics beyond the Standard Model

Physics beyond the Standard Model (**BSM**) refers to the theoretical developments needed to explain the deficiencies of the Standard Model, such as the origin of mass, the strong CP problem, neutrino oscillations, matter–antimatter asymmetry, and the nature of dark matter and dark energy.[1] Another problem lies within the mathematical framework of the Standard Model itself – the Standard Model is inconsistent with that of general relativity, to the point that one or both theories break down under certain conditions (for example within known space-time singularities like the Big Bang and black hole event horizons).

Theories that lie beyond the Standard Model include various extensions of the standard model through supersymmetry, such as the Minimal Supersymmetric Standard Model (MSSM) and Next-to-Minimal Supersymmetric Standard Model (NMSSM), or entirely novel explanations, such as string theory, M-theory and extra dimensions. As these theories tend to reproduce the entirety of current phenomena, the question of which theory is the right one, or at least the "best step" towards a Theory of Everything, can only be settled via experiments, and is one of the most active areas of research in both theoretical and experimental physics.

9.1 Problems with the Standard Model

Despite being the most successful theory of particle physics to date, the Standard Model is not perfect.[2] A large share of the published output of theoretical physicists consists of proposals for various forms of "Beyond the Standard Model" new physics proposals that would modify the Standard Model in ways subtle enough to be consistent with existing data, yet address its imperfections materially enough to predict non-Standard Model outcomes of new experiments that can be proposed.

9.1.1 Phenomena not explained

The Standard Model is inherently an incomplete theory. There are fundamental physical phenomena in nature that the Standard Model does not adequately explain:

- *Gravity*. The standard model does not explain gravity. The approach of simply adding a "graviton" (whose properties are the subject of considerable consensus among physicists if it exists) to the Standard Model does not recreate what is observed experimentally without other modifications, as yet undiscovered, to the Standard Model. Moreover, instead, the Standard Model is widely considered to be incompatible with the most successful theory of gravity to date, general relativity.[3]

- *Dark matter and dark energy*. Cosmological observations tell us the standard model explains about 5% of the energy present in the universe. About 26% should be dark matter, which would behave just like other matter, but which only interacts weakly (if at all) with the Standard Model fields. Yet, the Standard Model does not supply

9.1. PROBLEMS WITH THE STANDARD MODEL

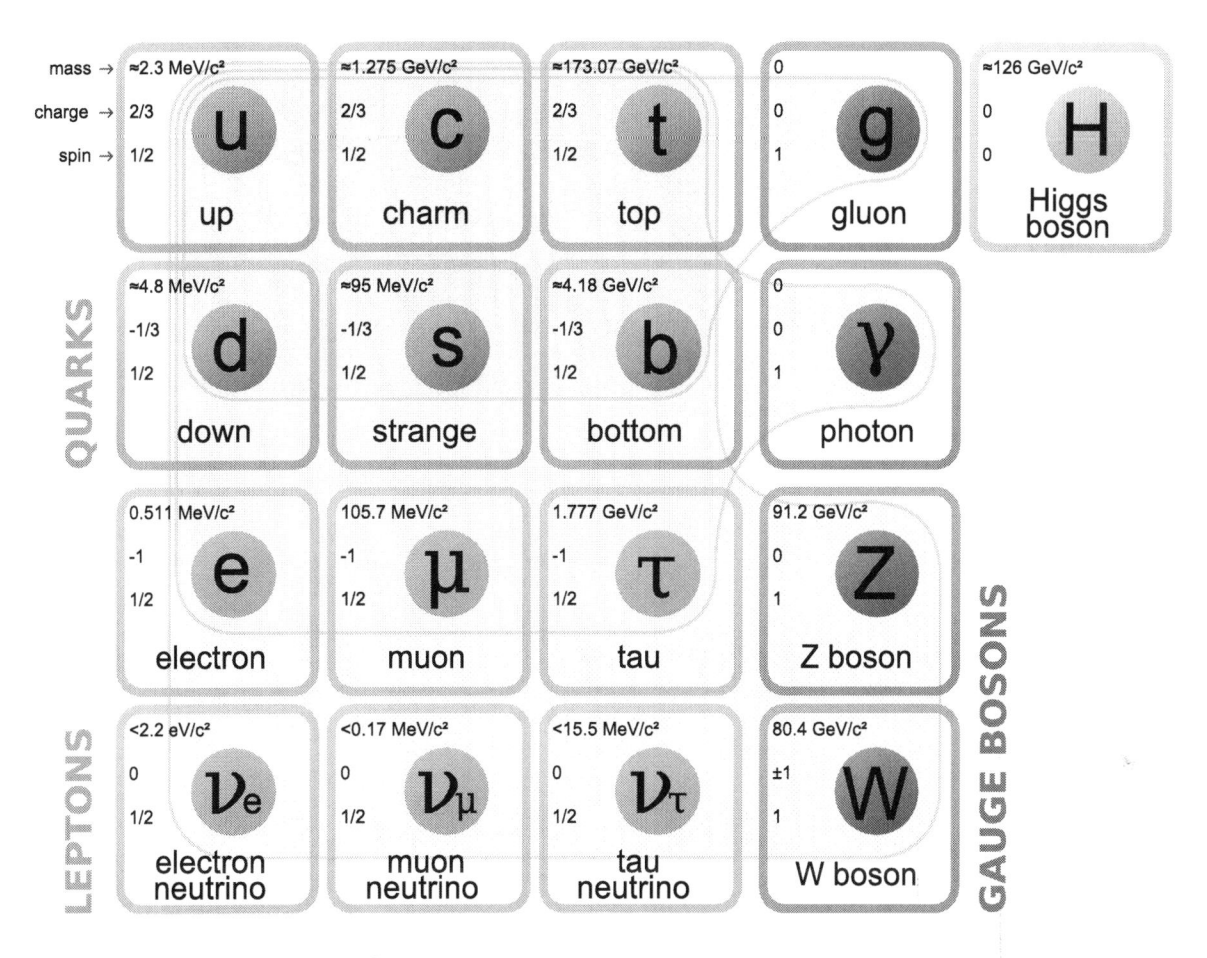

mass →	≈2.3 MeV/c²	≈1.275 GeV/c²	≈173.07 GeV/c²	0	≈126 GeV/c²
charge →	2/3	2/3	2/3	0	0
spin →	1/2 **u**	1/2 **c**	1/2 **t**	1 **g**	0 **H**
	up	charm	top	gluon	Higgs boson
QUARKS	≈4.8 MeV/c²	≈95 MeV/c²	≈4.18 GeV/c²	0	
	-1/3	-1/3	-1/3	0	
	1/2 **d**	1/2 **s**	1/2 **b**	1 **γ**	
	down	strange	bottom	photon	
	0.511 MeV/c²	105.7 MeV/c²	1.777 GeV/c²	91.2 GeV/c²	
	-1	-1	-1	0	
	1/2 **e**	1/2 **μ**	1/2 **τ**	1 **Z**	GAUGE BOSONS
	electron	muon	tau	Z boson	
LEPTONS	<2.2 eV/c²	<0.17 MeV/c²	<15.5 MeV/c²	80.4 GeV/c²	
	0	0	0	±1	
	1/2 ν_e	1/2 ν_μ	1/2 ν_τ	1 **W**	
	electron neutrino	muon neutrino	tau neutrino	W boson	

The Standard Model of elementary particles

any fundamental particles that are good dark matter candidates. The rest (69%) should be dark energy, a constant energy density for the vacuum. Attempts to explain dark energy in terms of vacuum energy of the standard model lead to a mismatch of 120 orders of magnitude.[4]

- *Neutrino masses*. According to the standard model, neutrinos are massless particles. However, neutrino oscillation experiments have shown that neutrinos do have mass. Mass terms for the neutrinos can be added to the standard model by hand, but these lead to new theoretical problems. For example, the mass terms need to be extraordinarily small and it is not clear if the neutrino masses would arise in the same way that the masses of other fundamental particles do in the Standard Model.

- *Matter-antimatter asymmetry*. The universe is made out of mostly matter. However, the standard model predicts that matter and antimatter should have been created in (almost) equal amounts if the initial conditions of the universe did not involve disproportionate matter relative to antimatter. Yet, no mechanism sufficient to explain this asymmetry exists in the Standard Model.

9.1.2 Experimental results not explained

No experimental result is widely accepted as contradicting the Standard Model at a level that definitively contradicts it at the "five sigma" (i.e. five standard deviation) level widely considered to be the threshold of a "discovery" in particle physics. But, because every experiment contains some degree of statistical and systemic uncertainty, and the theoretical predictions themselves are also almost never calculated exactly and are subject to uncertainties in measurements of the fundamental constants of the Standard Model (some of which are tiny and others of which are substantial,) it is mathematically expected that some of the hundreds of experimental tests of the Standard Model will deviate to some extent from the Standard Model even if there were no "new physics" beyond the Standard Model to be discovered.

At any given time there are a number of experimental results that are significantly different from the Standard Model expectation, although many of these have been found to be statistical flukes or experimental errors as more data has been collected. On the other hand, any "beyond the Standard Model" physics would necessarily first manifest experimentally as a statistically significant difference between an experiment and a Standard Model theoretical prediction.

In each case, physicists seek to determine if a result is a mere statistical fluke or experimental error on the one hand, or a sign of new physics on the other. More statistically significant results cannot be mere statistical flukes but can still result from experimental error or inaccurate estimates of experimental precision. Frequently, experiments are tailored to be more sensitive to experimental results that would distinguish the Standard Model from theoretical alternatives.

Some of the most notable examples include the following:

- *Muonic hydrogen* – the Standard Model makes precise theoretical predictions regarding the atomic radius size of ordinary hydrogen (a proton-electron system) and that of muonic hydrogen (a proton-muon system in which a muon is a "heavy" variant of an electron). However, the measured atomic radius of muonic hydrogen differs significantly from that of the radius predicted by the Standard Model using existing physical constant measurements by what appears to be as many as seven standard deviations.[5] Doubts about the accuracy of the error estimates in earlier experiments, which are still within 4% of each other in measuring a truly tiny distance, and a lack of a well motivated theory that could explain the discrepancy, have caused physicists to be hesitant to describe these results as contradicting the Standard Model despite the apparent statistical significance of the result and a lack of any clearly identified possible source of experimental error in the results.

- *BaBar Data Suggests Possible Flaws in the Standard Model* – results from a BaBar experiment may suggest a surplus over Standard Model predictions of a type of particle decay called "B to D-star-tau-nu." In this, an electron and positron collide, resulting in a B meson and an antimatter B-bar meson, which then decays into a D meson and a tau lepton as well as a smaller antineutrino. While the level of certainty of the excess (3.4 sigma in statistical language) is not enough to claim a break from the Standard Model, the results are a potential sign of something amiss and are likely to affect existing theories, including those attempting to deduce the properties of Higgs bosons.[6] However, results at LHCb have demonstrated no significant deviation from the Standard Model prediction of very nearly zero asymmetry.[7][8]

- Proton radius - radius measured using electrons is different from radius measured using muons[9]

9.1.3 Theoretical predictions not observed

Observation at particle colliders of all of the fundamental particles predicted by the Standard Model has been confirmed. The Higgs boson is predicted by the Standard Model's explanation of the Higgs mechanism, which describes how the weak SU(2) gauge symmetry is broken and how fundamental particles obtain mass; it was the last particle predicted by the Standard Model to be observed. On July 4, 2012, CERN scientists using the Large Hadron Collider announced the discovery of a particle consistent with the Higgs boson, with a mass of about 126 GeV/c^2. A Higgs boson was confirmed to exist on March 14, 2013, although efforts to confirm that it has all of the properties predicted by the Standard Model are ongoing.[10]

A few hadrons (i.e. composite particles made of quarks) whose existence is predicted by the Standard Model, which can be produced only at very high energies in very low frequencies have not yet been definitively observed, and "glueballs"[11] (i.e. composite particles made of gluons) have also not yet been definitively observed. Some very low frequency particle decays predicted by the Standard Model have also not yet been definitively observed because insufficient data is available to make a statistically significant observation.

9.1.4 Theoretical problems

Some features of the standard model are added in an ad hoc way. These are not problems per se (i.e. the theory works fine with these ad hoc features), but they imply a lack of understanding. These ad hoc features have motivated theorists

9.1. PROBLEMS WITH THE STANDARD MODEL

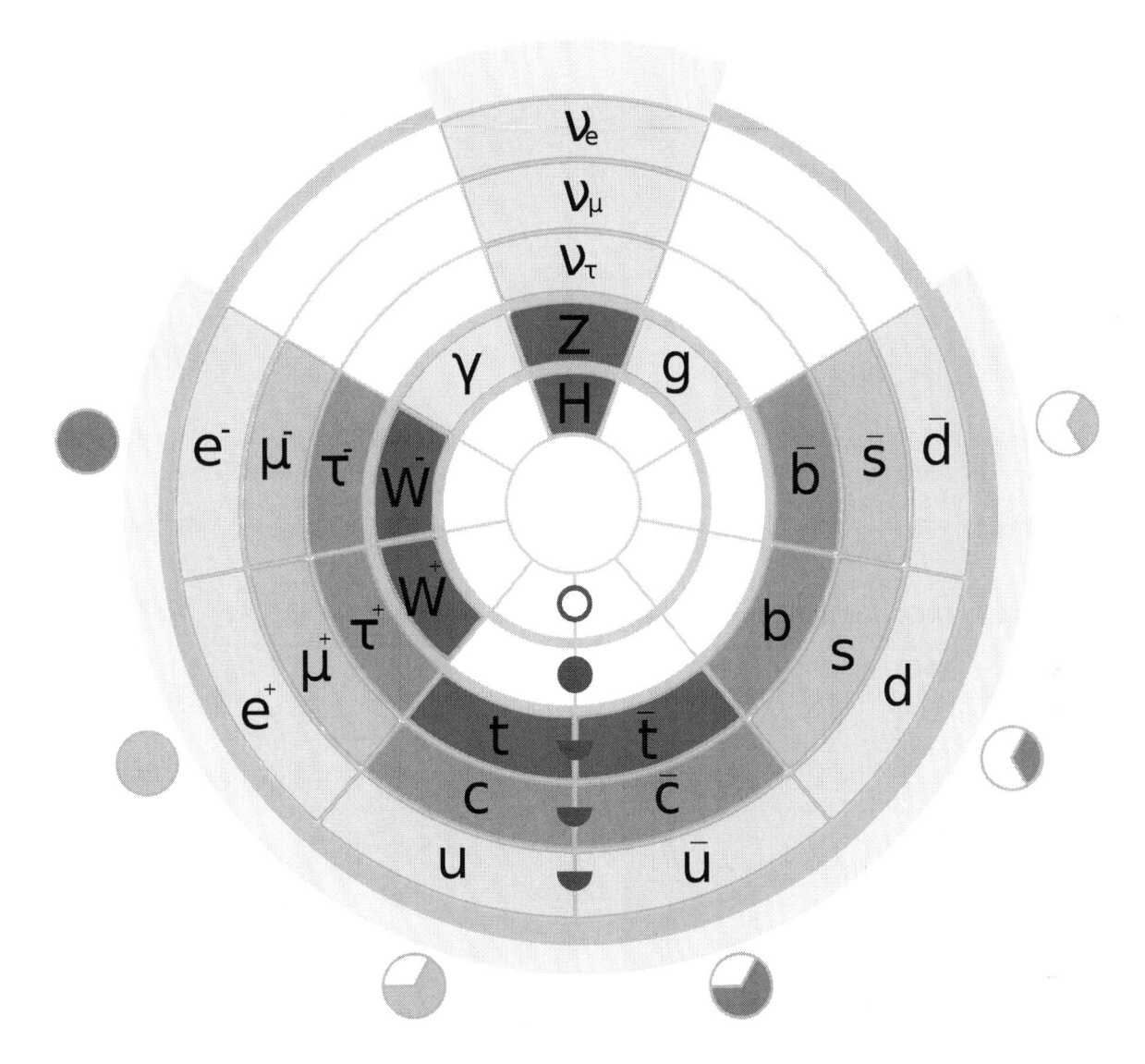

Masses of fundamental particles ----
more than 80 GeV/c^2
1-5 GeV/c^2
90-110 MeV/c^2
less than 16 MeV/c^2
Massless

to look for more fundamental theories with fewer parameters. Some of the ad hoc features are:

- *Hierarchy problem* – the standard model introduces particle masses through a process known as spontaneous symmetry breaking caused by the Higgs field. Within the standard model, the mass of the Higgs gets some very large quantum corrections due to the presence of virtual particles (mostly virtual top quarks). These corrections are much larger than the actual mass of the Higgs. This means that the bare mass parameter of the Higgs in the standard model must be fine tuned in such a way that almost completely cancels the quantum corrections. This level of fine-tuning is deemed unnatural by many theorists. There are also issues of Quantum triviality, which suggests that it may not be possible to create a consistent quantum field theory involving elementary scalar particles.

- *Strong CP problem* – theoretically it can be argued that the standard model should contain a term that breaks CP symmetry —relating matter to antimatter— in the strong interaction sector. Experimentally, however, no such violation has been found, implying that the coefficient of this term is very close to zero. This fine tuning is also considered unnatural.

- *Number of parameters* – the standard model depends on 19 numerical parameters. Their values are known from experiment, but the origin of the values is unknown. Some theorists have tried to find relations between different parameters, for example, between the masses of particles in different generations.

9.2 Grand unified theories

Main article: Grand Unified Theory

The standard model has three gauge symmetries; the colour SU(3), the weak isospin SU(2), and the hypercharge U(1) symmetry, corresponding to the three fundamental forces. Due to renormalization the coupling constants of each of these symmetries vary with the energy at which they are measured. Around 10^{16} GeV these couplings become approximately equal. This has led to speculation that above this energy the three gauge symmetries of the standard model are unified in one single gauge symmetry with a simple group gauge group, and just one coupling constant. Below this energy the symmetry is spontaneously broken to the standard model symmetries.[12] Popular choices for the unifying group are the special unitary group in five dimensions SU(5) and the special orthogonal group in ten dimensions SO(10).[13]

Theories that unify the standard model symmetries in this way are called Grand Unified Theories (or GUTs), and the energy scale at which the unified symmetry is broken is called the GUT scale. Generically, grand unified theories predict the creation of magnetic monopoles in the early universe,[14] and instability of the proton.[15] Neither of which have been observed, and this absence of observation puts limits on the possible GUTs.

9.3 Supersymmetry

Main article: Supersymmetry

Supersymmetry extends the Standard Model by adding another class of symmetries to the Lagrangian. These symmetries exchange fermionic particles with bosonic ones. Such a symmetry predicts the existence of *supersymmetric particles*, abbreviated as *sparticles*, which include the sleptons, squarks, neutralinos and charginos. Each particle in the Standard Model would have a superpartner whose spin differs by 1/2 from the ordinary particle. Due to the breaking of supersymmetry, the sparticles are much heavier than their ordinary counterparts; they are so heavy that existing particle colliders may not be powerful enough to produce them.

9.4 Neutrinos

In the standard model, neutrinos have exactly zero mass. This is a consequence of the standard model containing only left-handed neutrinos. With no suitable right-handed partner, it is impossible to add a renormalizable mass term to the standard model.[16] Measurements however indicated that neutrinos spontaneously change flavour, which implies that neutrinos have a mass. These measurements only give the relative masses of the different flavours. The best constraint on the absolute mass of the neutrinos comes from precision measurements of tritium decay, providing an upper limit 2 eV, which makes them at least five orders of magnitude lighter than the other particles in the standard model.[17] This necessitates an extension of the standard model, which not only needs to explain how neutrinos get their mass, but also why the mass is so small.[18]

One approach to add masses to the neutrinos, the so-called seesaw mechanism, is to add right-handed neutrinos and have these couple to left-handed neutrinos with a Dirac mass term. The right-handed neutrinos have to be sterile, meaning that they do not participate in any of the standard model interactions. Because they have no charges, the right-handed neutrinos can act as their own anti-particles, and have a Majorana mass term. Like the other Dirac masses in the standard model, the neutrino Dirac mass is expected to be generated through the Higgs mechanism, and is therefore unpredictable. The standard model fermion masses differ by many orders of magnitude; the Dirac neutrino mass has at least the same uncertainty. On the other hand, the Majorana mass for the right-handed neutrinos does not arise from the Higgs mechanism, and is therefore expected to be tied to some energy scale of new physics beyond the standard model, for example the Planck scale.[19] Therefore, any process involving right-handed neutrinos will be suppressed at low energies. The correction due to these suppressed processes effectively gives the left-handed neutrinos a mass that is inversely proportional to the right-handed Majorana mass, a mechanism known as the see-saw.[20] The presence of heavy right-handed neutrinos thereby explains both the small mass of the left-handed neutrinos and the absence of the right-handed neutrinos in observations. However, due to the uncertainty in the Dirac neutrino masses, the right-handed neutrino masses can lie anywhere. For example, they could be as light as keV and be dark matter,[21] they can have a mass in the LHC energy range[22][23] and lead to observable lepton number violation,[24] or they can be near the GUT scale, linking the right-handed neutrinos to the possibility of a grand unified theory.[25][26]

The mass terms mix neutrinos of different generations. This mixing is parameterized by the PMNS matrix, which is the neutrino analogue of the CKM quark mixing matrix. Unlike the quark mixing, which is almost minimal, the mixing of the neutrinos appears to be almost maximal. This has led to various speculations of symmetries between the various generations that could explain the mixing patterns.[27] The mixing matrix could also contain several complex phases that break CP invariance, although there has been no experimental probe of these. These phases could potentially create a surplus of leptons over anti-leptons in the early universe, a process known as leptogenesis. This asymmetry could then at a later stage be converted in an excess of baryons over anti-baryons, and explain the matter-antimatter asymmetry in the universe.[13]

The light neutrinos are disfavored as an explanation for the observation of dark matter, due to considerations of large-scale structure formation in the early universe. Simulations of structure formation show that they are too hot—i.e. their kinetic energy is large compared to their mass—while formation of structures similar to the galaxies in our universe requires cold dark matter. The simulations show that neutrinos can at best explain a few percent of the missing dark matter. The heavy sterile right-handed neutrinos are however a possible candidate for a dark matter WIMP.[28]

9.5 Preon Models

Several preon models have been proposed to address the unsolved problem concerning the fact that there are three generations of quarks and leptons. Preon models generally postulate some additional new particles which are further postulated to be able to combine to form the quarks and leptons of the standard model. One of the earliest preon models was the Rishon model.[29][30][31]

To date, no preon model is widely accepted or fully verified.

9.6 Theories of everything

9.6.1 Theory of everything

Main article: Theory of everything

Theoretical physics continues to strive toward a theory of everything, a theory that fully explains and links together all known physical phenomena, and predicts the outcome of any experiment that could be carried out in principle. In practical terms the immediate goal in this regard is to develop a theory which would unify the Standard Model with General Relativity in a theory of quantum gravity. Additional features, such as overcoming conceptual flaws in either theory or accurate prediction of particle masses, would be desired. The challenges in putting together such a theory are not just conceptual - they include the experimental aspects of the very high energies needed to probe exotic realms.

Several notable attempts in this direction are supersymmetry, string theory, and loop quantum gravity.

9.6.2 String theory

Main article: String theory

Extensions, revisions, replacements, and reorganizations of the Standard Model exist in attempt to correct for these and other issues. String theory is one such reinvention, and many theoretical physicists think that such theories are the next theoretical step toward a true Theory of Everything. Theories of quantum gravity such as loop quantum gravity and others are thought by some to be promising candidates to the mathematical unification of quantum field theory and general relativity, requiring less drastic changes to existing theories.[32] However recent work places stringent limits on the putative effects of quantum gravity on the speed of light, and disfavours some current models of quantum gravity.[33]

Among the numerous variants of string theory, M-theory, whose mathematical existence was first proposed at a String Conference in 1995, is believed by many to be a proper "ToE" candidate, notably by physicists Brian Greene and Stephen Hawking. Though a full mathematical description is not yet known, solutions to the theory exist for specific cases.[34] Recent works have also proposed alternate string models, some of which lack the various harder-to-test features of M-theory (e.g. the existence of Calabi–Yau manifolds, many extra dimensions, etc.) including works by well-published physicists such as Lisa Randall.[35][36]

9.7 See also

- *A New Kind of Science*
- Antimatter tests of Lorentz violation
- Fundamental physical constants in the standard model
- Higgsless model
- Holographic principle
- Little Higgs
- Lorentz-violating neutrino oscillations
- Minimal Supersymmetric Standard Model
- Peccei–Quinn theory
- Preon
- Standard-Model Extension
- Supergravity
- Seesaw mechanism
- Supersymmetry
- Superfluid vacuum theory
- String theory
- Technicolor (physics)
- Theory of everything
- Unsolved problems in physics
- Unparticle physics

9.8 References

[1] Womersley, J. (February 2005). "Beyond the Standard Model" (PDF). *Symmetry Magazine*. Retrieved 2010-11-23.

[2] Lykken, J. D. (2010). "Beyond the Standard Model". *CERN Yellow Report*. CERN. pp. 101–109. arXiv:1005.1676. CERN-2010-002.

[3] Sushkov, A. O.; Kim, W. J.; Dalvit, D. A. R.; Lamoreaux, S. K. (2011). "New Experimental Limits on Non-Newtonian Forces in the Micrometer Range". *Physical Review Letters* **107** (17): 171101. arXiv:1108.2547. Bibcode:2011PhRvL.107q1101S. doi:10.1103/PhysRevLett.107.171101. It is remarkable that two of the greatest successes of 20th century physics, general relativity and the standard model, appear to be fundamentally incompatible. But see also Donoghue, John F. (2012). "The effective field theory treatment of quantum gravity". *AIP Conference Proceedings* **1473**: 73. arXiv:1209.3511. doi:10.1063/1.4756964. One can find thousands of statements in the literature to the effect that "general relativity and quantum mechanics are incompatible". These are completely outdated and no longer relevant. Effective field theory shows that general relativity and quantum mechanics work together perfectly normally over a range of scales and curvatures, including those relevant for the world that we see around us. However, effective field theories are only valid over some range of scales. General relativity certainly does have problematic issues at extreme scales. There are important problems which the effective field theory does not solve because they are beyond its range of validity. However, this means that the issue of quantum gravity is not what we thought it to be. Rather than a fundamental incompatibility of quantum mechanics and gravity, we are in the more familiar situation of needing a more complete theory beyond the range of their combined applicability. The usual marriage of general relativity and quantum mechanics is fine at ordinary energies, but we now seek to uncover the modifications that must be present in more extreme conditions. This is the modern view of the problem of quantum gravity, and it represents progress over the outdated view of the past."

[4] Krauss, L. (2009). *A Universe from Nothing*. AAI Conference.

[5] Randolf Pohl, Ronald Gilman, Gerald A. Miller, Krzysztof Pachucki, "Muonic hydrogen and the proton radius puzzle" (May 30, 2013) http://arxiv.org/abs/1301.0905 in print Annu. Rev. Nucl. Part. Sci. Vol 63 (2013) 10.1146/annurev-nucl-102212-170627 ("The recent determination of the proton radius using the measurement of the Lamb shift in the muonic hydrogen atom startled the physics world. The obtained value of 0.84087(39) fm differs by about 4% or 7 standard deviations from the CODATA value of 0.8775(51) fm. The latter is composed from the electronic hydrogenate atom value of 0.8758(77) fm and from a similar value with larger uncertainties determined by electron scattering.")

[6] Lees, J. P.; et al. (BaBar Collaboration) (1970). "Evidence for an excess of $B \to D^{(*)}\tau^-\tau\nu$ decays". *Physical Review Letters* **109** (10). arXiv:1205.5442. Bibcode:2012PhRvL.109j1802L. doi:10.1103/PhysRevLett.109.101802.

[7] Article on LHCb results

[8] 2012 LHCb paper

[9] http://arxiv.org/pdf/1502.05314.pdf

[10] O'Luanaigh, C. (14 March 2013). "New results indicate that new particle is a Higgs boson". CERN.

[11] Marco Frasca, "What is a Glueball?" (March 31, 2009) http://marcofrasca.wordpress.com/2009/03/31/what-is-a-glueball-2/

[12] Peskin, M. E.; Schroeder, D. V. (1995). *An introduction to quantum field theory*. Addison-Wesley. pp. 786–791. ISBN 978-0-201-50397-5.

[13] Buchmüller, W. (2002). "Neutrinos, Grand Unification and Leptogenesis". arXiv:hep-ph/0204288 [hep-ph].

[14] Milstead, D.; Weinberg, E.J. (2009). "Magnetic Monopoles" (PDF). Particle Data Group. Retrieved 2010-12-20.

[15] P., Nath; P. F., Perez (2006). "Proton stability in grand unified theories, in strings, and in branes". *Physics Reports* **441** (5–6): 191–317. arXiv:hep-ph/0601023. Bibcode:2007PhR...441..191N. doi:10.1016/j.physrep.2007.02.010.

[16] Peskin, M. E.; Schroeder, D. V. (1995). *An introduction to quantum field theory*. Addison-Wesley. pp. 713–715. ISBN 978-0-201-50397-5.

[17] Nakamura, K.; et al. (Particle Data Group) (2010). "Neutrino Properties". Particle Data Group. Retrieved 2010-12-20.

[18] Mohapatra, R. N.; Pal, P. B. (2007). *Massive neutrinos in physics and astrophysics*. Lecture Notes in Physics **72** (3rd ed.). World Scientific. ISBN 978-981-238-071-5.

[19] Senjanovic, G. (2011). "Probing the Origin of Neutrino Mass: from GUT to LHC". arXiv:1107.5322 [hep-ph].

[20] Grossman, Y. (2003). "TASI 2002 lectures on neutrinos". arXiv:hep-ph/0305245v1 [hep-ph].

[21] Dodelson, S.; Widrow, L. M. (1993). "Sterile neutrinos as dark matter". *Physical Review Letters* **72**: 17. arXiv:hep-ph/9303287. Bibcode:1994PhRvL..72...17D. doi:10.1103/PhysRevLett.72.17.

[22] Minkowski, P. (1977). "$\mu \to e\gamma$ at a Rate of One Out of 10_9Muon Decays?".*Physics Letters B***67**(4): 421.Bibcode:1977PhLB...6 7..42doi:10.1016/0370-2693(77)90435-X.

[23] Mohapatra, R. N.; Senjanovic, G. (1980). "Neutrino mass and spontaneous parity nonconservation". *Physical Review Letters* **44** (14): 912. Bibcode:1980PhRvL..44..912M. doi:10.1103/PhysRevLett.44.912.

[24] Keung, W.-Y.; Senjanovic, G. (1983). "Majorana Neutrinos And The Production Of The Right-handed Charged Gauge Boson". *Physical Review Letters* **50** (19): 1427. Bibcode:1983PhRvL..50.1427K. doi:10.1103/PhysRevLett.50.1427.

[25] Gell-Mann, M.; Ramond, P.; Slansky, R. (1979). P. van Nieuwenhuizen; D. Freedman, eds. *Supergravity*. North Holland.

[26] Glashow, S. L. (1979). M. Levy, ed. *Proceedings of the 1979 Cargèse Summer Institute on Quarks and Leptons*. Plenum Press.

[27] Altarelli, G. (2007). "Lectures on Models of Neutrino Masses and Mixings". arXiv:0711.0161 [hep-ph].

[28] Murayama, H. (2007). "Physics Beyond the Standard Model and Dark Matter". arXiv:0704.2276 [hep-ph].

[29] Harari, H. (1979). "A Schematic Model of Quarks and Leptons". Physics Letters B 86 (1): 83-86.

[30] Shupe, M. A. (1979). "A Composite Model of Leptons and Quarks". Physics Letters B 86 (1): 87-92.

[31] Zenczykowski, P. (2008). "The Harari-Shupe preon model and nonrelativistic quantum phase space". Physics Letters B 660 (5): 567-572.

[32] Smolin, L. (2001). *Three Roads to Quantum Gravity*. Basic Books. ISBN 0-465-07835-4.

[33] Abdo, A. A.; et al. (Fermi GBM/LAT Collaborations) (2009). "A limit on the variation of the speed of light arising from quantum gravity effects". *Nature* **462** (7271): 331–4. arXiv:0908.1832. Bibcode:2009Natur.462..331A. doi:10.1038/nature08574. PMID 19865083.

[34] Maldacena, J.; Strominger, A.; Witten, E. (1997). "Black hole entropy in M-Theory". *Journal of High Energy Physics* **1997** (12): 2. arXiv:hep-th/9711053. Bibcode:1997JHEP...12..002M. doi:10.1088/1126-6708/1997/12/002.

[35] Randall, L.; Sundrum, R. (1999). "Large Mass Hierarchy from a Small Extra Dimension". *Physical Review Letters* **83** (17): 3370. arXiv:hep-ph/9905221. Bibcode:1999PhRvL..83.3370R. doi:10.1103/PhysRevLett.83.3370.

[36] Randall, L.; Sundrum, R. (1999). "An Alternative to Compactification". *Physical Review Letters* **83** (23): 4690. arXiv:hep-th/9906064. Bibcode:1999PhRvL..83.4690R. doi:10.1103/PhysRevLett.83.4690.

9.9 Further reading

- Lisa Randall (2005). *Warped Passages: Unraveling the Mysteries of the Universe's Hidden Dimensions*. HarperCollins. ISBN 0-06-053108-8.

9.10 External resources

- Standard Model Theory @ SLAC
- Scientific American Apr 2006
- LHC. Nature July 2007
- Open Questions
- Working group - schedule
- Les Houches Conference, Summer 2005

Chapter 10

Structure formation

In physical cosmology, **structure formation** refers to the formation of galaxies, galaxy clusters and larger structures from small early density fluctuations. The Universe, as is now known from observations of the cosmic microwave background radiation, began in a hot, dense, nearly uniform state approximately 13.8 billion years ago.[1] However, looking in the sky today, we see structures on all scales, from stars and planets to galaxies and, on still larger scales still, galaxy clusters and sheet-like structures of galaxies separated by enormous voids containing few galaxies. Structure formation attempts to model how these structures formed by gravitational instability of small early density ripples.[2][3][4][5]

The modern Lambda-CDM model is successful at predicting the observed large-scale distribution of galaxies, clusters and voids; but on the scale of individual galaxies there are many complications due to highly nonlinear processes involving baryonic physics, gas heating and cooling, star formation and feedback. Understanding the processes of galaxy formation is a major topic of modern cosmology research, both via observations such as the Hubble Ultra-Deep Field and via large computer simulations.

10.1 Overview

Under present models, the structure of the visible universe was formed in the following stages:

- **The very early Universe**
  In this stage, some mechanism, such as cosmic inflation, was responsible for establishing the initial conditions of the Universe: homogeneity, isotropy, and flatness.[3][6] Cosmic inflation also would have amplified minute quantum fluctuations (pre-inflation) into slight density ripples of overdensity and underdensity (post-inflation).

- **Growth of structure**
  The early universe was dominated by radiation; in this case density fluctuations larger than the cosmic horizon grow proportional to the scale factor, as the gravitational potential fluctuations remain constant. Structures smaller than the horizon remained essentially frozen due to radiation domination impeding growth. As the universe expanded, the density of radiation drops faster than matter (due to redshifting of photon energy); this led to a crossover called matter-radiation equality at ~ 50,000 years after the Big Bang. After this all dark matter ripples could grow freely, forming seeds into which the baryons could later fall. The size of the universe at this epoch forms a turnover in the matter power spectrum which can be measured in large redshift surveys.

- **Recombination**
  The Universe was dominated by radiation for most of this stage, and due to the intense heat and radiation, the primordial hydrogen and helium were fully ionized into nuclei and free electrons. In this hot and dense situation, the radiation (photons) could not travel far before Thomson scattering off an electron. The Universe was very hot and dense, but expanding rapidly and therefore cooling. Finally, at a little less than 400,000 years after the 'bang', it become cool enough (around 3000 K) for the protons to capture negatively charged electrons, forming neutral hydrogen atoms. (Helium atoms formed somewhat earlier due to their larger binding energy). Once nearly

all the charged particles were bound in neutral atoms, the photons no longer interacted with them and were free to propagate for the next 13.8 billion years; we currently detect those photons redshifted by a factor 1090 down to 2.725 K as the Cosmic Microwave Background Radiation (CMB) filling today's Universe. Several remarkable space-based missions (COBE,WMAP,Planck),have detected very slight variations in the density and temperature of the CMB. These variations were subtle, and the CMB appears very nearly uniformly the same in every direction. However, the slight temperature variations of order a few parts in 100,000 are of enormous importance, for they essentially were early "seeds" from which all subsequent complex structures in the universe ultimately developed.

The theory of what happened after the universe's first 400,000 years is one of hierarchical structure formation: the smaller gravitationally bound structures such as matter peaks containing the first stars and stellar clusters formed first, and these subsequently merged together with gas and dark matter to form galaxies, followed by groups, clusters and superclusters of galaxies.

10.2 Very early Universe

The very early Universe is still a poorly understood epoch, from the viewpoint of fundamental physics. The prevailing theory, cosmic inflation, does a good job explaining the observed flatness, homogeneity and isotropy of the Universe, as well as the absence of exotic relic particles (such as magnetic monopoles). In addition, it has made a crucial prediction that has been borne out by observation: that the primordial Universe would have tiny perturbations which seed the formation of structure in the later Universe. These fluctuations, while they form the foundation for all structure in the Universe, appear most clearly as tiny temperature fluctuations at one part in 100,000. (To put this in perspective, the same level of fluctuations on a topographic map of the United States would show no feature higher than a few centimeters high.) These fluctuations are critical, because they provide the seeds from which the largest structures within the Universe can grow and eventually collapse to form galaxies and stars. COBE (Cosmic Background Explorer) provided the first detection of the intrinsic fluctuations in the cosmic microwave background radiation in the 1990s.

These perturbations are thought to have a very specific character: they form a Gaussian random field whose covariance function is diagonal and nearly scale-invariant. The observed fluctuations appear to have exactly this form, and in addition the *spectral index* measured by WMAP—the spectral index measures the deviation from a scale-invariant (or Harrison-Zel'dovich) spectrum—is very nearly the value predicted by the simplest and most robust models of inflation. Another important property of the primordial perturbations, that they are adiabatic (or isentropic between the various kinds of matter that compose the Universe), is predicted by cosmic inflation and has been confirmed by observations.

Other theories of the very early Universe, which are claimed to make very similar predictions, have been proposed, such as the brane gas cosmology, cyclic model, pre-big bang model and holographic universe, but they remain in their nascency and are not as widely accepted. Some theories, such as cosmic strings, have largely been refuted by increasingly precise data.

10.2.1 The horizon problem

An extremely important concept in the theory of structure formation is the notion of the Hubble radius, often called simply the *horizon* as it is closely related to the particle horizon. The Hubble radius, which is related to the Hubble parameter H as $R = c/H$, where c is the speed of light, defines, roughly speaking, the volume of the nearby universe that has recently (in the last expansion time) been in causal contact with an observer. Since the Universe is continually expanding, its energy density is continually decreasing (in the absence of truly exotic matter such as phantom energy). The Friedmann equation relates the energy density of the Universe to the Hubble parameter, and shows that the Hubble radius is continually increasing.

The horizon problem of the big bang cosmology says that, without inflation, perturbations were never in causal contact before they entered the horizon and thus the homogeneity and isotropy of, for example, the large scale galaxy distributions cannot be explained. This is because, in an ordinary Friedmann–Lemaître–Robertson–Walker cosmology, the Hubble radius increases more rapidly than space expands, so perturbations are only ever entering the Hubble radius, and they are not being pushed out by the expansion of space. This paradox is resolved by cosmic inflation, which suggests that there

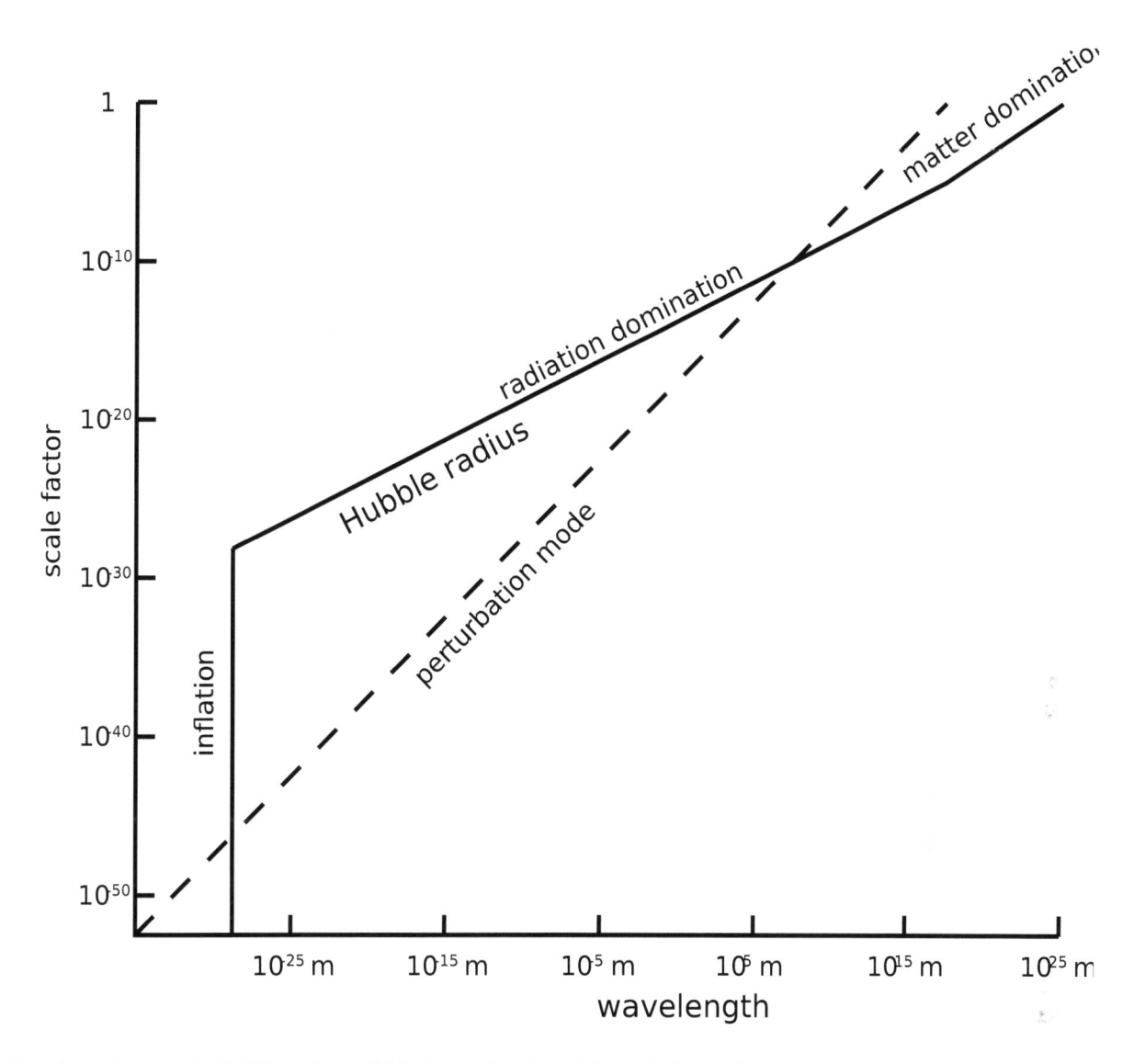

The physical size of the Hubble radius (solid line) as a function of the scale factor of the Universe. The physical wavelength of a perturbation mode (dashed line) is shown as well. The plot illustrates how the perturbation mode exits the horizon during cosmic inflation in order to reenter during radiation domination. If cosmic inflation never happened, and radiation domination continued back until a gravitational singularity, then the mode would never have exited the horizon in the very early Universe.

was a phase of very rapid expansion in the early Universe in which the Hubble radius was very nearly constant. Thus, the large scale isotropy that we see today is due to quantum fluctuations produced during cosmic inflation being pushed outside the horizon.

10.3 Primordial plasma

The end of inflation is called reheating, when the inflation particles decay into a hot, thermal plasma of other particles. In this epoch, the energy content of the Universe is entirely radiation, with standard model particles having relativistic velocities. As the plasma cools, baryogenesis and leptogenesis are thought to occur, as the quark–gluon plasma cools, electroweak symmetry breaking occurs and the Universe becomes principally composed of ordinary protons, neutrons and electrons. As the Universe cools further, big bang nucleosynthesis occurs and small quantities of deuterium, helium and lithium nuclei are created. As the Universe cools and expands, the energy in photons begins to redshift away, particles become non-relativistic and ordinary matter begins to dominate the Universe. Eventually, atoms begin to form as free electrons bind to nuclei. This suppresses Thomson scattering of photons. Combined with the rarefaction of the Universe (and consequent increase in the mean free path of photons), this makes the Universe transparent and the cosmic microwave background is emitted at recombination (the *surface of last scattering*).

10.3.1 Acoustic oscillations

Main article: baryon acoustic oscillations

The primordial plasma would have had very slight overdensities of matter, thought to have derived from the enlargement of quantum fluctuations during inflation. Whatever the source, these overdensities gravitationally attract matter. But the intense heat of the near constant photon-matter interactions of this epoch rather forcefully seeks thermal equilibrium, which creates a large amount of outward pressure. These counteracting forces of gravity and pressure create oscillations, analogous to sound waves created in air by pressure differences.

These perturbations are important, as they are responsible for the subtle physics that result in the cosmic microwave background anisotropy. In this epoch, the amplitude of perturbations that enter the horizon oscillate sinusoidally, with dense regions becoming more rarefied and then becoming dense again, with a frequency which is related to the size of the perturbation. If the perturbation oscillates an integral or half-integral number of times between coming into the horizon and recombination, it appears as an acoustic peak of the cosmic microwave background anisotropy. (A half-oscillation, in which a dense region becomes a rarefied region or vice versa, appears as a peak because the anisotropy is displayed as a *power spectrum*, so underdensities contribute to the power just as much as overdensities.) The physics that determines the detailed peak structure of the microwave background is complicated, but these oscillations provide the essence.[7][8][9][10][11]

10.4 Linear structure

One of the key realizations made by cosmologists in the 1970s and 1980s was that the majority of the matter content of the Universe was composed not of atoms, but rather a mysterious form of matter known as dark matter. Dark matter interacts through the force of gravity, but it is not composed of baryons and it is known with very high accuracy that it does not emit or absorb radiation. It may be composed of particles that interact through the weak interaction, such as neutrinos, but it cannot be composed entirely of the three known kinds of neutrinos (although some have suggested it is a sterile neutrino). Recent evidence suggests that there is about five times as much dark matter as baryonic matter, and thus the dynamics of the Universe in this epoch are dominated by dark matter.

Dark matter plays a key role in structure formation because it feels only the force of gravity: the gravitational Jeans instability which allows compact structures to form is not opposed by any force, such as radiation pressure. As a result, dark matter begins to collapse into a complex network of dark matter halos well before ordinary matter, which is impeded by pressure forces. Without dark matter, the epoch of galaxy formation would occur substantially later in the Universe than is observed.

The physics of structure formation in this epoch is particularly simple, as dark matter perturbations with different wavelengths evolve independently. As the Hubble radius grows in the expanding Universe, it encompasses larger and larger perturbations. During matter domination, all causal dark matter perturbations grow through gravitational clustering. However, the shorter-wavelength perturbations that are encompassed during radiation domination have their growth retarded until matter domination. At this stage, luminous, baryonic matter is expected to simply mirror the evolution of the dark matter, and their distributions should closely trace one another.

It is a simple matter to calculate this "linear power spectrum" and, as a tool for cosmology, it is of comparable importance to the cosmic microwave background. The power spectrum has been measured by galaxy surveys, such as the Sloan Digital Sky Survey, and by surveys of the Lyman-α forest. Since these surveys observe radiation emitted from galaxies and quasars, they do not directly measure the dark matter, but the large scale distribution of galaxies (and of absorption lines in the Lyman-α forest) is expected to closely mirror the distribution of dark matter. This depends on the fact that galaxies will be larger and more numerous in denser parts of the Universe, whereas they will be comparatively scarce in rarefied regions.

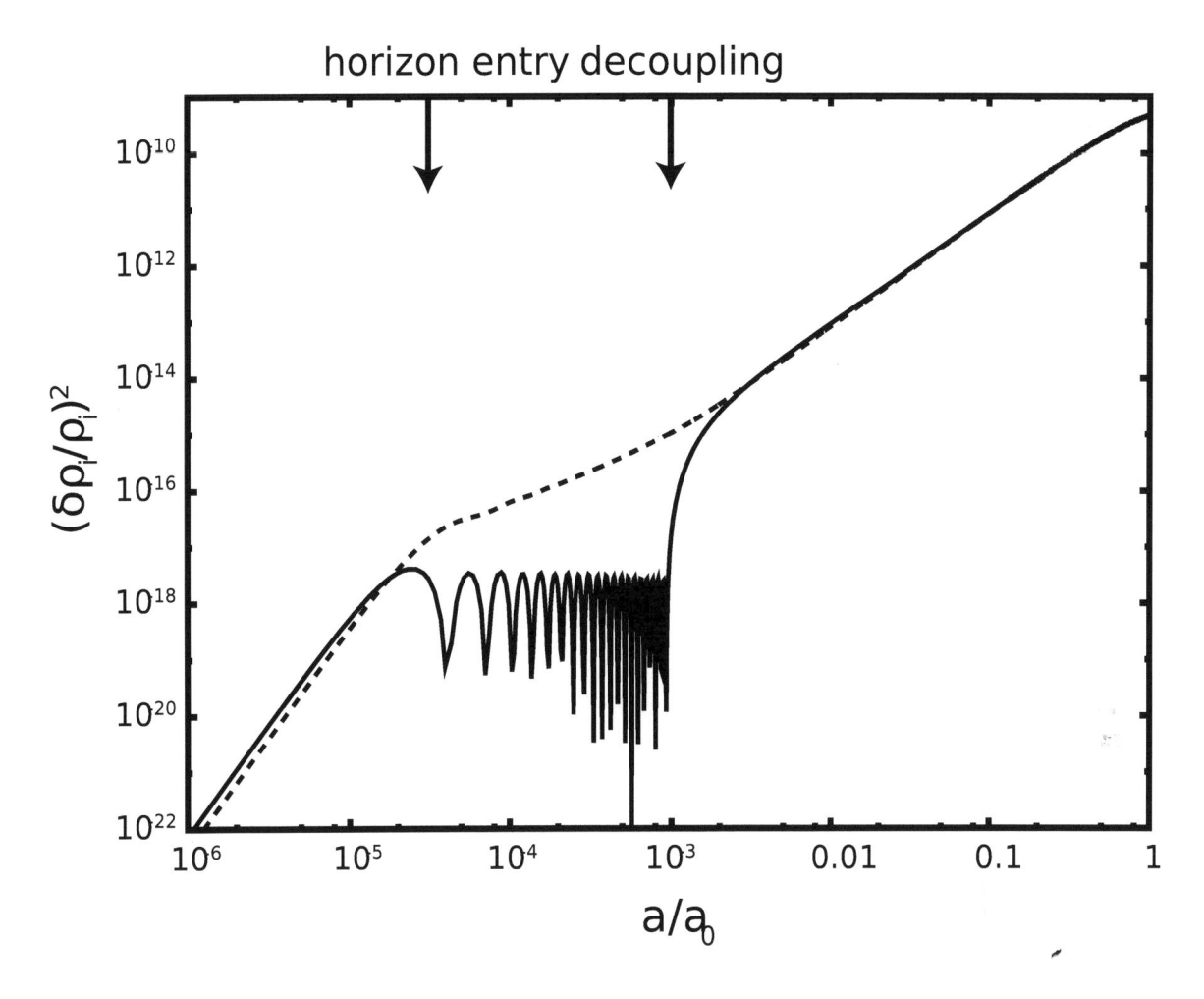

Evolution of two perturbations to the ΛCDM homogeneous big bang model. Between entering the horizon and decoupling, the dark matter perturbation (dashed line) grows logarithmically, before the growth accelerates in matter domination. On the other hand, between entering the horizon and decoupling, the perturbation in the baryon–photon fluid (solid line) oscillates rapidly. After decoupling, it grows rapidly to match the dominant matter perturbation, the dark matter mode.

10.5 Nonlinear structure

When the perturbations have grown sufficiently, a small region might become substantially denser than the mean density of the Universe. At this point, the physics involved becomes substantially more complicated. When the deviations from homogeneity are small, the dark matter may be treated as a pressureless fluid and evolves by very simple equations. In regions which are significantly denser than the background, the full Newtonian theory of gravity must be included. (The Newtonian theory is appropriate because the masses involved are much less than those required to form a black hole, and the speed of gravity may be ignored as the light-crossing time for the structure is still smaller than the characteristic dynamical time.) One sign that the linear and fluid approximations become invalid is that dark matter starts to form caustics in which the trajectories of adjacent particles cross, or particles start to form orbits. These dynamics are generally best understood using N-body simulations (although a variety of semi-analytic schemes, such as the Press–Schechter formalism, can be used in some cases). While in principle these simulations are quite simple, in practice they are very difficult to implement, as they require simulating millions or even billions of particles. Moreover, despite the large number of particles, each particle typically weighs 10^9 solar masses and discretization effects may become significant. The largest such simulation as of 2005 is the Millennium simulation.[12]

The result of N-body simulations suggests that the Universe is composed largely of voids, whose densities might be as low as one tenth the cosmological mean. The matter condenses in large filaments and haloes which have an intricate web-like structure. These form galaxy groups, clusters and superclusters. While the simulations appear to agree broadly with observations, their interpretation is complicated by the understanding of how dense accumulations of dark matter spur galaxy formation. In particular, many more small haloes form than we see in astronomical observations as dwarf galaxies and globular clusters. This is known as the galaxy bias problem, and a variety of explanations have been proposed. Most account for it as an effect in the complicated physics of galaxy formation, but some have suggested that it is a problem with our model of dark matter and that some effect, such as warm dark matter, prevents the formation of the smallest haloes.

10.6 Gas evolution

See also: galaxy formation and evolution and stellar evolution

The final stage in evolution comes when baryons condense in the centres of galaxy haloes to form galaxies, stars and quasars. A paradoxical aspect of structure formation is that while dark matter greatly accelerates the formation of dense haloes, because dark matter does not have radiation pressure, the formation of smaller structures from dark matter is impossible because dark matter cannot dissipate angular momentum, whereas ordinary baryonic matter can collapse to form dense objects by dissipating angular momentum through radiative cooling. Understanding these processes is an enormously difficult computational problem, because they can involve the physics of gravity, magnetohydrodynamics, atomic physics, nuclear reactions, turbulence and even general relativity. In most cases, it is not yet possible to perform simulations that can be compared quantitatively with observations, and the best that can be achieved are approximate simulations that illustrate the main qualitative features of a process such as star formation.

10.7 Modelling structure formation

10.7.1 Cosmological perturbations

Main article: cosmological perturbation theory

Much of the difficulty, and many of the disputes, in understanding the large-scale structure of the Universe can be resolved by better understanding the choice of gauge in general relativity. By the scalar-vector-tensor decomposition, the metric includes four scalar perturbations, two vector perturbations, and one tensor perturbation. Only the scalar perturbations are significant: the vectors are exponentially suppressed in the early Universe, and the tensor mode makes only a small (but important) contribution in the form of primordial gravitational radiation and the B-modes of the cosmic microwave background polarization. Two of the four scalar modes may be removed by a physically meaningless coordinate transformation. Which modes are eliminated determine the infinite number of possible gauge fixings. The most popular gauge is Newtonian gauge (and the closely related conformal Newtonian gauge), in which the retained scalars are the Newtonian potentials Φ and Ψ, which correspond exactly to the Newtonian potential energy from Newtonian gravity. Many other gauges are used, including synchronous gauge, which can be an efficient gauge for numerical computation (it is used by CMBFAST). Each gauge still includes some unphysical degrees of freedom. There is a so-called gauge-invariant formalism, in which only gauge invariant combinations of variables are considered.

10.7.2 Inflation and initial conditions

The initial conditions for the Universe are thought to arise from the scale invariant quantum mechanical fluctuations of cosmic inflation. The perturbation of the background energy density at a given point $\rho(\mathbf{x}, t)$ in space is then given by an isotropic, homogeneous Gaussian random field of mean zero. This means that the spatial Fourier transform of $\rho - \hat{\rho}(\mathbf{k}, t)$ has the following correlation functions

10.7. MODELLING STRUCTURE FORMATION

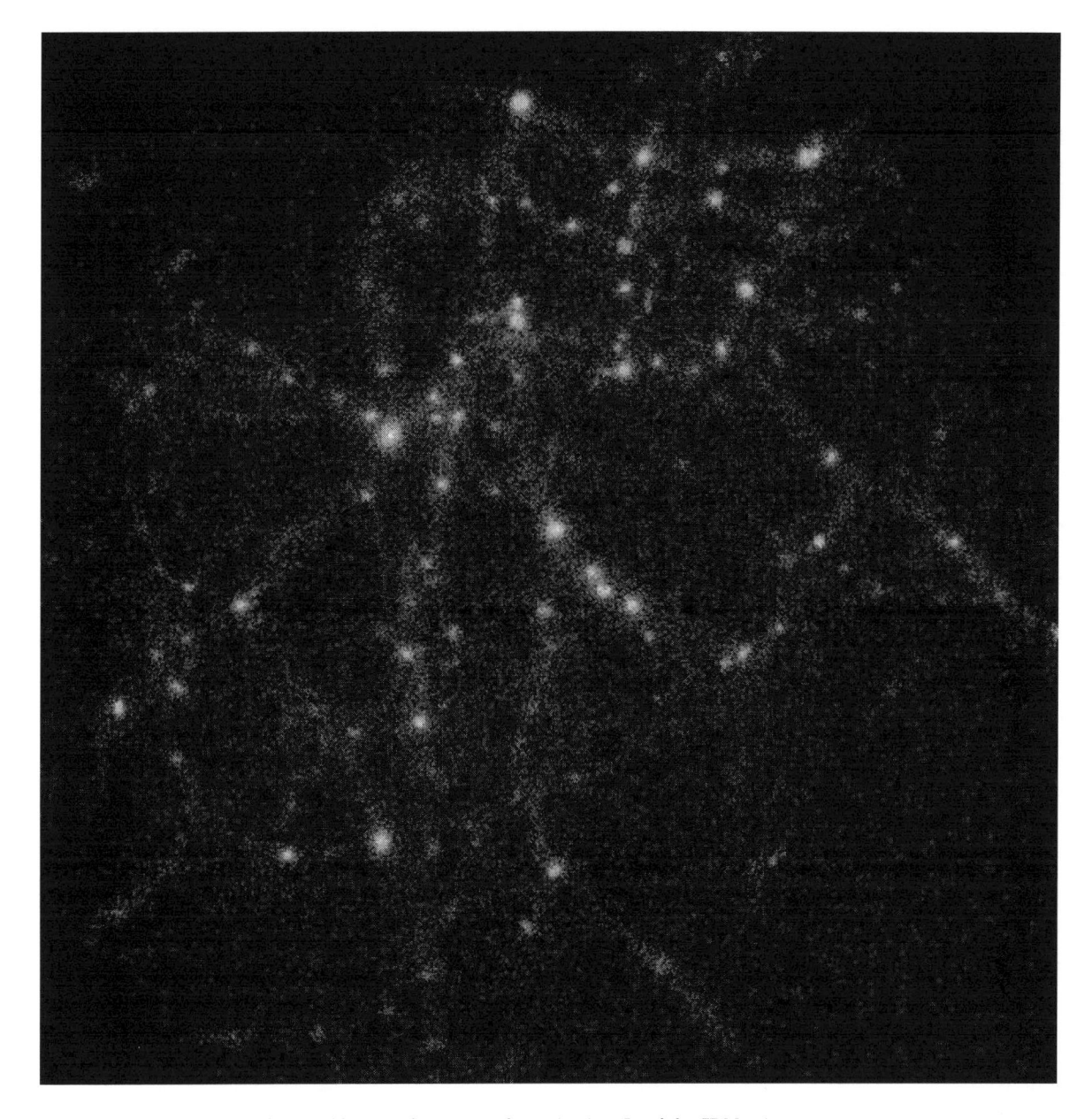

Snapshot from a computer simulation of large scale structure formation in a Lambda-CDM universe.

$$\langle \hat{\rho}(\mathbf{k},t)\hat{\rho}(\mathbf{k}',t)\rangle = f(k)\delta^{(3)}(\mathbf{k}-\mathbf{k}')$$

where $\delta^{(3)}$ is the three-dimensional Dirac delta function and $k = |\mathbf{k}|$ is the length of $\mathbf{k}$. Moreover, the spectrum predicted by inflation is nearly scale invariant, which means

$$\langle \hat{\rho}(\mathbf{k},t)\hat{\rho}(\mathbf{k}',t)\rangle = k^{n_s-1}\delta^{(3)}(\mathbf{k}-\mathbf{k}')$$

where $n_s - 1$ is a small number. Finally, the initial conditions are adiabatic or isentropic, which means that the fractional perturbation in the entropy of each species of particle is equal.

10.8 See also

- Big Bang
- Chronology of the universe
- Galaxy formation and evolution
- Illustris project
- Stellar evolution
- Timeline of the Big Bang

10.9 References

[1] "Cosmic Detectives". The European Space Agency (ESA). 2013-04-02. Retrieved 2013-04-15.

[2] Dodelson, Scott (2003). *Modern Cosmology*. Academic Press. ISBN 0-12-219141-2.

[3] Liddle, Andrew; David Lyth (2000). *Cosmological Inflation and Large-Scale Structure*. Cambridge. ISBN 0-521-57598-2.

[4] Padmanabhan, T. (1993). *Structure formation in the universe*. Cambridge University Press. ISBN 0-521-42486-0.

[5] Peebles, P. J. E. (1980). *The Large-Scale Structure of the Universe*. Princeton University Press. ISBN 0-691-08240-5.

[6] Kolb, Edward; Michael Turner (1988). *The Early Universe*. Addison-Wesley. ISBN 0-201-11604-9.

[7] Harrison, E. R. (1970). "Fluctuations at the threshold of classical cosmology". *Phys. Rev.* **D1**(10): 2726. Bibcode:1970PhRvD...1 doi:10.1103/PhysRevD.1.2726.

[8] Peebles, P. J. E.; Yu, J. T. (1970). "Primeval adiabatic perturbation in an expanding universe". *Astrophysical Journal* **162**: 815. Bibcode:1970ApJ...162..815P. doi:10.1086/150713.

[9] Zel'dovich, Yaa B. (1972). "A hypothesis, unifying the structure and entropy of the Universe". *Monthly Notices of the Royal Astronomical Society* **160**: 1P. Bibcode:1972MNRAS.160P...1Z. doi:10.1093/mnras/160.1.1p.

[10] R. A. Sunyaev, "Fluctuations of the microwave background radiation", in *Large Scale Structure of the Universe* ed. M. S. Longair and J. Einasto, 393. Dordrecht: Reidel 1978.

[11] U. Seljak & M. Zaldarriaga (1996). "A line-of-sight integration approach to cosmic microwave background anisotropies". *Astrophysics J.* **469**: 437–444. arXiv:astro-ph/9603033. Bibcode:1996ApJ...469..437S. doi:10.1086/177793.

[12] Springel, V. et al. (2005). "Simulations of the formation, evolution and clustering of galaxies and quasars". *Nature* **435** (7042): 629–636. arXiv:astro-ph/0504097. Bibcode:2005Natur.435..629S. doi:10.1038/nature03597. PMID 15931216.

Chapter 11

Gravitational binding energy

A **gravitational binding energy** is the energy that must be exported from a system for the system to enter a gravitationally bound state at a negative level of energy. Negative energy is called "potential energy".[1] A gravitationally bound system has a lower (*i.e.*, more negative) gravitational potential energy than the sum of its parts—this is what keeps the system aggregated in accordance with the minimum total potential energy principle. Therefore, a system's gravitational binding energy is the system's gravitational synergy.

For a spherical mass of uniform density, the gravitational binding energy U is given by the formula[2][3]

$$U = \frac{3GM^2}{5R}$$

where G is the gravitational constant, M is the mass of the sphere, and R is its radius. The energy required to separate to infinity the two hemispheres of a spherical mass is

$$U = \frac{3GM^2}{25R}$$

Assuming that the Earth is a uniform sphere (which is not correct, but is close enough to get an order-of-magnitude estimate) with $M = 5.97 \cdot 10^{24}$kg and $r = 6.37 \cdot 10^6$m, U is $2.24 \cdot 10^{32}$J. This is roughly equal to one week of the Sun's total energy output. It is 37.5 MJ/kg, 60% of the absolute value of the potential energy per kilogram at the surface.

The actual depth-dependence of density, inferred from seismic travel times (see Adams–Williamson equation), is given in the Preliminary Reference Earth Model (PREM).[4] Using this, the real gravitational binding energy of Earth can be calculated numerically to U = $2.487 \cdot 10^{32}$ J

According to the virial theorem, the gravitational binding energy of a star is about two times its internal thermal energy.[2]

11.1 Derivation for a uniform sphere

The gravitational binding energy of a sphere with Radius R is found by imagining that it is pulled apart by successively moving spherical shells to infinity, the outermost first, and finding the total energy needed for that.

If we assume a constant density ρ then the masses of a shell and the sphere inside it are:

$$m_{\text{shell}} = 4\pi r^2 \rho \, dr \text{ and } m_{\text{interior}} = \tfrac{4}{3}\pi r^3 \rho$$

The required energy for a shell is the negative of the gravitational potential energy:

$$dU = -G\frac{m_{\text{shell}} m_{\text{interior}}}{r}$$

Integrating over all shells we get:

$$U = -G \int_0^R \frac{(4\pi r^2 \rho)(\frac{4}{3}\pi r^3 \rho)}{r} dr = -G\frac{16}{3}\pi^2 \rho^2 \int_0^R r^4 dr = -G\frac{16}{15}\pi^2 \rho^2 R^5$$

Remembering that ρ is simply equal to the mass of the whole divided by its volume for objects with uniform density we get:

$$\rho = \frac{M}{\frac{4}{3}\pi R^3}$$

And finally, plugging this into our result we get:

$$U = -G\frac{16}{15}\pi^2 R^5 \left(\frac{M}{\frac{4}{3}\pi R^3}\right)^2 = -\frac{3GM^2}{5R}$$

11.2 Non-uniform spheres

Planets and stars have radial density gradients from their lower density surfaces to their much larger density compressed cores. Degenerate matter objects (white dwarfs; neutron star pulsars) have radial density gradients plus relativistic corrections.

Neutron star relativistic equations of state provided by Jim Lattimer include a graph of radius vs. mass for various models.[5] The most likely radii for a given neutron star mass are bracketed by models AP4 (smallest radius) and MS2 (largest radius). BE is the ratio of gravitational binding energy mass equivalent to observed neutron star gravitational mass of "M" kilograms with radius "R" meters,

$$BE = \frac{0.60\,\beta}{1-\frac{\beta}{2}} \quad \beta = GM/Rc^2$$

Given current values

$$G = 6.6742 \times 10^{-11}\, m^3 kg^{-1} sec^{-2}\,[6]$$

$c^2 = 8.98755 \times 10^{16}\, m^2 sec^{-2}$

$M_{solar} = 1.98844 \times 10^{30}\, kg$

and star masses "M" commonly reported as multiples of one solar mass,

$$M_x = \frac{M}{M_\odot}$$

then the relativistic fractional binding energy of a neutron star is

$$BE = \frac{885.975\, M_x}{R - 738.313\, M_x}$$

11.3 See also

- Nordtvedt effect

11.4 References

[1] Why is the Potential Energy Negative? *HyperPhysics*

[2] Chandrasekhar, S. 1939, *An Introduction to the Study of Stellar Structure* (Chicago: U. of Chicago; reprinted in New York: Dover), section 9, eqs. 90-92, p. 51 (Dover edition)

[3] Lang, K. R. 1980, *Astrophysical Formulae* (Berlin: Springer Verlag), p. 272

[4] Dziewonski, A. M.; Anderson, D. L.. "Preliminary Reference Earth Model". *Physics of the Earth and Planetary Interiors* **25**: 297–356. Bibcode:1981PEPI...25..297D. doi:10.1016/0031-9201(81)90046-7.

[5] Neutron Star Masses and Radii, p. 9/20, bottom

[6] Measurement of Newton's Constant Using a Torsion Balance with Angular Acceleration Feedback , Phys. Rev. Lett. 85(14) 2869 (2000)

Chapter 12

Galaxy formation and evolution

The study of **galaxy formation and evolution** is concerned with the processes that formed a heterogeneous universe from a homogeneous beginning, the formation of the first galaxies, the way galaxies change over time, and the processes that have generated the variety of structures observed in nearby galaxies.

Galaxy formation is hypothesized to occur, from structure formation theories, as a result of tiny quantum fluctuations in the aftermath of the Big Bang. The simplest model for this that is in general agreement with observed phenomena is the Λ-Cold Dark Matter cosmology; that is to say that clustering and merging is how galaxies gain in mass, and can also determine their shape and structure.

12.1 Commonly observed properties of galaxies

Some notable observed features of galaxy structure (including our own Milky Way) that astronomers wish to explain with galactic formation theories, include (but are certainly not limited to) the following:

- Spiral galaxies and the galactic disk are quite thin, dense, and rotate relatively fast. (Our Milky Way galaxy is believed to be a barred spiral.)

- The majority of mass in galaxies is made up of dark matter, a substance which is not directly observable, and might not interact through any means except gravity.

- Halo stars are typically much older and have much lower metallicities (that is to say, they are almost exclusively composed of hydrogen and helium) than disk stars.

- Many disk galaxies have a puffed up outer disk (often called the "thick disk") that is composed of old stars.

- Globular clusters are typically old and metal-poor as well, but there are a few that are not nearly as metal-poor as most, or have some younger stars.

- High-velocity clouds, clouds of neutral hydrogen are "raining" down on the galaxy, and presumably have been from the beginning (this would be the necessary source of a gas disk from which the disk stars formed).

- Galaxies come in a great variety of shapes and sizes (see the Hubble sequence), from giant, featureless blobs of old stars (called elliptical galaxies) to thin disks with gas and stars arranged in highly ordered spirals.

- The majority of giant galaxies contain a supermassive black hole in their centers, ranging in mass from millions to billions of times the mass of our Sun. The black hole mass is tied to properties of its host galaxy.

- Many of the properties of galaxies (including the galaxy color–magnitude diagram) indicate that there are fundamentally two types of galaxies. These groups divide into blue star-forming galaxies that are more like spiral types, and red non-star forming galaxies that are more like elliptical galaxies.

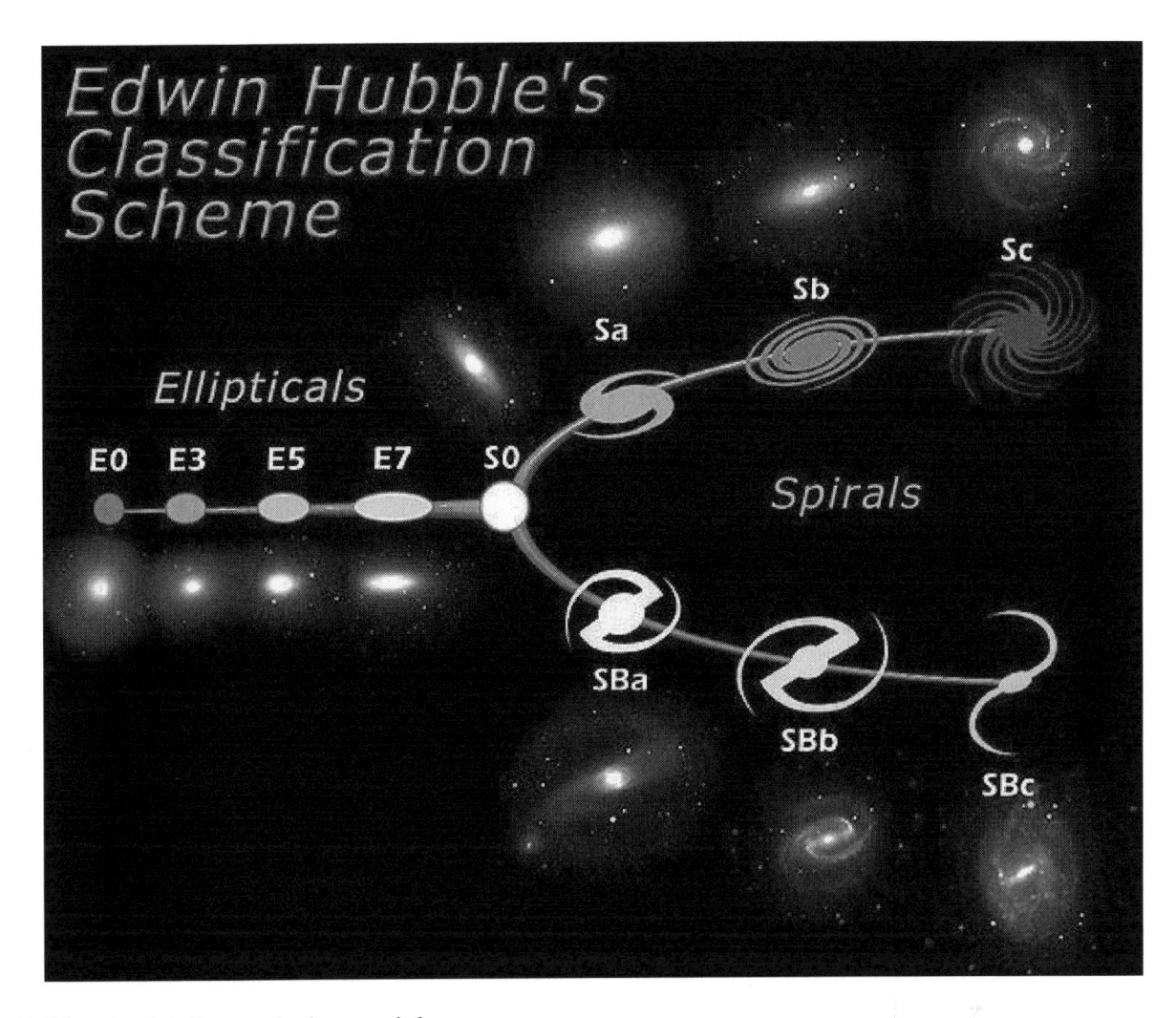

Hubble tuning fork diagram of galaxy morphology

12.2 Formation of disk galaxies

The key properties of disk galaxies, which are also commonly called spiral galaxies, are that they are very thin, rotate rapidly, and often show spiral structure. One of the main challenges to galaxy formation is the great number of thin disk galaxies in the local universe. The problem is that disks are very fragile, and mergers with other galaxies can quickly destroy thin disks.

Olin Eggen, Donald Lynden-Bell, and Allan Sandage[1] in 1962, proposed a theory that disk galaxies form through a monolithic collapse of a large gas cloud. As the cloud collapses the gas settles into a rapidly rotating disk. Known as a top-down formation scenario, this theory is quite simple yet no longer widely accepted because observations of the early universe strongly suggest that objects grow from bottom-up (i.e. smaller objects merging to form larger ones). It was first proposed by Leonard Searle and Robert Zinn[2] that galaxies form by the coalescence of smaller progenitors.

More recent theories include the clustering of dark matter halos in the bottom-up process. Essentially early on in the universe galaxies were composed mostly of gas and dark matter, and thus, there were fewer stars. As a galaxy gained mass (by accreting smaller galaxies) the dark matter stays mostly on the outer parts of the galaxy. This is because the dark matter can only interact gravitationally, and thus will not dissipate. The gas, however, can quickly contract, and as it does so it rotates faster, until the final result is a very thin, very rapidly rotating disk.

Astronomers do not currently know what process stops the contraction. In fact, theories of disk galaxy formation are not successful at producing the rotation speed and size of disk galaxies. It has been suggested that the radiation from bright

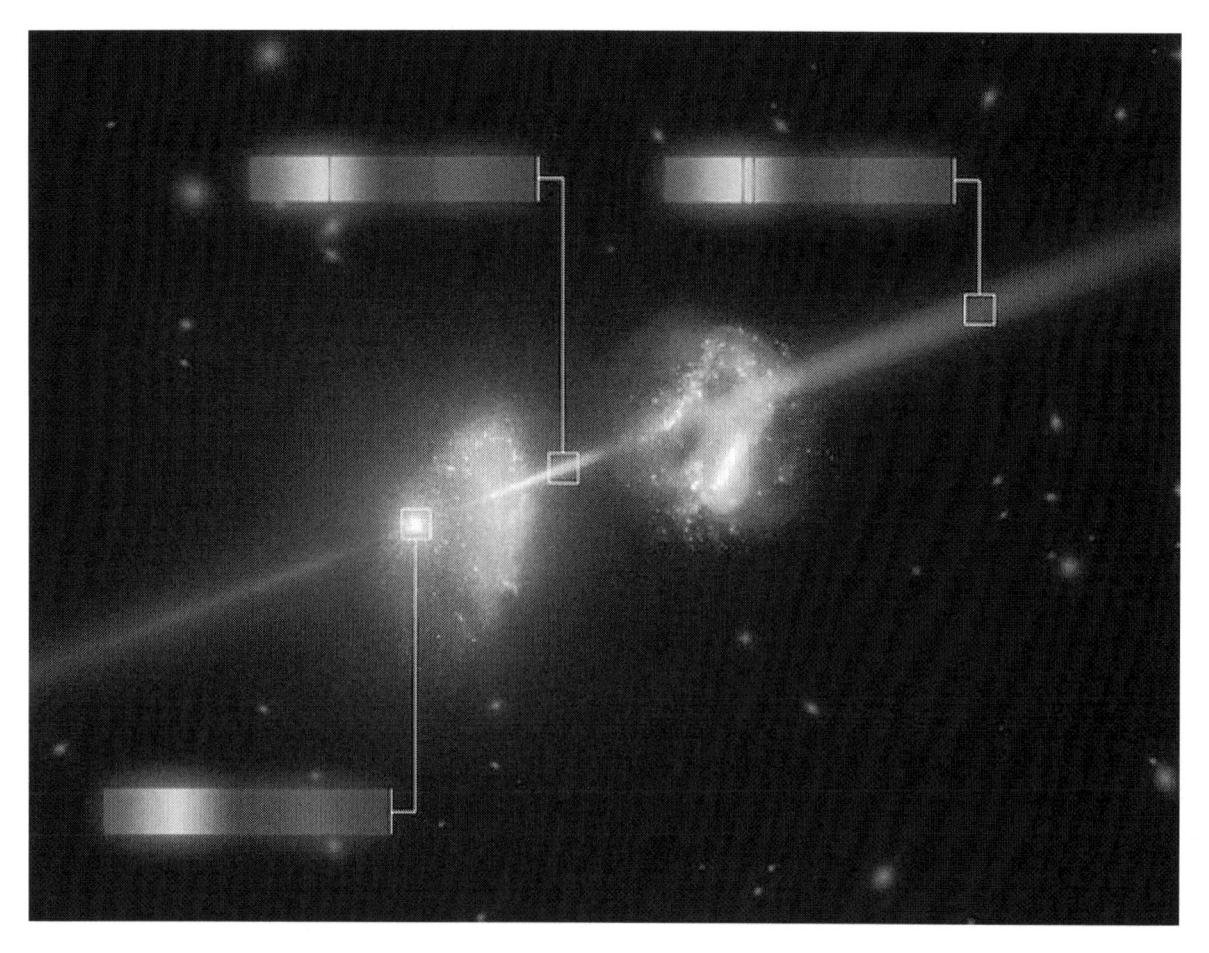

This artist's impression shows two galaxies in the early universe. The brilliant explosion on the left is a gamma-ray burst. As the light from the burst passes through the two galaxies on the way to Earth (outside the frame to the right), some colours are absorbed by the cool gas in the galaxies, leaving characteristic dark lines in the spectrum. Careful study of these spectra has allowed astronomers to discover that these two galaxies are remarkably rich in heavier chemical elements.

newly formed stars, or from an active galactic nuclei can slow the contraction of a forming disk. It has also been suggested that the dark matter halo can pull the galaxy, thus stopping disk contraction.

In recent years, a great deal of focus has been put on understanding merger events in the evolution of galaxies. Our own galaxy (the Milky Way) has a tiny satellite galaxy (the Sagittarius Dwarf Elliptical Galaxy) which is currently gradually being ripped up and "eaten" by the Milky Way. It is thought these kinds of events may be quite common in the evolution of large galaxies. The Sagittarius dwarf galaxy is orbiting our galaxy at almost a right angle to the disk. It is currently passing through the disk; stars are being stripped off of it with each pass and joining the halo of our galaxy. There are other examples of these minor accretion events, and it is likely a continual process for many galaxies. Such mergers provide "new" gas, stars, and dark matter to galaxies. Evidence for this process is often observable as warps or streams coming out of galaxies.

The Lambda-CDM model of galaxy formation underestimates the number of thin disk galaxies in the universe.[4] The reason is that these galaxy formation models predict a large number of mergers. If disk galaxies merge with another galaxy of comparable mass (at least 15 percent of its mass) the merger will likely destroy, or at a minimum greatly disrupt the disk, yet the resulting galaxy is not expected to be a disk galaxy. While this remains an unsolved problem for astronomers, it does not necessarily mean that the Lambda-CDM model is completely wrong, but rather that it requires further refinement to accurately reproduce the population of galaxies in the universe.

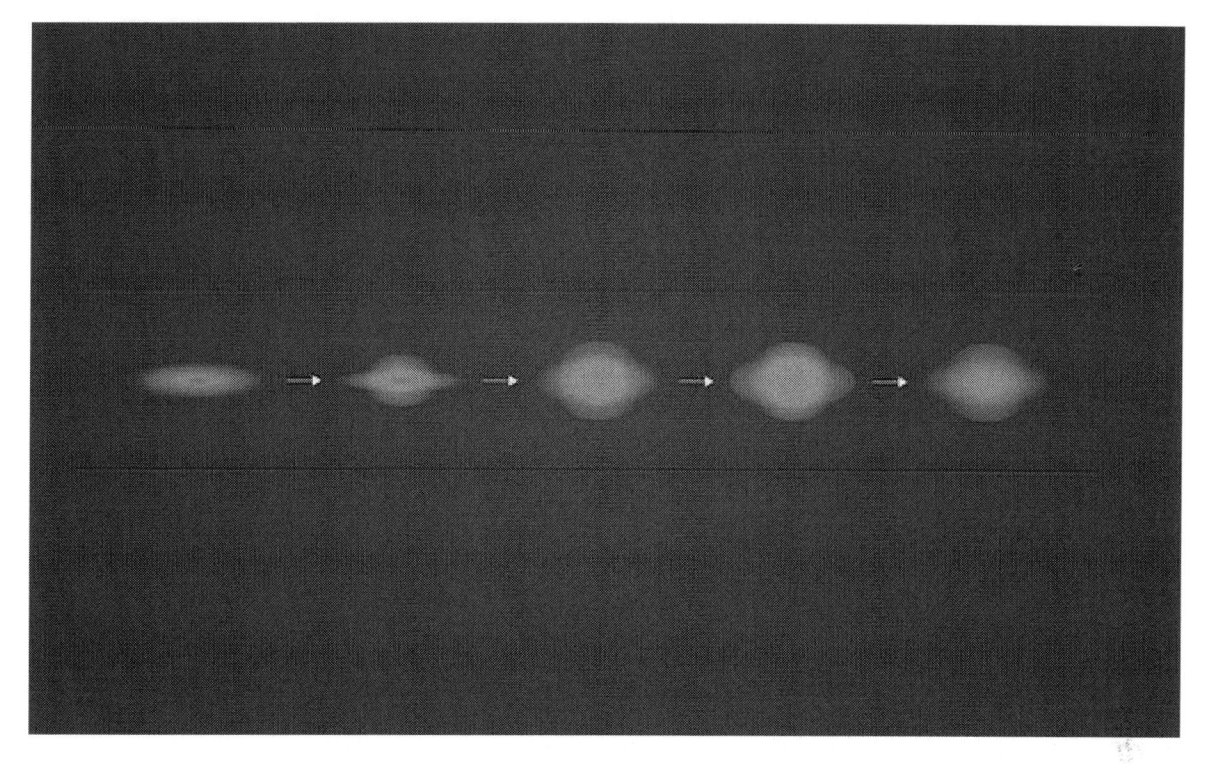

Star formation in what are now "dead" galaxies sputtered out billions of years ago.[3]

12.3 Galaxy mergers and the formation of elliptical galaxies

Main article: Galaxy merger

The most massive galaxies in the sky are giant elliptical galaxies. Their stars are on orbits that are randomly oriented within the galaxy (i.e. they are not rotating like disk galaxies). They are composed of old stars and have little to no dust. All elliptical galaxies probed so far have supermassive black holes in their center, and the mass of these black holes is correlated with the mass of the elliptical galaxy. They are also correlated to a property called sigma which is the speed of the stars at the far edge of the elliptical galaxies. Elliptical galaxies do not have disks around them, although some bulges of disk galaxies look similar to elliptical galaxies. One is more likely to find elliptical galaxies in more crowded regions of the universe (such as galaxy clusters).

Astronomers now see elliptical galaxies as some of the most evolved systems in the universe. It is widely accepted that the main driving force for the evolution of elliptical galaxies is mergers of smaller galaxies. These mergers can be extremely violent; galaxies often collide at speeds of 500 kilometers per second.

Many galaxies in the universe are gravitationally bound to other galaxies, that is to say they will never escape the pull of the other galaxy. If the galaxies are of similar size, the resultant galaxy will appear similar to neither of the two galaxies merging,[5] but would instead be an elliptical galaxy.

In the Local Group, the Milky Way and M31 (the Andromeda Galaxy) are gravitationally bound, and currently approaching each other at high speed. If the two galaxies do meet they will pass through each other, with gravity distorting both galaxies severely and ejecting some gas, dust and stars into intergalactic space. They will travel apart, slow down, and then again be drawn towards each other, and again collide. Eventually both galaxies will have merged completely, streams of gas and dust will be flying through the space near the newly formed giant elliptical galaxy. M31 is actually already distorted: the edges are warped. This is probably because of interactions with its own galactic companions, as well as possible mergers with dwarf spheroidal galaxies in the recent past - the remnants of which are still visible in the disk populations.

In our epoch, large concentrations of galaxies (clusters and superclusters) are still assembling.

Artist image of a firestorm of star birth deep inside core of young, growing elliptical galaxy

NGC 4676 (Mice Galaxies) is an example of a present merger

While scientists have learned a great deal about ours and other galaxies, the most fundamental questions about formation and evolution remain only tentatively answered.

Antennae Galaxies are a pair of colliding galaxies - the bright, blue knots are young stars that have recently ignited as a result of the merger

12.4 See also

- Big Bang
- Bulge (astronomy)
- Chronology of the universe
- Cosmology
- Disc (galaxy)
- Formation and evolution of the Solar System
- Galactic coordinate system

ESO 325-G004, a typical elliptical galaxy

- Galactic corona
- Galactic halo
- Galaxy rotation problem
- Illustris project
- Mass segregation
- Metallicity distribution function
- Pea galaxy

- Stellar formation
- Structure formation
- Zeldovich pancake
- List of galaxies

12.5 Further reading

- Ho, Houjun; van den Bosch, Frank; White, Simon (June 2010), *Galaxy Formation and Evolution* (1 ed.), Cambridge University Press, ISBN 978-0521857932

12.6 References

[1] Eggen, O. J.; Lynden-Bell, D.; Sandage, A. R. (1962). "Evidence from the motions of old stars that the Galaxy collapsed". *The Astrophysical Journal* **136**: 748. Bibcode:1962ApJ...136..748E. doi:10.1086/147433.

[2] Searle, L.; Zinn, R. (1978). "Compositions of halo clusters and the formation of the galactic halo". *The Astrophysical Journal* **225**: 357–379. Bibcode:1978ApJ...225..357S. doi:10.1086/156499.

[3] "Giant Galaxies Die from the Inside Out". *www.eso.org*. European Southern Observatory. Retrieved 21 April 2015.

[4] Steinmetz, M.; Navarro, J.F. (2002). "The hierarchical origin of galaxy morphologies". *New Astronomy* **7** (4): 155–160. arXiv:astro-ph/0202466. Bibcode:2002NewA....7..155S. doi:10.1016/S1384-1076(02)00102-1.

[5] Barnes,J. Nature, vol. 338, March 9, 1989, p. 123-126

12.7 External links

- NOAO gallery of galaxy images
    - Image of Andromeda galaxy (M31)
- Javascript passive evolution calculator for early type (elliptical) galaxies
- Video on the evolution of galaxies by Canadian astrophysicist Doctor P

Chapter 13

Anisotropy

WMAP image of the (extremely tiny) anisotropies in the cosmic background radiation

Anisotropy /ˌænaɪˈsɒtrəpi/ is the property of being directionally dependent, as opposed to isotropy, which implies identical properties in all directions. It can be defined as a difference, when measured along different axes, in a material's physical or mechanical properties (absorbance, refractive index, conductivity, tensile strength, etc.) An example of anisotropy is the light coming through a polarizer. Another is wood, which is easier to split along its grain than against it.

13.1 Fields of interest

13.1.1 Computer graphics

In the field of computer graphics, an anisotropic surface changes in appearance as it rotates about its geometric normal, as is the case with velvet.

Anisotropic filtering (AF) is a method of enhancing the image quality of textures on surfaces that are far away and steeply angled with respect to the point of view. Older techniques, such as bilinear and trilinear filtering, do not take into account the angle a surface is viewed from, which can result in aliasing or blurring of textures. By reducing detail in one direction more than another, these effects can be reduced.

13.1.2 Chemistry

A chemical anisotropic filter, as used to filter particles, is a filter with increasingly smaller interstitial spaces in the direction of filtration so that the proximal regions filter out larger particles and distal regions increasingly remove smaller particles, resulting in greater flow-through and more efficient filtration.

In NMR spectroscopy, the orientation of nuclei with respect to the applied magnetic field determines their chemical shift. In this context, anisotropic systems refer to the electron distribution of molecules with abnormally high electron density, like the pi system of benzene. This abnormal electron density affects the applied magnetic field and causes the observed chemical shift to change.

In fluorescence spectroscopy, the fluorescence anisotropy, calculated from the polarization properties of fluorescence from samples excited with plane-polarized light, is used, e.g., to determine the shape of a macromolecule. Anisotropy measurements reveal the average angular displacement of the fluorophore that occurs between absorption and subsequent emission of a photon.

13.1.3 Real-world imagery

Images of a gravity-bound or man-made environment are particularly anisotropic in the orientation domain, with more image structure located at orientations parallel with or orthogonal to the direction of gravity (vertical and horizontal).

13.1.4 Physics

Physicists from University of California, Berkeley reported about their detection of the cosine anisotropy in cosmic microwave background radiation in 1977. Their experiment demonstrated the Doppler shift caused by the movement of the earth with respect to the early Universe matter, the source of the radiation.[1] Cosmic anisotropy has also been seen in the alignment of galaxies' rotation axes and polarisation angles of quasars.

Physicists use the term anisotropy to describe direction-dependent properties of materials. Magnetic anisotropy, for example, may occur in a plasma, so that its magnetic field is oriented in a preferred direction. Plasmas may also show "filamentation" (such as that seen in lightning or a plasma globe) that is directional.

An *anisotropic liquid* has the fluidity of a normal liquid, but has an average structural order relative to each other along the molecular axis, unlike water or chloroform, which contain no structural ordering of the molecules. Liquid crystals are examples of anisotropic liquids.

Some materials conduct heat in a way that is isotropic, that is independent of spatial orientation around the heat source. Heat conduction is more commonly anisotropic, which implies that detailed geometric modeling of typically diverse materials being thermally managed is required. The materials used to transfer and reject heat from the heat source in electronics are often anisotropic.

Many crystals are anisotropic to light ("optical anisotropy"), and exhibit properties such as birefringence. Crystal optics describes light propagation in these media. An "axis of anisotropy" is defined as the axis along which isotropy is broken (or an axis of symmetry, such as normal to crystalline layers). Some materials can have multiple such optical axes.

13.1.5 Geology and geophysics

Seismic anisotropy is the variation of seismic wavespeed with direction. Seismic anisotropy is an indicator of long range order in a material, where features smaller than the seismic wavelength (e.g., crystals, cracks, pores, layers or inclusions) have a dominant alignment. This alignment leads to a directional variation of elasticity wavespeed. Measuring the effects of anisotropy in seismic data can provide important information about processes and mineralogy in the Earth; indeed, significant seismic anisotropy has been detected in the Earth's crust, mantle and inner core.

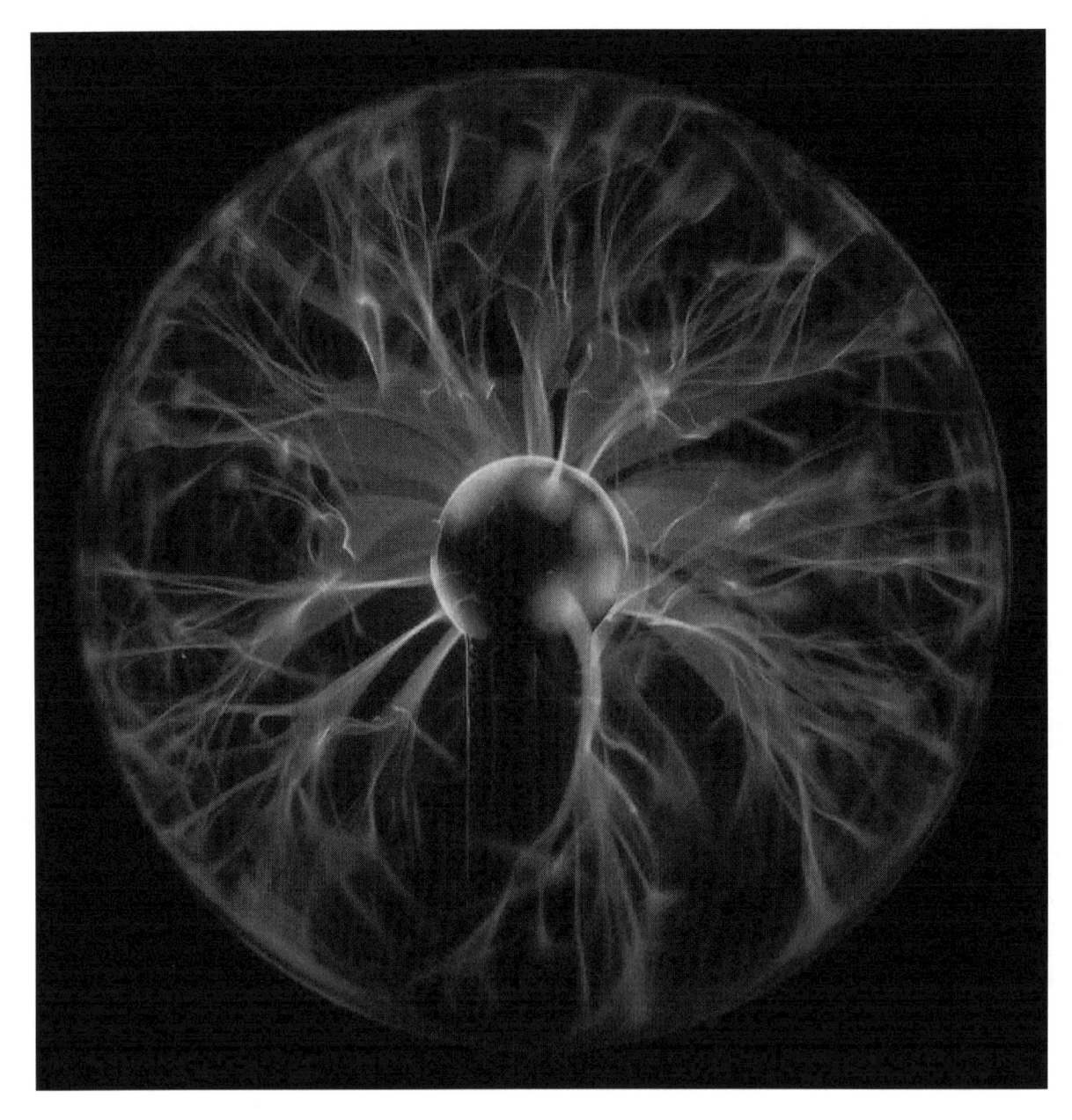

A plasma lamp displaying the nature of plasmas, in this case, the phenomenon of "filamentation"

Geological formations with distinct layers of sedimentary material can exhibit electrical anisotropy; electrical conductivity in one direction (e.g. parallel to a layer), is different from that in another (e.g. perpendicular to a layer). This property is used in the gas and oil exploration industry to identify hydrocarbon-bearing sands in sequences of sand and shale. Sand-bearing hydrocarbon assets have high resistivity (low conductivity), whereas shales have lower resistivity. Formation evaluation instruments measure this conductivity/resistivity and the results are used to help find oil and gas in wells.

The hydraulic conductivity of aquifers is often anisotropic for the same reason. When calculating groundwater flow to drains[2] or to wells,[3] the difference between horizontal and vertical permeability must be taken into account, otherwise the results may be subject to error.

Most common rock-forming minerals are anisotropic, including quartz and feldspar. Anisotropy in minerals is most reliably seen in their optical properties. An example of an isotropic mineral is garnet.

13.1.6 Medical acoustics

Anisotropy is also a well-known property in medical ultrasound imaging describing a different resulting echogenicity of soft tissues, such as tendons, when the angle of the transducer is changed. Tendon fibers appear hyperechoic (bright) when the transducer is perpendicular to the tendon, but can appear hypoechoic (darker) when the transducer is angled obliquely. This can be a source of interpretation error for inexperienced practitioners.

13.1.7 Material science and engineering

Anisotropy, in Material Science, is a material's directional dependence of a physical property. Most materials exhibit anisotropic behavior. An example would be the dependence of Young's modulus on the direction of load.[4] Anisotropy in polycrystalline materials can also be due to certain texture patterns often produced during manufacturing of the material. In the case of rolling, "stringers" of texture are produced in the direction of rolling, which can lead to vastly different properties in the rolling and transverse directions. Some materials, such as wood and fibre-reinforced composites are very anisotropic, being much stronger along the grain/fibre than across it. Metals and alloys tend to be more isotropic, though they can sometimes exhibit significant anisotropic behaviour. This is especially important in processes such as deep-drawing.

Wood is a naturally anisotropic (transversely isotropic) material. Its properties vary widely when measured with or against the growth grain. For example, wood's strength and hardness is different for the same sample measured in different orientations.

13.1.8 Microfabrication

Anisotropic etching techniques (such as deep reactive ion etching) are used in microfabrication processes to create well defined microscopic features with a high aspect ratio. These features are commonly used in MEMS and microfluidic devices, where the anisotropy of the features is needed to impart desired optical, electrical, or physical properties to the device. Anisotropic etching can also refer to certain chemical etchants used to etch a certain material preferentially over certain crystallographic planes (e.g., KOH etching of silicon [100] produces pyramid-like structures)

13.1.9 Neuroscience

Diffusion tensor imaging is an MRI technique that involves measuring the fractional anisotropy of the random motion (Brownian motion) of water molecules in the brain. Water molecules located in fiber tracts are more likely to be anisotropic, since they are restricted in their movement (they move more in the dimension parallel to the fiber tract rather than in the two dimensions orthogonal to it), whereas water molecules dispersed in the rest of the brain have less restricted movement and therefore display more isotropy. This difference in fractional anisotropy is exploited to create a map of the fiber tracts in the brains of the individual.

13.1.10 Atmospheric Radiative Transfer

Radiance fields (see BRDF) from a reflective surface are often not isotropic in nature. This makes calculations of the total energy being reflected from any scene a difficult quantity to calculate. In remote sensing applications, anisotropy functions can be derived for specific scenes, immensely simplifying the calculation of the net reflectance or (thereby) the net irradiance of a scene. For example, let the BRDF be $\gamma(\Omega_i, \Omega_v)$ where 'i' denotes incident direction and 'v' denotes viewing direction (as if from a satellite or other instrument). And let P be the Planar Albedo, which represents the total reflectance from the scene.

$$P(\Omega_i) = \int_{\Omega_v} \gamma(\Omega_i, \Omega_v) \hat{n} \cdot d\hat{\Omega}_v$$

$$A(\Omega_i, \Omega_v) = \frac{\gamma(\Omega_i, \Omega_v)}{P(\Omega_i)}$$

It is of interest because, with knowledge of the anisotropy function as defined, a measurement of the BRDF from a single viewing direction (say, Ω_v) yields a measure of the total scene reflectance (Planar Albedo) for that specific incident geometry (say, Ω_i).

13.2 References

[1] Smoot G. F., Gorenstein M. V., and Muller R. A. (5 October 1977). "Detection of Anisotropy in the Cosmic Blackbody Radiation" (PDF). Lawrence Berkeley Laboratory and Space Sciences Laboratory, University of California, Berkeley. Retrieved 15 September 2013.

[2] R.J.Oosterbaan, 1997, The energy balance of groundwater flow applied to subsurface drainage in anisotropic soils by pipes or ditches with entrance resistance. On line: . The corresponding free EnDrain program can be downloaded from: .

[3] R.J.Oosterbaan, 2002, Subsurface drainage by (tube)wells, 9 pp. On line: . The corresponding free WellDrain program can be downloaded from:

[4] Kocks, U.F. (2000). *Texture and Anisotropy: Preferred Orientations in Polycrystals and their effect on Materials Properties*. Cambridge. ISBN 9780521794206.

13.3 External links

- "Gauge, and knitted fabric generally, is an anisotropic phenomenon"
- "Overview of Anisotropy"
- DoITPoMS Teaching and Learning Package: "Introduction to Anisotropy"

Chapter 14

Baryon

Not to be confused with Baryonyx.

A **baryon** is a composite subatomic particle made up of three quarks (as distinct from mesons, which are composed of one quark and one antiquark). Baryons and mesons belong to the hadron family of particles, which are the quark-based particles. The name "baryon" comes from the Greek word for "heavy" (βαρύς, *barys*), because, at the time of their naming, most known elementary particles had lower masses than the baryons.

As quark-based particles, baryons participate in the strong interaction, whereas leptons, which are not quark-based, do not. The most familiar baryons are the protons and neutrons that make up most of the mass of the visible matter in the universe. Electrons (the other major component of the atom) are leptons.

Each baryon has a corresponding antiparticle (antibaryon) where quarks are replaced by their corresponding antiquarks. For example, a proton is made of two up quarks and one down quark; and its corresponding antiparticle, the antiproton, is made of two up antiquarks and one down antiquark.

14.1 Background

Baryons are strongly interacting fermions that is, they experience the strong nuclear force and are described by Fermi–Dirac statistics, which apply to all particles obeying the Pauli exclusion principle. This is in contrast to the bosons, which do not obey the exclusion principle.

Baryons, along with mesons, are hadrons, meaning they are particles composed of quarks. Quarks have baryon numbers of $B = \frac{1}{3}$ and antiquarks have baryon number of $B = -\frac{1}{3}$. The term "baryon" usually refers to *triquarks*—baryons made of three quarks ($B = \frac{1}{3} + \frac{1}{3} + \frac{1}{3} = 1$).

Other exotic baryons have been proposed, such as pentaquarks—baryons made of four quarks and one antiquark ($B = \frac{1}{3} + \frac{1}{3} + \frac{1}{3} + \frac{1}{3} - \frac{1}{3} = 1$), but their existence is not generally accepted. In theory, heptaquarks (5 quarks, 2 antiquarks), nonaquarks (6 quarks, 3 antiquarks), etc. could also exist. Until recently, it was believed that some experiments showed the existence of pentaquarks—baryons made of four quarks and one antiquark.[1][2] The particle physics community as a whole did not view their existence as likely in 2006,[3] and in 2008, considered evidence to be overwhelmingly against the existence of the reported pentaquarks.[4] However, in July 2015, the LHCb experiment reported observations consistent with pentaquark formation along the Λ0
b → J/ψK−
p decay, with a statistical significance of 9-σ.[5]

14.2 Baryonic matter

Nearly all matter that may be encountered or experienced in everyday life is baryonic matter, which includes atoms of any sort, and provides those with the quality of mass. Non-baryonic matter, as implied by the name, is any sort of matter that is not composed primarily of baryons. Those might include neutrinos or free electrons, dark matter, such as supersymmetric particles, axions, or black holes.

The very existence of baryons is also a significant issue in cosmology because it is assumed that the Big Bang produced a state with equal amounts of baryons and antibaryons. The process by which baryons came to outnumber their antiparticles is called baryogenesis.

14.3 Baryogenesis

Main article: Baryogenesis

Experiments are consistent with the number of quarks in the universe being a constant and, to be more specific, the number of baryons being a constant ; in technical language, the total baryon number appears to be *conserved*. Within the prevailing Standard Model of particle physics, the number of baryons may change in multiples of three due to the action of sphalerons, although this is rare and has not been observed under experiment. Some grand unified theories of particle physics also predict that a single proton can decay, changing the baryon number by one; however, this has not yet been observed under experiment. The excess of baryons over antibaryons in the present universe is thought to be due to non-conservation of baryon number in the very early universe, though this is not well understood.

14.4 Properties

14.4.1 Isospin and charge

Main article: Isospin

The concept of isospin was first proposed by Werner Heisenberg in 1932 to explain the similarities between protons and neutrons under the strong interaction.[6] Although they had different electric charges, their masses were so similar that physicists believed they were actually the same particle. The different electric charges were explained as being the result of some unknown excitation similar to spin. This unknown excitation was later dubbed *isospin* by Eugene Wigner in 1937.[7]

This belief lasted until Murray Gell-Mann proposed the quark model in 1964 (containing originally only the u, d, and s quarks).[8] The success of the isospin model is now understood to be the result of the similar masses of the u and d quarks. Since the u and d quarks have similar masses, particles made of the same number then also have similar masses. The exact specific u and d quark composition determines the charge, as u quarks carry charge $+2/3$ while d quarks carry charge $-1/3$. For example the four Deltas all have different charges (Δ++ (uuu), Δ+ (uud), Δ0 (udd), Δ− (ddd)), but have similar masses (~1,232 MeV/c^2) as they are each made of a combination of three u and d quarks. Under the isospin model, they were considered to be a single particle in different charged states.

The mathematics of isospin was modeled after that of spin. Isospin projections varied in increments of 1 just like those of spin, and to each projection was associated a "charged state". Since the "Delta particle" had four "charged states", it was said to be of isospin $I = 3/2$. Its "charged states" Δ++, Δ+, Δ0, and Δ−, corresponded to the isospin projections $I_3 = +3/2$, $I_3 = +1/2$, $I_3 = -1/2$, and $I_3 = -3/2$, respectively. Another example is the "nucleon particle". As there were two nucleon "charged states", it was said to be of isospin $1/2$. The positive nucleon N+ (proton) was identified with $I_3 = +1/2$ and the neutral nucleon N0 (neutron) with $I_3 = -1/2$.[9] It was later noted that the isospin projections were related to the up and down quark content of particles by the relation:

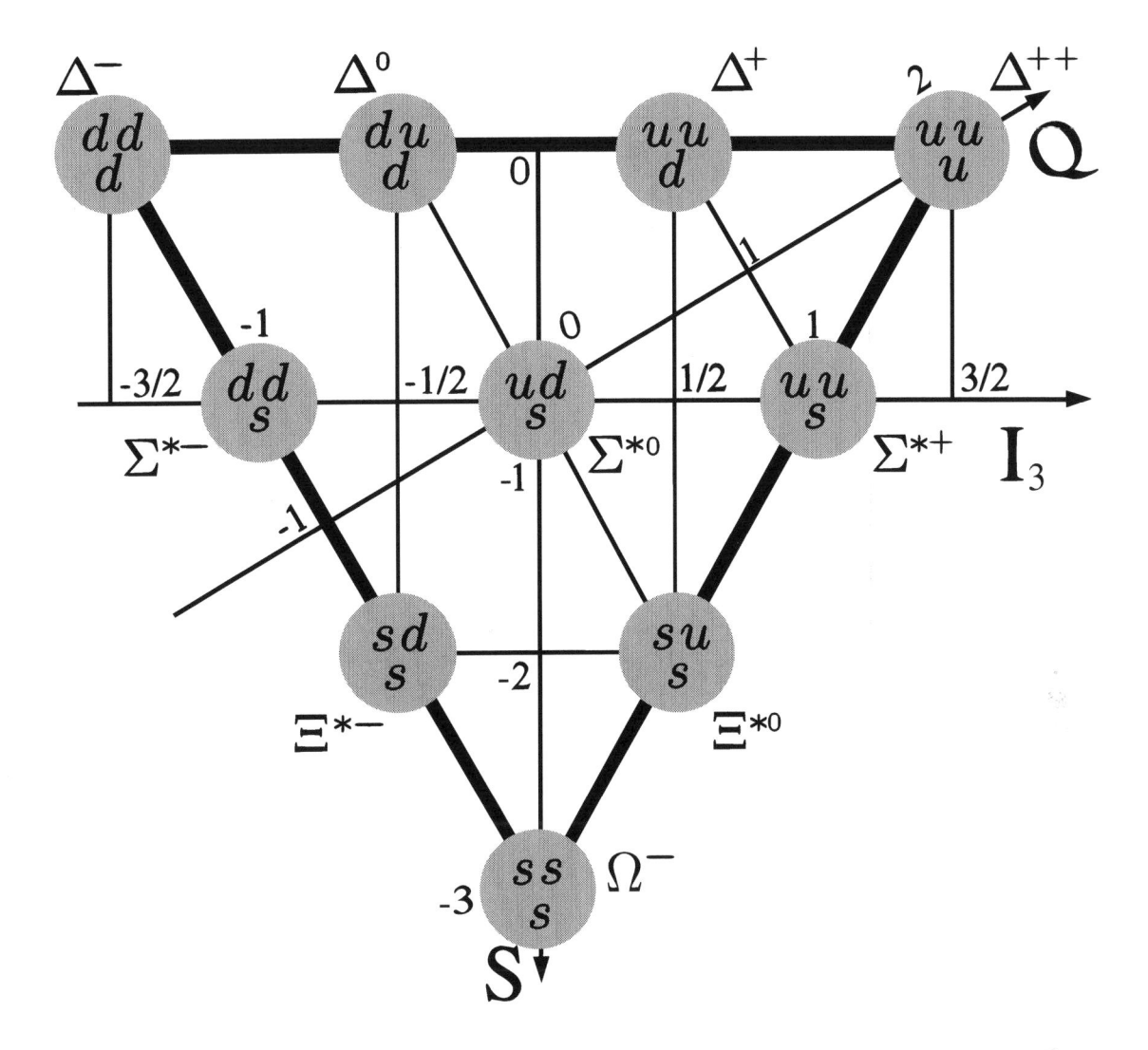

*Combinations of three **u**, **d** or **s** quarks forming baryons with a spin-$3/2$ form the* uds baryon decuplet

$$I_3 = \frac{1}{2}[(n_u - n_{\bar{u}}) - (n_d - n_{\bar{d}})],$$

where the *n*'s are the number of up and down quarks and antiquarks.

In the "isospin picture", the four Deltas and the two nucleons were thought to be the different states of two particles. However in the quark model, Deltas are different states of nucleons (the N^{++} or N^- are forbidden by Pauli's exclusion principle). Isospin, although conveying an inaccurate picture of things, is still used to classify baryons, leading to unnatural and often confusing nomenclature.

14.4.2 Flavour quantum numbers

Main article: Flavour (particle physics) § Flavour quantum numbers

The strangeness flavour quantum number S (not to be confused with spin) was noticed to go up and down along with particle mass. The higher the mass, the lower the strangeness (the more s quarks). Particles could be described with

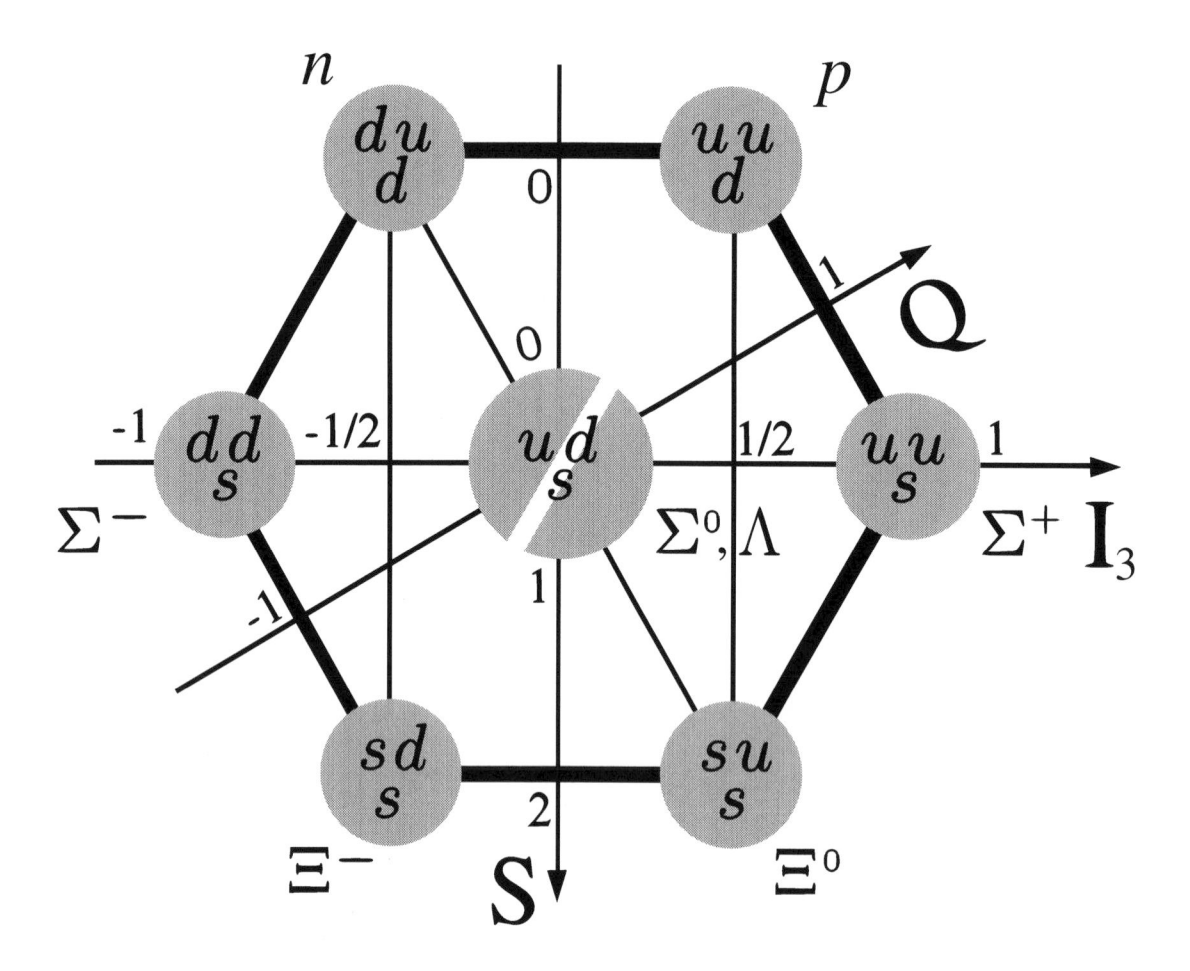

*Combinations of three **u, d** or s quarks forming baryons with a spin-$\frac{1}{2}$ form the* uds baryon octet

isospin projections (related to charge) and strangeness (mass) (see the uds octet and decuplet figures on the right). As other quarks were discovered, new quantum numbers were made to have similar description of udc and udb octets and decuplets. Since only the u and d mass are similar, this description of particle mass and charge in terms of isospin and flavour quantum numbers works well only for octet and decuplet made of one u, one d, and one other quark, and breaks down for the other octets and decuplets (for example, ucb octet and decuplet). If the quarks all had the same mass, their behaviour would be called *symmetric*, as they would all behave in exactly the same way with respect to the strong interaction. Since quarks do not have the same mass, they do not interact in the same way (exactly like an electron placed in an electric field will accelerate more than a proton placed in the same field because of its lighter mass), and the symmetry is said to be broken.

It was noted that charge (Q) was related to the isospin projection (I_3), the baryon number (B) and flavour quantum numbers (S, C, B', T) by the Gell-Mann–Nishijima formula:[9]

$$Q = I_3 + \frac{1}{2}(B + S + C + B' + T),$$

where S, C, B', and T represent the strangeness, charm, bottomness and topness flavour quantum numbers, respectively. They are related to the number of strange, charm, bottom, and top quarks and antiquark according to the relations:

$$S = -(n_s - n_{\bar{s}}),$$
$$C = +(n_c - n_{\bar{c}}),$$
$$B' = -(n_b - n_{\bar{b}}),$$
$$T = +(n_t - n_{\bar{t}}),$$

meaning that the Gell-Mann–Nishijima formula is equivalent to the expression of charge in terms of quark content:

$$Q = \frac{2}{3}[(n_u - n_{\bar{u}}) + (n_c - n_{\bar{c}}) + (n_t - n_{\bar{t}})] - \frac{1}{3}[(n_d - n_{\bar{d}}) + (n_s - n_{\bar{s}}) + (n_b - n_{\bar{b}})].$$

14.4.3 Spin, orbital angular momentum, and total angular momentum

Main articles: Spin (physics), Angular momentum operator, Quantum numbers and Clebsch–Gordan coefficients

Spin (quantum number S) is a vector quantity that represents the "intrinsic" angular momentum of a particle. It comes in increments of $\frac{1}{2}$ ℏ (pronounced "h-bar"). The ℏ is often dropped because it is the "fundamental" unit of spin, and it is implied that "spin 1" means "spin 1 ℏ". In some systems of natural units, ℏ is chosen to be 1, and therefore does not appear anywhere.

Quarks are fermionic particles of spin $\frac{1}{2}$ ($S = \frac{1}{2}$). Because spin projections vary in increments of 1 (that is 1 ℏ), a single quark has a spin vector of length $\frac{1}{2}$, and has two spin projections ($S_z = +\frac{1}{2}$ and $S_z = -\frac{1}{2}$). Two quarks can have their spins aligned, in which case the two spin vectors add to make a vector of length $S = 1$ and three spin projections ($S_z = +1$, $S_z = 0$, and $S_z = -1$). If two quarks have unaligned spins, the spin vectors add up to make a vector of length $S = 0$ and has only one spin projection ($S_z = 0$), etc. Since baryons are made of three quarks, their spin vectors can add to make a vector of length $S = \frac{3}{2}$, which has four spin projections ($S_z = +\frac{3}{2}$, $S_z = +\frac{1}{2}$, $S_z = -\frac{1}{2}$, and $S_z = -\frac{3}{2}$), or a vector of length $S = \frac{1}{2}$ with two spin projections ($S_z = +\frac{1}{2}$, and $S_z = -\frac{1}{2}$).[10]

There is another quantity of angular momentum, called the orbital angular momentum, (azimuthal quantum number L), that comes in increments of 1 ℏ, which represent the angular moment due to quarks orbiting around each other. The total angular momentum (total angular momentum quantum number J) of a particle is therefore the combination of intrinsic angular momentum (spin) and orbital angular momentum. It can take any value from $J = |L - S|$ to $J = |L + S|$, in increments of 1.

Particle physicists are most interested in baryons with no orbital angular momentum ($L = 0$), as they correspond to ground states—states of minimal energy. Therefore the two groups of baryons most studied are the $S = \frac{1}{2}$; $L = 0$ and $S = \frac{3}{2}$; $L = 0$, which corresponds to $J = \frac{1}{2}^+$ and $J = \frac{3}{2}^+$, respectively, although they are not the only ones. It is also possible to obtain $J = \frac{3}{2}^+$ particles from $S = \frac{1}{2}$ and $L = 2$, as well as $S = \frac{3}{2}$ and $L = 2$. This phenomenon of having multiple particles in the same total angular momentum configuration is called *degeneracy*. How to distinguish between these degenerate baryons is an active area of research in baryon spectroscopy.[11][12]

14.4.4 Parity

Main article: Parity (physics)

If the universe were reflected in a mirror, most of the laws of physics would be identical—things would behave the same way regardless of what we call "left" and what we call "right". This concept of mirror reflection is called *intrinsic parity* or *parity* (P). Gravity, the electromagnetic force, and the strong interaction all behave in the same way regardless of whether or not the universe is reflected in a mirror, and thus are said to conserve parity (P-symmetry). However, the weak interaction *does* distinguish "left" from "right", a phenomenon called parity violation (P-violation).

Based on this, one might think that, if the wavefunction for each particle (in more precise terms, the quantum field for each particle type) were simultaneously mirror-reversed, then the new set of wavefunctions would perfectly satisfy the laws of physics (apart from the weak interaction). It turns out that this is not quite true: In order for the equations to be satisfied, the wavefunctions of certain types of particles have to be multiplied by −1, in addition to being mirror-reversed. Such particle types are said to have *negative* or *odd* parity ($P = -1$, or alternatively $P = -$), while the other particles are said to have *positive* or *even* parity ($P = +1$, or alternatively $P = +$).

For baryons, the parity is related to the orbital angular momentum by the relation:[13]

$$P = (-1)^L.$$

As a consequence, baryons with no orbital angular momentum ($L = 0$) all have even parity ($P = +$).

14.5 Nomenclature

Baryons are classified into groups according to their isospin (I) values and quark (q) content. There are six groups of baryons—nucleon (N), Delta (Δ), Lambda (Λ), Sigma (Σ), Xi (Ξ), and Omega (Ω). The rules for classification are defined by the Particle Data Group. These rules consider the up (u), down (d) and strange (s) quarks to be *light* and the charm (c), bottom (b), and top (t) quarks to be *heavy*. The rules cover all the particles that can be made from three of each of the six quarks, even though baryons made of t quarks are not expected to exist because of the t quark's short lifetime. The rules do not cover pentaquarks.[14]

- Baryons with three u and/or d quarks are N's ($I = 1/2$) or Δ's ($I = 3/2$).

- Baryons with two u and/or d quarks are Λ's ($I = 0$) or Σ's ($I = 1$). If the third quark is heavy, its identity is given by a subscript.

- Baryons with one u or d quark are Ξ's ($I = 1/2$). One or two subscripts are used if one or both of the remaining quarks are heavy.

- Baryons with no u or d quarks are Ω's ($I = 0$), and subscripts indicate any heavy quark content.

- Baryons that decay strongly have their masses as part of their names. For example, Σ^0 does not decay strongly, but Δ^{++}(1232) does.

It is also a widespread (but not universal) practice to follow some additional rules when distinguishing between some states that would otherwise have the same symbol.[9]

- Baryons in total angular momentum $J = 3/2$ configuration that have the same symbols as their $J = 1/2$ counterparts are denoted by an asterisk (*).

- Two baryons can be made of three different quarks in $J = 1/2$ configuration. In this case, a prime (′) is used to distinguish between them.
  - *Exception*: When two of the three quarks are one up and one down quark, one baryon is dubbed Λ while the other is dubbed Σ.

Quarks carry charge, so knowing the charge of a particle indirectly gives the quark content. For example, the rules above say that a Λ+
c contains a c quark and some combination of two u and/or d quarks. The c quark has a charge of ($Q = +2/3$), therefore the other two must be a u quark ($Q = +2/3$), and a d quark ($Q = -1/3$) to have the correct total charge ($Q = +1$).

14.6 See also

- Eightfold way
- List of baryons
- List of particles
- Meson
- Timeline of particle discoveries

14.7 Notes

[1] H. Muir (2003)

[2] K. Carter (2003)

[3] W.-M. Yao *et al.* (2006): Particle listings – Θ^+

[4] C. Amsler *et al.* (2008): Pentaquarks

[5] LHCb (14 July 2015). "Observation of particles composed of five quarks, pentaquark-charmonium states, seen in $\Lambda_b^0 \to J/\psi pK^-$ decays.". *CERN website*. Retrieved 2015-07-14.

[6] W. Heisenberg (1932)

[7] E. Wigner (1937)

[8] M. Gell-Mann (1964)

[9] S.S.M. Wong (1998a)

[10] R. Shankar (1994)

[11] H. Garcilazo *et al.* (2007)

[12] D.M. Manley (2005)

[13] S.S.M. Wong (1998b)

[14] C. Amsler *et al.* (2008): Naming scheme for hadrons

14.8 References

- C. Amsler *et al.* (Particle Data Group) (2008). "Review of Particle Physics". *Physics Letters B* **667** (1): 1–1340. Bibcode:2008PhLB..667....1P. doi:10.1016/j.physletb.2008.07.018.

- H. Garcilazo, J. Vijande, and A. Valcarce (2007). "Faddeev study of heavy-baryon spectroscopy". *Journal of Physics G* **34** (5): 961–976. doi:10.1088/0954-3899/34/5/014.

- K. Carter (2006). "The rise and fall of the pentaquark". Fermilab and SLAC. Retrieved 2008-05-27.

- W.-M. Yao *et al.*(Particle Data Group) (2006). "Review of Particle Physics". *Journal of Physics G* **33**: 1–1232. arXiv:astro-ph/0601168. Bibcode:2006JPhG...33....1Y. doi:10.1088/0954-3899/33/1/001.

- D.M. Manley (2005). "Status of baryon spectroscopy". *Journal of Physics: Conference Series* **5**: 230–237. Bibcode:2005JPhCS...9..230M. doi:10.1088/1742-6596/9/1/043.

- H. Muir (2003). "Pentaquark discovery confounds sceptics". New Scientist. Retrieved 2008-05-27.

- S.S.M. Wong (1998a). "Chapter 2—Nucleon Structure". *Introductory Nuclear Physics* (2nd ed.). New York (NY): John Wiley & Sons. pp. 21–56. ISBN 0-471-23973-9.

- S.S.M. Wong (1998b). "Chapter 3—The Deuteron". *Introductory Nuclear Physics* (2nd ed.). New York (NY): John Wiley & Sons. pp. 57–104. ISBN 0-471-23973-9.

- R. Shankar (1994). *Principles of Quantum Mechanics* (2nd ed.). New York (NY): Plenum Press. ISBN 0-306-44790-8.

- E. Wigner (1937). "On the Consequences of the Symmetry of the Nuclear Hamiltonian on the Spectroscopy of Nuclei". *Physical Review* **51** (2): 106–119. Bibcode:1937PhRv...51..106W. doi:10.1103/PhysRev.51.106.

- M. Gell-Mann (1964). "A Schematic of Baryons and Mesons".*Physics Letters***8**(3): 214–215.Bibcode:1964PhL.....8. doi:10.1016/S0031-9163(64)92001-3.

- W. Heisenberg (1932). "Über den Bau der Atomkerne I". *Zeitschrift für Physik* (in German) **77**: 1–11. Bibcode: doi:10.1007/BF01342433.

- W. Heisenberg (1932). "Über den Bau der Atomkerne II". *Zeitschrift für Physik* (in German) **78** (3–4): 156–164. Bibcode:1932ZPhy...78..156H. doi:10.1007/BF01337585.

- W. Heisenberg (1932). "Über den Bau der Atomkerne III". *Zeitschrift für Physik* (in German) **80** (9–10): 587–596. Bibcode:1933ZPhy...80..587H. doi:10.1007/BF01335696.

14.9 External links

- Particle Data Group—Review of Particle Physics (2008).

- Georgia State University—HyperPhysics

- Baryons made thinkable, an interactive visualisation allowing physical properties to be compared

Chapter 15

Weakly interacting massive particles

In particle physics and astrophysics, **weakly interacting massive particles**, or **WIMPs**, are among the leading hypothetical particle physics candidates for dark matter. The term "WIMP" is given to a dark matter particle that was produced by falling out of thermal equilibrium with the hot dense plasma of the early universe, although it is often used to refer to any dark matter candidate that interacts with standard particles via a force similar in strength to the weak nuclear force. Its name comes from the fact that obtaining the correct abundance of dark matter today via thermal production requires a self-annihilation cross section of $\langle \sigma v \rangle \simeq 3 \times 10^{-26}$cm^3 s^{-1}, which is roughly what is expected for a new particle in the 100 GeV mass range that interacts via the electroweak force. This apparent coincidence is known as the "WIMP miracle". Because supersymmetric extensions of the standard model of particle physics readily predict a new particle with these properties, a stable supersymmetric partner has long been a prime WIMP candidate.[1] However, recent null results from direct detection experiments including LUX and SuperCDMS, along with the failure to produce evidence of supersymmetry in the Large Hadron Collider (LHC) experiment[2][3] has cast doubt on the simplest WIMP hypothesis.[4] Experimental efforts to detect WIMPs include the search for products of WIMP annihilation, including gamma rays, neutrinos and cosmic rays in nearby galaxies and galaxy clusters; direct detection experiments designed to measure the collision of WIMPs with nuclei in the laboratory, as well as attempts to directly produce WIMPs in colliders, such as the LHC.

15.1 Theoretical framework and properties

WIMP-like particles are predicted by R-parity-conserving supersymmetry, a popular type of extension to the standard model of particle physics, although none of the large number of new particles in supersymmetry have been observed.[5] WIMP-like particles are also predicted by universal extra dimension and little Higgs.

The main theoretical characteristics of a WIMP are:

- Interactions only through the weak nuclear force and gravity, or possibly other interactions with cross-sections no higher than the weak scale;[6]

- Large mass compared to standard particles (WIMPs with sub-GeV masses may be considered to be light dark matter).

Because of their lack of electromagnetic interaction with normal matter, WIMPs would be dark and invisible through normal electromagnetic observations. Because of their large mass, they would be relatively slow moving and therefore "cold".[7] Their relatively low velocities would be insufficient to overcome the mutual gravitational attraction, and as a result WIMPs would tend to clump together.[8] WIMPs are considered one of the main candidates for cold dark matter, the others being massive compact halo objects (MACHOs) and axions. (These names were deliberately chosen for contrast, with MACHOs named later than WIMPs.[9]) Also, in contrast to MACHOs, there are no known stable particles within the standard model of particle physics that have all the properties of WIMPs. The particles that have little interaction with normal matter, such as neutrinos, are all very light, and hence would be fast moving, or "hot".

15.2 WIMPs as dark matter

Although the existence of WIMPs in nature is hypothetical at this point, it would resolve a number of astrophysical and cosmological problems related to dark matter. There is near consensus today among astronomers that most of the mass in the Universe is dark. Simulations of a universe full of cold dark matter produce galaxy distributions that are roughly similar to what is observed.[10][11] By contrast, hot dark matter would smear out the large-scale structure of galaxies and thus is not considered a viable cosmological model.

The WIMP fits the model of a relic dark matter particle from the early Universe, when all particles were in a state of thermal equilibrium. For sufficiently high temperatures, such as those existing in the early Universe, the dark matter particle and its antiparticle would have been both forming from and annihilating into lighter particles. As the Universe expanded and cooled, the average thermal energy of these lighter particles decreased and eventually became insufficient to form a dark matter particle-antiparticle pair. The annihilation of the dark matter particle-antiparticle pairs, however, would have continued, and the number density of dark matter particles would have begun to decrease exponentially.[6] Eventually, however, the number density would become so low that the dark matter particle and antiparticle interaction would cease, and the number of dark matter particles would remain (roughly) constant as the Universe continued to expand.[8] Particles with a larger interaction cross section would continue to annihilate for a longer period of time, and thus would have a smaller number density when the annihilation interaction ceases. Based on the current estimated abundance of dark matter in the Universe, if the dark matter particle is such a relic particle, the interaction cross section governing the particle-antiparticle annihilation can be no larger than the cross section for the weak interaction.[6] If this model is correct, the dark matter particle would have the properties of the WIMP.

15.3 Experimental detection

Because WIMPs may only interact through gravitational and weak forces, they are extremely difficult to detect. However, there are many experiments underway to attempt to detect WIMPs both directly and indirectly. *Direct detection* refers to the observation of the effects of a WIMP-nucleus collision as the dark matter passes through a detector in an Earth laboratory. *Indirect detection* refers to the observation of annihilation or decay products of WIMPs far away from Earth.

Indirect detection efforts typically focus on locations where WIMP dark matter is thought to accumulate the most: in the centers of galaxies and galaxy clusters, as well as in the smaller satellite galaxies of the Milky Way. These are particularly useful since they tend to contain very little baryonic matter, reducing the expected background from standard astrophysical processes. Typical indirect searches look for excess gamma rays, which are predicted both as final-state products of annihilation, or are produced as charged particles interact with ambient radiation via inverse Compton scattering. The spectrum and intensity of a gamma ray signal depends on the annihilation products, and must be computed on a model-by-model basis. Experiments that have placed bounds on WIMP annihilation, via the non-observation of an annihilation signal, include the Fermi-LAT gamma ray telescope[12] and the VERITAS ground-based gamma ray observatory.[13] Although the annihilation of WIMPs into standard model particles also predicts the production of high-energy neutrinos, their interaction rate is too low to reliably detect a dark matter signal at present. Future observations from the IceCube observatory in Antarctica may be able to differentiate WIMP-produced neutrinos from standard astrophysical neutrinos; however, at present, only 37 cosmological neutrinos have been observed,[14] making such a distinction impossible.

Another type of indirect WIMP signal could come from the Sun. Halo WIMPs may, as they pass through the Sun, interact with solar protons, helium nuclei as well as heavier elements. If a WIMP loses enough energy in such an interaction to fall below the local escape velocity, it would not have enough energy to escape the gravitational pull of the Sun and would remain gravitationally bound.[8] As more and more WIMPs thermalize inside the Sun, they begin to annihilate with each other, forming a variety of particles, including high-energy neutrinos.[15] These neutrinos may then travel to the Earth to be detected in one of the many neutrino telescopes, such as the Super-Kamiokande detector in Japan. The number of neutrino events detected per day at these detectors depends on the properties of the WIMP, as well as on the mass of the Higgs boson. Similar experiments are underway to detect neutrinos from WIMP annihilations within the Earth[16] and from within the galactic center.[17][18]

While most WIMP models indicate that a large enough number of WIMPs must be captured in large celestial bodies for these experiments to succeed, it remains possible that these models are either incorrect or only explain part of the dark

15.3. EXPERIMENTAL DETECTION

matter phenomenon. Thus, even with the multiple experiments dedicated to providing indirect evidence for the existence of cold dark matter, direct detection measurements are also necessary to solidify the theory of WIMPs.

Although most WIMPs encountering the Sun or the Earth are expected to pass through without any effect, it is hoped that a large number of dark matter WIMPs crossing a sufficiently large detector will interact often enough to be seen—at least a few events per year. The general strategy of current attempts to detect WIMPs is to find very sensitive systems that can be scaled up to large volumes. This follows the lessons learned from the history of the discovery and (by now) routine detection of the neutrino.

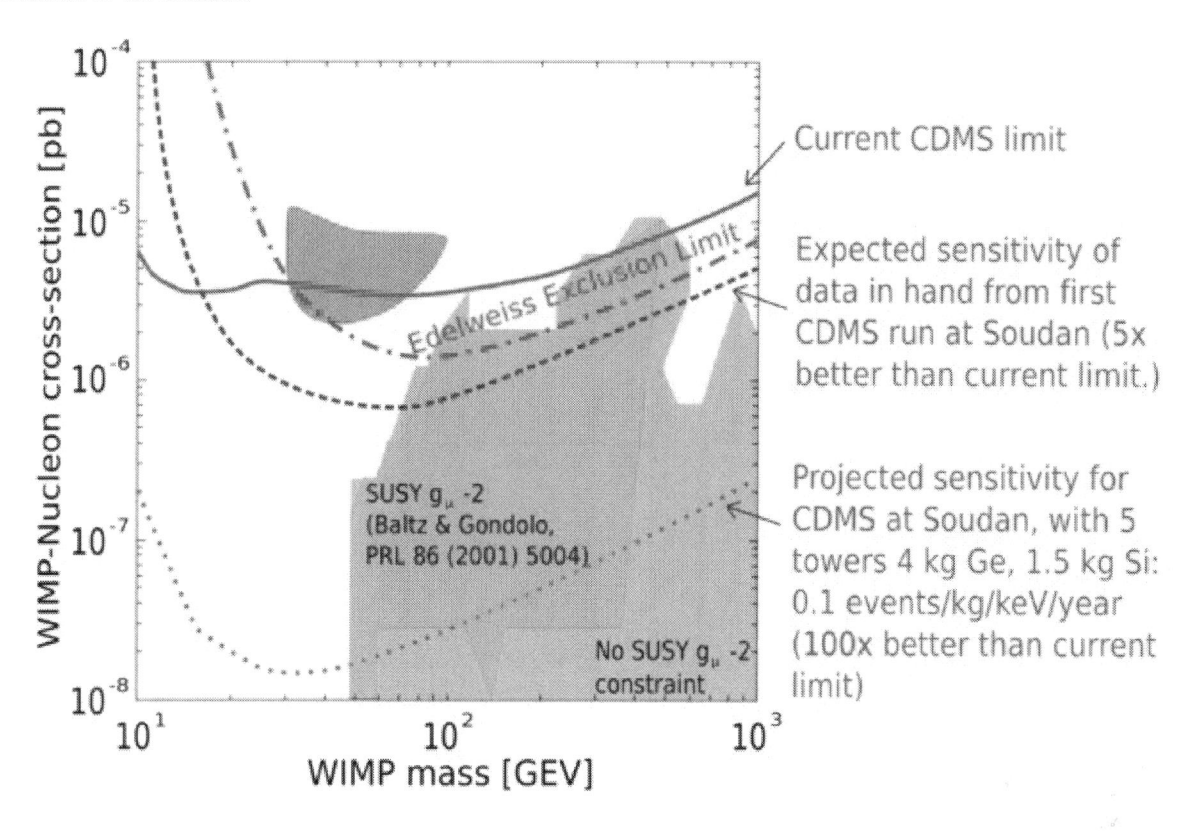

CDMS parameter space excluded as of 2004. DAMA result is located in green area and is disallowed.

15.3.1 Cryogenic Crystal Detectors

A technique used by the Cryogenic Dark Matter Search (CDMS) detector at the Soudan Mine relies on multiple very cold germanium and silicon crystals. The crystals (each about the size of a hockey puck) are cooled to about 50 mK. A layer of metal (aluminium and tungsten) at the surfaces is used to detect a WIMP passing through the crystal. This design hopes to detect vibrations in the crystal matrix generated by an atom being "kicked" by a WIMP. The tungsten transition edge sensors (TES) are held at the critical temperature so they are in the superconducting state. Large crystal vibrations will generate heat in the metal and are detectable because of a change in resistance. CRESST, CoGeNT, and EDELWEISS run similar setups.

15.3.2 Noble Gas Scintillators

Another way of detecting atoms "knocked about" by a WIMP is to use scintillating material, so that light pulses are generated by the moving atom and detected, often with PMTs. Experiments such as DEAP at SNOLAB, DarkSide, or WARP at the LNGS plan to instrument a very large target mass of liquid argon for sensitive WIMP searches. ZEPLIN, and XENON used xenon to exclude WIMPs at higher sensitivity until superseded in sensitivity by LUX in 2013. Larger ton-scale expansions of these xenon detectors have been approved for construction. PandaX is also using xenon. Neon may used in future studies.

15.3.3 Crystal Scintillators

Instead of a liquid noble gas, an in principle simpler approach is the use of a scintillating crystal such as NaI(Tl). This approach is taken by DAMA/LIBRA, an experiment that observed an annular modulation of the signal consistent with WIMP detection (see #Recent Limits). Several experiments are attempting to replicate those results, including ANAIS and DM-Ice, which is codeploying NaI crystals with the IceCube detector at the South Pole. KIMS is approaching the same problem using CsI(Tl) as a scintillator.

15.3.4 Bubble Chambers

The PICASSO (Project In Canada to Search for Supersymmetric Objects) experiment is a direct dark matter search experiment that is located at SNOLAB in Canada. It uses bubble detectors with Freon as the active mass. PICASSO is predominantly sensitive to spin-dependent interactions of WIMPs with the fluorine atoms in the Freon. COUPP, a similar experiment using trifluoroiodomethane(CF_3I), published limits for mass above 20 GeV in 2011.[19]

A bubble detector is a radiation sensitive device that uses small droplets of superheated liquid that are suspended in a gel matrix.[20] It uses the principle of a bubble chamber but, since only the small droplets can undergo a phase transition at a time, the detector can stay active for much longer periods. When enough energy is deposited in a droplet by ionizing radiation, the superheated droplet becomes a gas bubble. The bubble development is accompanied by an acoustic shock wave that is picked up by piezo-electric sensors. The main advantage of the bubble detector technique is that the detector is almost insensitive to background radiation. The detector sensitivity can be adjusted by changing the temperature, typically operated between 15 °C and 55 °C. There is another similar experiment using this technique in Europe called SIMPLE.

PICASSO reports results (November 2009) for spin-dependent WIMP interactions on ^{19}F, for masses of 24 Gev new stringent limits have been obtained on the spin-dependent cross section of 13.9 pb (90% CL). The obtained limits restrict recent interpretations of the DAMA/LIBRA annual modulation effect in terms of spin dependent interactions.[21]

PICO is an expansion of the concept planned in 2015.[22]

15.3.5 Other

Time Project Chambers (TPC) filled with low pressure gases are being studied for WIMP detection. The Directional Recoil Identification From Tracks (DRIFT) collaboration is attempting to utilize the predicted directionality of the WIMP signal. DRIFT uses a carbon disulfide target, that allows WIMP recoils to travel several millimetres, leaving a track of charged particles. This charged track is drifted to an MWPC readout plane that allows it to be reconstructed in three dimensions and determine the origin direction. DMTPC is a similar experiment with CF_4 gas.

15.4 Recent Limits

In February 2010, researchers at CDMS-II announced that they had observed two events that may have been caused by WIMP-nucleus collisions.[23][24][25] CoGeNT, a smaller detector using a single germanium puck, designed to sense WIMPs with smaller masses, reported hundreds of detection events in 56 days. Juan Collar, who presented the results to a conference at the University of California, was quoted: "If it's real, we're looking at a very beautiful dark-matter signal".[26][27]

Annual modulation is one of the predicted signatures of a WIMP signal,[28][29] and on this basis the DAMA collaboration has claimed a positive detection. Other groups, however, have not confirmed this result. The CDMS data made public in May 2004 exclude the entire DAMA signal region given certain standard assumptions about the properties of the WIMPs and the dark matter halo. CDMS and EDELWEISS would be expected to observe a significant number of WIMP-nucleus scatters if the DAMA signal were in fact caused by WIMPs.

Current limits from LUX and other searches are in disagreement with any WIMP interpretation of these results. With 370 kilograms of xenon it is more sensitive than XENON or CDMS.[30] First results from October 2013 report that no signals were seen, appearing to refute results obtained from less sensitive instruments.[31]

15.5 See also

- Higgs boson
- Massive compact halo object (MACHO)
- Micro black hole
- Robust associations of massive baryonic objects (RAMBOs)

15.5.1 Theoretical candidates

- Lightest Supersymmetric Particle (LSP)
- Neutralino
- Majorana fermion
- Sterile neutrino

15.6 References

[1] Jungman, Kamionkowski and Griest, Supersymmetric dark matter, Physics Reports, 1996

[2] LHC discovery maims supersymmetry again, Discovery News

[3] Nathaniel Craig, The State of Supersymmetry after Run I of the LHC

[4] Patrick J. Fox, Gabriel Jung, Peter Sorensen and Neal Weiner, Dark Matter in Light of LUX, Physical Review D, 2014

[5] H.V. Klapdor-Kleingrothaus, Double Beta Decay and Dark Matter Search - Window to New Physics now, and in future (GENIUS), 4 Feb 1998

[6] M. Kamionkowski, WIMP and Axion Dark Matter, 24 Oct 1997

[7] V. Zacek, Dark Matter Proc. of the 2007 Lake Louise Winter Institute, March 2007

[8] K. Griest, The Search for Dark Matter: WIMPs and MACHOs, 13 Mar 1993

[9] Griest, Kim (1991). "Galactic Microlensing as a Method of Detecting Massive Compact Halo Objects". *The Astrophysical Journal* **366**: 412–421. Bibcode:1991ApJ...366..412G. doi:10.1086/169575.

[10] C. Conroy, R. H. Wechsler, A. V. Kravtsov, Modeling Luminosity-Dependent Galaxy Clustering Through Cosmic Time, 21 Feb 2006.

[11] The Millennium Simulation Project, Introduction: The Millennium Simulation The Millennium Run used more than 10 billion particles to trace the evolution of the matter distribution in a cubic region of the Universe over 2 billion light-years on a side.

[12] The Fermi-LAT Collaboration Dark Matter Constraints from Observations of 25 Milky Way Satellite Galaxies with the Fermi Large Area Telescope, Physical Review D 89, 042001 (2014)

[13] Grube et al. VERITAS Limits on Dark Matter Annihilation from Dwarf Galaxies, Proceedings of the 5th International Symposium on High-Energy Gamma-Ray Astronomy, Heidelberg, July 9–13, 2012

[14] IceCube Collaboration, Observation of High-Energy Astrophysical Neutrinos in Three Years of IceCube Data, May 2014.

[15] F. Ferrer, L. Krauss, and S. Profumo, Indirect detection of light neutralino dark matter in the NMSSM. Phys.Rev. D74 (2006) 115007

[16] K. Freese, Can Scalar Neutrinos Or Massive Dirac Neutrinos Be the Missing Mass? . Phys.Lett.B167:295 (1986).

[17] Merritt, D.; Bertone, G. (2005). "Dark Matter Dynamics and Indirect Detection". *Modern Physics Letters A* **20** (14): 1021–1036. arXiv:astro-ph/0504422. Bibcode:2005MPLA...20.1021B. doi:10.1142/S0217732305017391.

[18] N. Fornengo, Status and perspectives of indirect and direct dark matter searches. 36th COSPAR Scientific Assembly, Beijing, China, 16–23 July 2006

[19] "Improved Limits on Spin-Dependent WIMP-Proton Interactions from a Two Liter CF3I Bubble Chamber". *PRL 106, 021303.* Jan 10, 2011.

[20] Bubble Technology Industries

[21] PICASSO Collaboration; Aubin, F.; Auger, M.; Behnke, E.; Beltran, B.; Clark, K.; Dai, X.; Davour, A. et al. (2009). "Dark Matter Spin-Dependent Limits for WIMP Interactions on ^{19}F by PICASSO". *Physics Letters B* **682** (2): 185. arXiv:0907.0307. Bibcode:2009PhLB..682..185A. doi:10.1016/j.physletb.2009.11.019.

[22] "Overview of non-liquid noble direct detection dark matter experiments". *Physics of the Dark Universe.* 28 October 2014. arXiv:1410.4960. Bibcode:2014PDU.....4...92C. doi:10.1016/j.dark.2014.10.005.

[23] "Key to the universe found on the Iron Range?". Retrieved December 18, 2009.

[24] CDMS Collaboration. "Results from the Final Exposure of the CDMS II Experiment" (PDF). See also a non-technical summary: CDMS Collaboration. "Latest Results in the Search for Dark Matter" (PDF)

[25] The CDMS II Collaboration (2010). "Dark Matter Search Results from the CDMS II Experiment" **327** (5973). Science. pp. 1619–21. arXiv:0912.3592. Bibcode:2010Sci...327.1619C. doi:10.1126/science.1186112. PMID 20150446.

[26] Eric Hand (2010-02-26). "A CoGeNT result in the hunt for dark matter". Nature News.

[27] C. E. Aalseth; CoGeNT collaboration (2011). "Results from a Search for Light-Mass Dark Matter with a P-type Point Contact Germanium Detector". *Physical Review Letters* **106** (13). arXiv:1002.4703. Bibcode:2011PhRvL.106m1301A. doi:10.1103/PhysRevLett.106.131301.

[28] A. Drukier, K. Freese, and D. Spergel, Detecting Cold Dark Matter Candidates, Phys.Rev.D33:3495-3508 (1986).(subscription required)

[29] K. Freese, J. Frieman, and A. Gould, Signal Modulation in Cold Dark Matter Detection, Phys.Rev.D37:3388 (1988).

[30] "New Experiment Torpedoes Lightweight Dark Matter Particles". 30 October 2013. Retrieved 6 May 2014.

[31] "First Results from LUX, the World's Most Sensitive Dark Matter Detector". Berkeley Lab News Center. 30 October 2013. Retrieved 6 May 2014.

15.7 Further reading

- Bertone, Gianfranco (2010). *Particle Dark Matter: Observations, Models and Searches.* Cambridge University Press. p. 762. ISBN 978-0-521-76368-4.

15.8 External links

- Particle Data Group review article on WIMP search
- Timothy J. Sumner, Experimental Searches for Dark Matter in Living Reviews in Relativity, Vol 5, 2002
- Portraits of darkness, New Scientist, August 31, 2013. Preview only.

Chapter 16

Cold dark matter

In cosmology and physics, **cold dark matter** (**CDM**) is a hypothetical form of matter (a kind of dark matter) whose particles moved slowly compared to the speed of light (the *cold* in CDM) since the universe was approximately one year old (a time when the cosmic particle horizon contained the mass of one typical galaxy); and interact very weakly with ordinary matter and electromagnetic radiation (the *dark* in CDM). It is believed that approximately 80% of matter in the Universe is dark matter, with only a small fraction being the ordinary baryonic matter that composes stars, planets and living organisms. Since the late 1980s or 1990s, most cosmologists favor the cold dark matter theory (specifically the modern Lambda-CDM model) as a description of how the Universe went from a smooth initial state at early times (as shown by the cosmic microwave background radiation) to the lumpy distribution of galaxies and their clusters we see today — the large-scale structure of the Universe. The theory sees the role that dwarf galaxies played as crucial, as they are thought to be natural building blocks that form larger structures, created by small-scale density fluctuations in the early Universe.[1]

The theory was originally published in 1982 by three independent groups of cosmologists; James Peebles at Princeton,[2] J. Richard Bond, Alex Szalay and Michael Turner;[3] and George Blumenthal, H. Pagels and Joel Primack.[4] An influential review article in 1984 by Blumenthal, Sandra Moore Faber, Primack and British scientist Martin Rees developed the details of the theory.[5]

In the cold dark matter theory, structure grows hierarchically, with small objects collapsing under their self-gravity first and merging in a continuous hierarchy to form larger and more massive objects. In the hot dark matter paradigm, popular in the early 1980s, structure does not form hierarchically (*bottom-up*), but rather forms by fragmentation (*top-down*), with the largest superclusters forming first in flat pancake-like sheets and subsequently fragmenting into smaller pieces like our galaxy the Milky Way. The predictions of the hot dark matter theory disagree with observations of large-scale structures, whereas the cold dark matter paradigm is in general agreement with the observations.

16.1 Composition

Dark matter is detected through its gravitational interactions with ordinary matter and radiation. As such, it is very difficult to determine what the constituents of cold dark matter are. The candidates fall roughly into three categories:

- Axions are very light particles with a specific type of self-interaction that makes them a suitable CDM candidate.[6][7] Axions have the theoretical advantage that their existence solves the Strong CP problem in QCD, but have not been detected.

- MACHOs or *Massive Compact Halo Objects* are large, condensed objects such as black holes, neutron stars, white dwarfs, very faint stars, or non-luminous objects like planets. The search for these consists of using gravitational lensing to see the effect of these objects on background galaxies. Most experts believe that the constraints from those searches rule out MACHOs as a viable dark matter candidate.[8][9][10][11][12][13]

- WIMPs: Dark matter is composed of *Weakly Interacting Massive Particles.* There is no currently known particle with the required properties, but many extensions of the standard model of particle physics predict such particles. The search for WIMPs involves attempts at direct detection by highly sensitive detectors, as well as attempts at production by particle accelerators. WIMPs are generally regarded as the most promising dark matter candidates.[9][11][13] The DAMA/NaI experiment and its successor DAMA/LIBRA have claimed to directly detect dark matter particles passing through the Earth, but many scientists remain skeptical, as no results from similar experiments seem compatible with the DAMA results.

16.2 Challenges

Several discrepancies between the predictions of the particle cold dark matter paradigm and observations of galaxies and their clustering have arisen:

- The cuspy halo problem: the density distributions of DM halos in cold dark matter simulations are much more peaked than what is observed in galaxies by investigating their rotation curves.[14]

- The missing satellites problem: cold dark matter simulations predict much larger numbers of small dwarf galaxies than are observed around galaxies like the Milky Way.[15]

- The disk of satellites problem: dwarf galaxies around the Milky Way and Andromeda galaxies are observed to be orbiting in thin, planar structures whereas the simulations predict that they should be distributed randomly about their parent galaxies.[16]

Some of these problems have proposed solutions but it remains unclear whether they can be solved without abandoning the CDM paradigm.[17]

16.3 See also

- Fuzzy cold dark matter
- Meta-cold dark matter
- Dark matter
  - Hot dark matter (HDM)
  - Warm dark matter (WDM)
  - Self-interacting dark matter
- Lambda-CDM model
- Modified Newtonian dynamics

16.4 References

[1] Battinelli, P.; S. Demers (2005-10-06). "The C star population of DDO 190: 1. Introduction" (PDF). Astronomy & Astrophysics. p. 1. Bibcode:2006A&A...447..473B. doi:10.1051/0004-6361:20052829. Archived from the original on 2005-10-06. Retrieved 2012-08-19. Dwarf galaxies play a crucial role in the CDM scenario for galaxy formation, having been suggested to be the natural building blocks from which larger structures are built up by merging processes. In this scenario dwarf galaxies are formed from small-scale density fluctuations in the primeval Universe.

[2] Peebles, P. J. E. (December 1982). "Large-scale background temperature and mass fluctuations due to scale-invariant primeval perturbations". *The Astrophysical Journal* **263**: L1. doi:10.1086/183911.

[3] . doi:10.1103/PhysRevLett.48.1636. Missing or empty |title= (help)

[4] Blumenthal, George R.; Pagels, Heinz; Primack, Joel R. (2 September 1982). "Galaxy formation by dissipationless particles heavier than neutrinos". *Nature* **299** (5878): 37–38. doi:10.1038/299037a0.

[5] Blumenthal, G. R.; Faber, S. M.; Primack, J. R.; Rees,, M. J. (1984). "Formation of galaxies and large-scale structure with cold dark matter". *Nature* **311** (517). Bibcode:1984Natur.311..517B. doi:10.1038/311517a0.

[6] e.g. M. Turner (2010). "Axions 2010 Workshop". U. Florida, Gainesville, USA.

[7] e.g. Pierre Sikivie (2008). "Axion Cosmology". Lect. Notes Phys. 741, 19-50.

[8] Carr, B. J. et al. (May 2010). "New cosmological constraints on primordial black holes". *Physical Review D* **81** (10): 104019. arXiv:0912.5297. Bibcode:2010PhRvD..81j4019C. doi:10.1103/PhysRevD.81.104019.

[9] Peter, A. H. G. (2012). "Dark Matter: A Brief Review". arXiv:1201.3942.

[10] Bertone, Gianfranco; Hooper, Dan; Silk, Joseph (January 2005). "Particle dark matter: evidence, candidates and constraints". *Physics Reports* **405** (5–6): 279–390. arXiv:hep-ph/0404175. Bibcode:2005PhR...405..279B. doi:10.1016/j.physrep.2004.08.031.

[11] Garrett, Katherine; Dūda, Gintaras. "Dark Matter: A Primer". *Advances in Astronomy* **2011**: 968283. arXiv:1006.2483. Bibcode:2011AdAst2011E...8G. doi:10.1155/2011/968283.. p. 3: "MACHOs can only account for a very small percentage of the nonluminous mass in our galaxy, revealing that most dark matter cannot be strongly concentrated or exist in the form of baryonic astrophysical objects. Although microlensing surveys rule out baryonic objects like brown dwarfs, black holes, and neutron stars in our galactic halo, can other forms of baryonic matter make up the bulk of dark matter? The answer, surprisingly, is no..."

[12] Gianfranco Bertone, "The moment of truth for WIMP dark matter," Nature 468, 389–393 (18 November 2010)

[13] Olive, Keith A (2003). "TASI Lectures on Dark Matter". *Physics* **54**: 21.

[14] Gentile, G.; P., Salucci (2004). "The cored distribution of dark matter in spiral galaxies". *Monthly Notices* **351**: 903–922. arXiv:astro-ph/0403154. Bibcode:2004MNRAS.351..903G. doi:10.1111/j.1365-2966.2004.07836.x.

[15] Klypin, Anatoly; Kravtsov, Andrey V.; Valenzuela, Octavio; Prada, Francisco (1999). "Where Are the Missing Galactic Satellites?". *ApJ* **522**: 82. arXiv:astro-ph/9901240. Bibcode:1999ApJ...522...82K. doi:10.1086/307643.

[16] Marcel Pawlowski et al., "Co-orbiting satellite galaxy structures are still in conflict with the distribution of primordial dwarf galaxies" MNRAS (2014) http://arxiv.org/abs/1406.1799

[17] Kroupa, P.; Famaey, B.; de Boer, Klaas S.; Dabringhausen, Joerg; Pawlowski, Marcel; Boily, Christian; Jerjen, Helmut; Forbes, Duncan; Hensler, Gerhard (2010). "Local-Group tests of dark-matter Concordance Cosmology: Towards a new paradigm for structure formation".*Astronomy and Astrophysics***523**: 32-54.arXiv:1006.1647.Bibcode:2010A&A...523A..36361/201014892.

16.5 Further reading

- Bertone, Gianfranco (2010). *Particle Dark Matter: Observations, Models and Searches*. Cambridge University Press. p. 762. ISBN 978-0-521-76368-4.

Chapter 17

Warm dark matter

Warm dark matter (**WDM**) is a hypothesized form of dark matter that has properties intermediate between those of hot dark matter and cold dark matter, causing structure formation to occur bottom-up from above their free-streaming scale, and top-down below their free streaming scale. The most common WDM candidates are sterile neutrinos and gravitinos. The WIMPs (weakly interacting massive particles), when produced non-thermally could be candidates for warm dark matter. In general, however the thermally produced WIMPs are cold dark matter candidates.

17.1 keVins and GeVins

One possible WDM candidate particle with a mass of a few keV comes from introducing two new, zero charge, zero lepton number fermions to the Standard Model of Particle Physics: "keV-mass inert fermions" (keVins) and "GeV-mass inert fermions" (GeVins). keVins are overproduced if they reach thermal equilibrium in the early universe, but in some scenarios the entropy production from the decays of unstable heavier particles can suppresses their abundance to the correct value. These particles are considered "inert" because they only have suppressed interactions with the Z boson. Sterile neutrinos with masses of a few keV are possible candidates for keVins. At temperatures below the electroweak scale their only interactions with standard model particles are weak interactions due to their mixing with ordinary neutrinos. Due to the smallness of the mixing angle they are not overproduced because they freeze out before reaching thermal equilibrium. Their properties are consistent with astrophysical bounds coming from structure formation and the Pauli principle if their mass is larger than 1-8 keV.

In February 2014, different analyses[1][2] have extracted from the spectrum of X-ray emissions observed by XMM-Newton, a monochromatic signal around 3.5 keV. This signal is coming from different galaxy cluster (like Perseus or Centaurus) and several scenarios of warm dark matter can justify such a line. We can cite for example a 3.5 keV candidate annihilating into 2 photons,[3] or a 7 keV dark matter particle decaying into photon and neutrino.[4]

17.2 See also

- Dark matter
  - Hot dark matter (HDM)
  - Cold dark matter (CDM)
- Lambda-CDM model
- Modified Newtonian Dynamics

17.3 References

[1] E. Bulbul *et al.* http://arxiv.org/abs/1402.2301 "Detection of An Unidentified Emission Line in the Stacked X-ray spectrum of Galaxy Clusters"

[2] A. Boyarski *et al.*: http://arxiv.org/abs/1402.4119, "An unidentified line in X-ray spectra of the Andromeda galaxy and Perseus galaxy cluster"

[3] E. Dudas, L. Heurtier, Y. Mambrini: http://arxiv.org/abs/arXiv:1404.1927, "Generating X-ray lines from annihilating dark matter"

[4] H. Ishida, K.S. Jeong, F. Takahashi : http://arxiv.org/abs/arXiv:1402.5837, "7 keV sterile neutrino dark matter from split flavor mechanism"

- Constraining warm dark matter candidates including sterile neutrinos and light gravitinos with WMAP and the Lyman-α forest

- King, S. & Merle, A. (2012) Warm Dark Matter from keVins. IOP Science, Journal of Cosmology and Astroparticle Physics, 1208 (2012) 016. doi: 10.1088/1475-7516/2012/08/016.

- The first star formation in WDM Universe

- W.B. Lin, D.H. Huang, X. Zhang, R. Brandenberger, Non-Thermal Production of WIMPs and the Sub-Galactic Structure of the Universe Phys. Rev. Lett. 86, 954, 2001.

- Millis, John. Warm Dark Matter. About.com. Retrieved 23 Jan., 2013. http://space.about.com/od/astronomydictionary/g/Warm-Dark-Matter.htm.

17.4 Further reading

- Bertone, Gianfranco (2010). *Particle Dark Matter: Observations, Models and Searches*. Cambridge University Press. p. 762. ISBN 978-0-521-76368-4.

Chapter 18

Hot dark matter

Hot dark matter is a form of dark matter which consists of particles that travel with ultrarelativistic velocities.

Dark matter is matter that does not interact with, and therefore cannot be detected by, electromagnetic radiation, hence *dark*. It is postulated to exist to explain how clusters and superclusters of galaxies formed after the big bang. Data from galaxy rotation curves indicate that around 90% of the mass of a galaxy cannot be seen. It can only be detected by its gravitational effect.

Hot dark matter cannot explain how individual galaxies formed from the big bang. The microwave background radiation as measured by the COBE satellite is very smooth and fast moving particles cannot clump together on this small scale from such a smooth initial clumping. Due to theory,[1] in order to explain small scale structure in the Universe, it is necessary to invoke cold dark matter (CDM) or warm dark matter. Hot dark matter as the sole explanation of dark matter is no longer viable,[1] therefore, it is nowadays considered only as part of a Mixed dark matter theory.

18.1 Neutrinos

The best example of a hot dark matter particle is the neutrino. Neutrinos have very small masses, and do not take part in two of the four fundamental forces, the electromagnetic interaction and the strong interaction. They *do* interact by the weak interaction, and gravity, but due to the feeble strength of these forces, they are difficult to detect. A number of projects, such as the Super-Kamiokande neutrino observatory, in Gifu, Japan are currently studying these neutrinos.

18.2 See also

- Lambda-CDM model
- Modified Newtonian Dynamics

18.3 References

[1] S. Frenk, Carlos; D. M. White, Simon (2012). "Dark matter and cosmic structure". arXiv:1210.0544.

18.4 Further reading

- Bertone, Gianfranco (2010). *Particle Dark Matter: Observations, Models and Searches*. Cambridge University Press. p. 762. ISBN 978-0-521-76368-4.

18.5 External links

- Hot dark matter by Berkeley
- Dark Matter

Chapter 19

Baryonic dark matter

In astronomy and cosmology, **baryonic dark matter** is dark matter (matter that is undetectable by its emitted radiation, but whose presence can be inferred from gravitational effects on visible matter) composed of baryons, i.e. protons and neutrons and combinations of these, such as non-emitting ordinary atoms. Candidates for baryonic dark matter include non-luminous gas, Massive Astrophysical Compact Halo Objects (MACHOs: condensed objects such as black holes, neutron stars, white dwarfs, very faint stars, or non-luminous objects like planets), and brown dwarfs.

The total amount of baryonic dark matter can be inferred from Big Bang nucleosynthesis, and observations of the cosmic microwave background. Both indicate that the amount of baryonic dark matter is much smaller than the total amount of dark matter.

In the case of big bang nucleosynthesis, the problem is that large amounts of ordinary matter means a denser early universe, more efficient conversion of matter to helium-4 and less unburned deuterium that can remain. If one assumes that all of the dark matter in the universe consists of baryons, then there is far too much deuterium in the universe. This could be resolved if there were some means of generating deuterium, but large efforts in the 1970s failed to come up with plausible mechanisms for this to occur. For instance, MACHOs, which include, for example, brown dwarfs (balls of hydrogen and helium with masses less than 0.08 M$\odot$ or 1.6×10^{29} kg), never begin nuclear fusion of hydrogen, but they do burn deuterium. Other possibilities that were examined include "Jupiters", which are similar to brown dwarfs but have masses ~ 0.001 M$\odot$ (2×10^{27} kg) and do not burn anything, and white dwarfs.[1][2]

19.1 See also

- Nonbaryonic dark matter

19.2 References

[1] G. Jungman, M. Kamionkowski, and K. Griest (1996). "Supersymmetric dark matter". *Physics Reports* **267**: 195. arXiv:hep-ph/9506380. Bibcode:1996PhR...267..195J. doi:10.1016/0370-1573(95)00058-5.

[2] M. S. Turner (1999). "Cosmological parameters". *AIP Conference Proceedings*. arXiv:astro-ph/9904051.Bibcode:1999AIPC..478..113T. doi:10.1063/1.59381.

This image shows the galaxy cluster Abell 1689, with the mass distribution of the dark matter in the gravitational lens overlaid (in purple). The mass in this lens is made up partly of normal (baryonic) matter and partly of dark matter. Distorted galaxies are clearly visible around the edges of the gravitational lens. The appearance of these distorted galaxies depends on the distribution of matter in the lens and on the relative geometry of the lens and the distant galaxies, as well as on the effect of dark energy on the geometry of the Universe.

Chapter 20

Massive compact halo object

"MACHO" redirects here. For other uses, see Macho (disambiguation).

Massive astrophysical compact halo object, or **MACHO**, is a general name for any kind of astronomical body that might explain the apparent presence of dark matter in galaxy halos. A MACHO is a body composed of normal baryonic matter, which emits little or no radiation and drifts through interstellar space unassociated with any planetary system. Since MACHOs would not emit any light of their own, they would be very hard to detect. MACHOs may sometimes be black holes or neutron stars as well as brown dwarfs or unassociated planets. White dwarfs and very faint red dwarfs have also been proposed as candidate MACHOs. The term was coined by astrophysicist Kim Griest, in contrast to WIMPs, another proposed form of dark matter.[1]

20.1 Detection

A MACHO may be detected when it passes in front of or nearly in front of a star and the MACHO's gravity bends the light, causing the star to appear brighter in an example of gravitational lensing known as gravitational microlensing. Several groups have searched for MACHOs by searching for the microlensing amplification of light. These groups have ruled out dark matter being explained by MACHOs with mass in the range 0.00000001 solar masses (0.3 lunar masses) to 100 solar masses. One group, the MACHO collaboration, claims to have found enough microlensing to predict the existence of many MACHOs with mass of about 0.5 solar masses, enough to make up perhaps 20% of the dark matter in the galaxy.[2] This suggests that MACHOs could be white dwarfs or red dwarfs which have similar masses. However, red and white dwarfs are not completely dark; they do emit some light, and so can be searched for with the Hubble Telescope and with proper motion surveys. These searches have ruled out the possibility that these objects make up a significant fraction of dark matter in our galaxy. Another group, the EROS2 collaboration does not confirm the signal claims by the MACHO group. They did not find enough microlensing effect with a sensitivity higher by a factor 2.[3] Observations using the Hubble Space Telescope's NICMOS instrument showed that less than one percent of the halo mass is composed of red dwarfs.[4][5] This corresponds to a negligible fraction of the dark matter halo mass. Therefore, the missing mass problem is not solved by MACHOs.

20.2 Types of MACHOs

MACHOs may sometimes be considered to include black holes. Black holes are truly black in that they emit no light and any light shone upon them is absorbed and not reflected. It is thought possible that there is a halo of black holes surrounding the galaxy. A black hole can sometimes be detected by the halo of bright gas and dust that forms around it as an accretion disc being pulled in by the black hole's gravity. Such a disk can generate jets of gas that are shot out away from the black hole because it cannot be absorbed quickly enough. An isolated black hole, however, would not have an accretion disk and would only be detectable by gravitational lensing. Cosmologists doubt they make up a majority of dark matter because the black holes are at isolated points of the galaxy. The largest contributor to the missing mass must be spread throughout the galaxy to balance the gravity. A minority of physicists, including Chapline and Laughlin, believe that the widely accepted model of the black hole is wrong and needs to be replaced by a new model, the dark-energy star; in the general case for the suggested new model, the cosmological distribution of dark energy would be slightly lumpy and dark-energy stars of primordial type might be a possible candidate for MACHOs.

Neutron stars are somewhat like black holes, but are not heavy enough to collapse completely, instead forming into a material rather like that of an atomic nucleus (sometimes informally called neutronium). After sufficient time these stars could radiate away enough energy to become cold enough that they would be too faint to see. Likewise, old white dwarfs may also become cold and dead, eventually becoming black dwarfs, although the universe is not thought to be old enough for any stars to have reached this stage.

The next candidate for MACHOs are the brown dwarfs mentioned above. Brown dwarfs are sometimes called "failed stars" as they do not have enough mass for nuclear fusion to begin and simply glow a dull brown. Hence, their only source of energy is released through their own gravitational contraction, and may therefore be faintly visible in some circumstances. Brown dwarfs are about thirteen to seventy-five times the mass of Jupiter.

20.3 Theoretical considerations

Theoretical work simultaneously also showed that ancient MACHOs are not likely to account for the large amounts of dark matter now thought to be present in the universe.[6] The Big Bang as it is currently understood could not have produced enough baryons and still be consistent with the observed elemental abundances,[7] including the abundance of deuterium.[8] Furthermore, separate observations of baryon acoustic oscillations, both in the cosmic microwave background and large-scale structure of galaxies, set limits on the ratio of baryons to the total amount of matter. These observations show that a large fraction of non-baryonic matter is necessary regardless of the presence or absence of MACHOs.

20.4 See also

- Weakly Interacting Massive Particles (WIMPS) (*An alternate theory of Dark Matter*)
- Robust Associations of Massive Baryonic Objects (RAMBOs)

20.5 References

[1] Croswell, Ken (2002). *The Universe at Midnight*. Simon and Schuster. p. 165.

[2] C. Alcock et al., The MACHO Project: Microlensing Results from 5.7 Years of LMC Observations. Astrophys.J. 542 (2000) 281-307

[3] P. Tisserand et al., Limits on the Macho Content of the Galactic Halo from the EROS-2 Survey of the Magellanic Clouds, 2007, Astron. Astrophys. 469, 387-404

[4] David Graff and Katherine Freese, , Analysis of a hubble space telescope search for red dwarfs: limits on baryonic matter in the galactic halo, Astrophys.J.456:L49,1996.

[5] J. Najita, G. Tiede, and S. Carr, From Stars to Superplanets: The Low-Mass Initial Mass Function in the Young Cluster IC 348. The Astrophysical Journal 541, 1 (2000), 977–1003

[6] Katherine Freese, Brian Fields, and David Graff, Limits on stellar objects as the dark matter of our halo: nonbaryonic dark matter seems to be required.

[7] Brian Fields, Katherine Freese, and David Graff,Chemical abundance constraints on white dwarfs as halo dark matter, Astrophys. J.534:265- 276,2000.

[8] Arnon Dar, Dark Matter and Big Bang Nucleosynthesis. Astrophys. J., 449 (1995) 550

Chapter 21

Neutralino

In supersymmetry, the **neutralino**[1] is a hypothetical particle. There are four neutralinos that are fermions and are electrically neutral, the lightest of which is typically stable. They are typically labeled Ñ0
1 (the lightest), Ñ0
2, Ñ0
3 and Ñ0
4 (the heaviest) although sometimes $\tilde{\chi}_1^0, \ldots, \tilde{\chi}_4^0$ is also used when $\tilde{\chi}_i^\pm$ is used to refer to charginos. These four states are mixtures of the bino and the neutral wino (which are the neutral electroweak gauginos), and the neutral higgsinos. As the neutralinos are Majorana fermions, each of them is identical to its antiparticle. Because these particles only interact with the weak vector bosons, they are not directly produced at hadron colliders in copious numbers. They would primarily appear as particles in cascade decays of heavier particles (decays that happen in multiple steps) usually originating from colored supersymmetric particles such as squarks or gluinos.

In R-parity conserving models, the lightest neutralino is stable and all supersymmetric cascade-decays end up decaying into this particle which leaves the detector unseen and its existence can only be inferred by looking for unbalanced momentum in a detector.

The heavier neutralinos typically decay through a neutral Z boson to a lighter neutralino or through a charged W boson to a light chargino:[2]

The mass splittings between the different neutralinos will dictate which patterns of decays are allowed.

Up to present, neutralinos have never been observed or detected in an experiment.

21.1 Origins in supersymmetric theories

In supersymmetry models, all Standard Model particles have partner particles with the same quantum numbers except for the quantum number spin, which differs by 1/2 from its partner particle. Since the superpartners of the Z boson (zino), the photon (photino) and the neutral higgs (higgsino) have the same quantum numbers, they can mix to form four eigenstates of the mass operator called "neutralinos". In many models the lightest of the four neutralinos turns out to be the lightest supersymmetric particle (LSP), though other particles may also take on this role.

21.2 Phenomenology

The exact properties of each neutralino will depend on the details of the mixing[1] (e.g. whether they are more higgsino-like or gaugino-like), but they tend to have masses at the weak scale (100 GeV – 1 TeV) and couple to other particles with strengths characteristic of the weak interaction. In this way they are phenomenologically similar to neutrinos, and so are not directly observable in particle detectors at accelerators.

In models in which R-parity is conserved and the lightest of the four neutralinos is the LSP, the lightest neutralino is stable and is eventually produced in the decay chain of all other superpartners.[3] In such cases supersymmetric processes at accelerators are characterized by a large discrepancy in energy and momentum between the visible initial and final state particles, with this energy being carried off by a neutralino which departs the detector unnoticed.[4][5] This is an important signature to discriminate supersymmetry from Standard Model backgrounds.

21.3 Relationship to dark matter

As a heavy, stable particle, the lightest neutralino is an excellent candidate to form the universe's cold dark matter.[6][7][8] In many models the lightest neutralino can be produced thermally in the hot early universe and leave approximately the right relic abundance to account for the observed dark matter. A lightest neutralino of roughly 10–10000 GeV is the leading weakly interacting massive particle (WIMP) dark matter candidate.[9]

Neutralino dark matter could be observed experimentally in nature either indirectly or directly. For indirect observation, gamma ray and neutrino telescopes look for evidence of neutralino annihilation in regions of high dark matter density such as the galactic or solar centre.[4] For direct observation, special purpose experiments such as the Cryogenic Dark Matter Search (CDMS) seek to detect the rare impacts of WIMPs in terrestrial detectors. These experiments have begun to probe interesting supersymmetric parameter space, excluding some models for neutralino dark matter, and upgraded experiments with greater sensitivity are under development.

21.4 See also

- Lightest Supersymmetric Particle
- Real neutral particle

21.5 Notes

[1] Martin, pp. 71–74

[2] J.-F. Grivaz & the Particle Data Group (2010). "Supersymmetry, Part II (Experiment)" (PDF). *Journal of Physics G* **37** (7): 1309–1319.

[3] Martin, p. 83

[4] Feng, Jonathan L (2010). "Dark Matter Candidates from Particle Physics and Methods of Detection". *Annual Review of Astronomy and Astrophysics* **48**: 495–545. arXiv:1003.0904. Bibcode:2010ARA&A..48..495F. doi:10.1146/annurev-astro-082708-101659. |chapter= ignored (help)

[5] Ellis, John; Olive, Keith A. (2010). "Supersymmetric Dark Matter Candidates". arXiv:1001.3651 [astro-ph]. Also published as Chapter 8 in Bertone

[6] M. Drees; G. Gerbier & the Particle Data Group (2010). "Dark Matter" (PDF). *Journal of Physics G* **37** (7A): 255–260.

[7] Martin, p. 99

[8] Bertone, p. 8

[9] Martin, p. 124

21.6 References

- Martin, Stephen P. (2008). "A Supersymmetry Primer". v5. arXiv:hep-ph/9709356 [hep-ph]. Also published as Chapter 1 in Kane, Gordon L, ed. (2010). *Perspectives on Supersymmetry II*. World Scientific. p. 604. ISBN 978-981-4307-48-2.

- Bertone, Gianfranco, ed. (2010). *Particle Dark Matter: Observations, Models and Searches*. Cambridge University Press. p. 762. ISBN 978-0-521-76368-4.

Chapter 22

Axion

For other uses, see Axion (disambiguation).

The **axion** is a hypothetical elementary particle postulated by the Peccei–Quinn theory in 1977 to resolve the strong CP problem in quantum chromodynamics (QCD). If axions exist and have low mass within a specific range, they are of interest as a possible component of cold dark matter.

22.1 History

22.1.1 Prediction

As shown by Gerardus 't Hooft, strong interactions of the standard model, QCD, possess a non-trivial vacuum structure that in principle permits violation of the combined symmetries of charge conjugation and parity, collectively known as CP. Together with effects generated by weak interactions, the effective periodic strong CP violating term, Θ, appears as a Standard Model input – its value is not predicted by the theory, but must be measured. However, large CP violating interactions originating from QCD would induce a large electric dipole moment (EDM) for the neutron. Experimental constraints on the currently unobserved EDM implies CP violation from QCD must be extremely tiny and thus Θ must itself be extremely small. Since a priori Θ could have any value between 0 and 2π, this presents a naturalness problem for the standard model. Why should this parameter find itself so close to 0? (Or, why should QCD find itself CP-preserving?) This question constitutes what is known as the strong CP problem.

One simple solution exists: if at least one of the quarks of the standard model is massless, Θ becomes unobservable. However, empirical evidence strongly suggests that none of the quarks are massless.

In 1977, Roberto Peccei and Helen Quinn postulated a more elegant solution to the strong CP problem, the Peccei–Quinn mechanism. The idea is to effectively promote Θ to a field. This is accomplished by adding a new global symmetry (called a Peccei–Quinn symmetry) that becomes spontaneously broken. This results in a new particle, as shown by Frank Wilczek and Steven Weinberg, that fills the role of Θ—naturally relaxing the CP violation parameter to zero. This hypothesized new particle is called the axion. (On a more technical note, the axion is the would-be Nambu–Goldstone boson that results from the spontaneously broken Peccei–Quinn symmetry. However, the non-trivial QCD vacuum effects (e.g. instantons) spoil the Peccei–Quinn symmetry explicitly and provide a small mass for the axion. Hence, the axion is actually a pseudo-Nambu–Goldstone boson.) The original Weinberg-Wilczek axion was ruled out. Current literature discusses the mechanism as the 'invisible axion' which has two forms: KSVZ [1][2] and DFSZ.[3][4]

22.1.2 Searches

It had been thought that the invisible axion solves the strong CP problem without being amenable to verification by experiment. Axion models choose coupling that does not appear in any of the prior experiments. The very weakly coupled axion is also very light because axion couplings and mass are proportional. The situation changed when it was shown that a very light axion is overproduced in the early universe and therefore excluded.[5][6][7] The critical mass is of order 10^{-11} times the electron mass, where axions may account for the dark matter. The axion is thus a dark matter candidate as well as a solution to the strong CP problem. Furthermore, in 1983, Pierre Sikivie wrote down the modification of Maxwell's equations from a light stable axion [8] and showed axions can be detected on Earth by converting them to photons with a strong magnetic field, the principle of the ADMX. Solar axions may be converted to x-rays, as in CAST. Many experiments are searching laser light for signs of axions.[9]

22.2 Experiments

The Italian PVLAS experiment searches for polarization changes of light propagating in a magnetic field. The concept was first put forward in 1986 by Luciano Maiani, Roberto Petronzio and Emilio Zavattini.[10] A rotation claim[11] in 2006 was excluded by an upgraded setup.[12] An optimized search began in 2014.

Another technique is so called "light shining through walls",[13] where light passes through an intense magnetic field to convert photons into axions, that pass through metal. Experiments by BFRS and a team led by Rizzo ruled out an axion cause.[14] GammeV saw no events in a 2008 PRL. ALPS-I conducted similar runs,[15] setting new constraints in 2010; ALPS-II will run in 2014. OSQAR found no signal, limiting coupling[16] and will continue.

Several experiments search for astrophysical axions by the Primakoff effect, which converts axions to photons and vice versa in electromagnetic fields. Axions can be produced in the Sun's core when x-rays scatter in strong electric fields. The CAST solar telescope is underway, and has set limits on coupling to photons and electrons. ADMX searches the galactic dark matter halo[17] for resonant axions with a cold microwave cavity and has excluded optimistic axion models in the 1.9-3.53 μeV range.[18][19][20] It is amidst a series of upgrades and is taking new data, including at 4.9-6.2 μeV.

Resonance effects may be evident in Josephson junctions[21] from a supposed high flux of axions from the galactic halo with mass of 0.11 meV and density $0.05 GeV/cm^3$ [22] compared to the implied dark matter density $(0.3 \pm 0.1) GeV/cm^3$, indicating said axions would only partially compose dark matter.

Dark matter cryogenic detectors have searched for electron recoils that would indicate axions. CDMS published in 2009 and EDELWEISS set coupling and mass limits in 2013. UORE and XMASS also set limits on solar axions in 2013. XENON100 used a 225-day run to set the best copling liming to date and exclude some parameters.[23]

Axion-like bosons could have a signature in astrophysical settings. In particular, several recent works have proposed axion-like particles as a solution to the apparent transparency of the Universe to TeV photons.[24][25] It has also been demonstrated in a few recent works that, in the large magnetic fields threading the atmospheres of compact astrophysical objects (e.g., magnetars), photons will convert much more efficiently. This would in turn give rise to distinct absorption-like features in the spectra detectable by current telescopes.[26] A new promising means is looking for quasi-particle refraction in systems with strong magnetic gradients. In particular, the refraction will lead to beam splitting in the radio light curves of highly magnetized pulsars and allow much greater sensitivities than currently achievable.[27] The International Axion Observatory (IAXO) is a proposed fourth generation helioscope.[28]

22.3 Possible detection

Axions may have been detected through irregularities in X-ray emission due to interaction of the Earth's magnetic field with radiation streaming from the Sun. Studying 15 years of data by the European Space Agency's XMM-Newton observatory, a research group at Leicester University noticed a seasonal variation for which no conventional explanation could be found. One potential explanation for the variation, described as "plausible" by the senior author of the paper, was X-rays produced by axions from the Sun's core.[29]

A term analogous to the one that must be added to Maxwell's equations[30] also appears in recent theoretical models for topological insulators.[31] This term leads to several interesting predicted properties at the interface between topological and normal insulators.[32] In this situation the field θ describes something very different from its use in high-energy physics.[32] In 2013, Christian Beck suggested that axions might be detectable in Josephson junctions; and in 2014, he argued that a signature, consistent with a mass ~110μeV, had in fact been observed in several preexisting experiments.[33]

22.4 Properties

22.4.1 Predictions

One theory of axions relevant to cosmology had predicted that they would have no electric charge, a very small mass in the range from 10^{-6} to 1 eV/c^2, and very low interaction cross-sections for strong and weak forces. Because of their properties, axions would interact only minimally with ordinary matter. Axions would change to and from photons in magnetic fields.

Supersymmetry

In supersymmetric theories the axion has both a scalar and a fermionic superpartner. The fermionic superpartner of the axion is called the axino, the scalar superpartner is called the saxion or dilaton. They are all bundled up in a chiral superfield.

The axino has been predicted to be the lightest supersymmetric particle in such a model.[34] In part due to this property, it is considered a candidate for dark matter.[35]

22.4.2 Cosmological implications

Theory suggests that axions were created abundantly during the Big Bang.[36] Because of a unique coupling to the instanton field of the primordial universe (the "misalignment mechanism"), an effective dynamical friction is created during the acquisition of mass following cosmic inflation. This robs all such primordial axions of their kinetic energy.

If axions have low mass, thus preventing other decay modes, theories predict that the universe would be filled with a very cold Bose–Einstein condensate of primordial axions. Hence, axions could plausibly explain the dark matter problem of physical cosmology.[37] Observational studies are underway, but they are not yet sufficiently sensitive to probe the mass regions if they are the solution to the dark matter problem. High mass axions of the kind searched for by Jain and Singh (2007)[38] would not persist in the modern universe. Moreover, if axions exist, scatterings with other particles in the thermal bath of the early universe unavoidably produce a population of hot axions.[39]

Low mass axions could have additional structure at the galactic scale. As they continuously fell into a galaxy from the intergalactic medium, they would be denser in "caustic" rings, just as the stream of water in a continuously-flowing fountain is thicker at its peak.[40] The gravitational effects of these rings on galactic structure and rotation might then be observable.[41] Other cold dark matter theoretical candidates, such as WIMPs and MACHOs, could also form such rings, but because such candidates are fermionic and thus experience friction or scattering among themselves, the rings would be less pronounced.

Axions would also have stopped interaction with normal matter at a different moment than other more massive dark particles. The lingering effects of this difference could perhaps be calculated and observed astronomically. Axions may hold the key to the Solar corona heating problem.[42]

22.5 References

22.5.1 Notes

[1] Kim, J.E. (1979). "Weak-Interaction Singlet and Strong CP Invariance". *Phys. Rev. Lett.* **43**(2): 103–107. Bibcode:1979PhRvL. doi:10.1103/PhysRevLett.43.103.

[2] Shifman, M.; Vainshtein, A.; Zakharov, V. (1980). "Can confinement ensure natural CP invariance of strong interactions?". *Nucl. Phys.* **B166**: 493. Bibcode:1980NuPhB.166..493S. doi:10.1016/0550-3213(80)90209-6.

[3] Dine, M.; Fischler, W.; Srednicki, M. (1981). "A simple solution to the strong CP problem with a harmless axion". *Phys. Lett.* **B104**: 199. Bibcode:1981PhLB..104..199D. doi:10.1016/0370-2693(81)90590-6.

[4] Zhitnitsky, A. (1980). "On possible suppression of the axion-hadron interactions". *Sov. J. Nucl. Phys.* **31**: 260.

[5] Preskill, J.; Wise, M.; Wilczek, F. (1983). "Cosmology of the invisible axion". *Phys. Lett.* **B120**: 127. Bibcode:1983PhLB..120. doi:10.1016/0370-2693(83)90637-8.

[6] Abbott, L.; Sikivie, P. (1983). "A cosmological bound on the invisible axion". *Phys.Lett.* **B120**: 133.

[7] Dine, M.; Fischler, W. (1983). "The not-so-harmless axion". *Phys.Lett.* **B120**: 137. Bibcode:1983PhLB..120..137D. doi:1-2693(83)90639-1.

[8] Sikivie, P. (1983). "Experimental Tests of the "Invisible" Axion". *Phys. Rev. Lett.* **51**(16): 1413. Bibcode:1983PhRvL..51.14S. doi:10.1103/physrevlett.51.1415.

[9] http://home.web.cern.ch/about/experiments/osqar

[10] Maiani, L.; Petronzio, R.; Zavattini, E. (1986). "Effects of nearly massless, spin-zero particles on light propagation in a magnetic field". *Phys. Lett.* **175** (3): 359–363. Bibcode:1986PhLB..175..359M. doi:10.1016/0370-2693(86)90869-5.

[11] Steve Reucroft, John Swain. Axion signature may be QED CERN Courier, 2006-10-05

[12] Zavattini, E.; Zavattini, G.; Ruoso, G.; Polacco, E.; Milotti, E.; Karuza, M.; Gastaldi, U.; Di Domenico, G.; Della Valle, F.; Cimino, R.; Carusotto, S.; Cantatore, G.; Bregant, M.; Pvlas, Collaboration (2006). "Experimental Observation of Optical Rotation Generated in Vacuum by a Magnetic Field". *Physical Review Letters* **96** (11): 110406. arXiv:hep-ex/0507107. Bibcode:2006PhRvL..96k0406Z. doi:10.1103/PhysRevLett.96.110406. PMID 16605804.

[13] Ringwald, A. (2003). "Electromagnetic Probes of Fundamental Physics - Proceedings of the Workshop". invited talk *"Fundamental Physics at an X-Ray Free Electron Laser"* at *Workshop on Electromagnetic Probes of Fundamental Physics*. The Science and Culture Series - Physics (Erice, Italy): 63–74. arXiv:hep-ph/0112254. doi:10.1142/9789812704214_0007. ISBN 9789812385666.

[14] Robilliard, C.; Battesti, R.; Fouche, M.; Mauchain, J.; Sautivet, A.-M.; Amiranoff, F.; Rizzo, C. (2007). "No "Light Shining through a Wall": Results from a Photoregeneration Experiment". *Physical Review Letters* **99** (19): 190403. arXiv:0707.1296. Bibcode:2007PhRvL..99s0403R. doi:10.1103/PhysRevLett.99.190403. PMID 18233050.

[15] Ehret, Klaus; Frede, Maik; Ghazaryan, Samvel; Hildebrandt, Matthias; Knabbe, Ernst-Axel; Kracht, Dietmar; Lindner, Axel; List, Jenny; Meier, Tobias; Meyer, Niels; Notz, Dieter; Redondo, Javier; Ringwald, Andreas; Wiedemann, Günter; Willke, Benno (May 2010). "New ALPS results on hidden-sector lightweights". *Phys Lett B* **689** (4–5): 149–155. arXiv:1004.1313. Bibcode:2010PhLB..689..149E. doi:10.1016/j.physletb.2010.04.066.

[16] Pugnat, P.; Ballou, R.; Schott, M.; Husek, T.; Sulc, M.; Deferne, G.; Duvillaret, L.; Finger, M.; Finger, M.; Flekova, L.; Hosek, J.; Jary, V.; Jost, R.; Kral, M.; Kunc, S.; MacUchova, K.; Meissner, K. A.; Morville, J.; Romanini, D.; Siemko, A.; Slunecka, M.; Vitrant, G.; Zicha, J. (Aug 2014). "Search for weakly interacting sub-eV particles with the OSQAR laser-based experiment: results and perspectives". *Eur Phys J C* **74** (8): 3027. arXiv:1306.0443. Bibcode:2014EPJC...74.3027P. doi:10.1140/epjc/s10052-014-3027-8.

[17] Duffy, L. D.; Sikivie, P.; Tanner, D. B.; Bradley, R. F.; Hagmann, C.; Kinion, D.; Rosenberg, L. J.; Van Bibber, K.; Yu, D. B.; Bradley, R. F. (2006). "High resolution search for dark-matter axions". *Physical Review D* **74**: 12006. arXiv:astro-ph/0603108. Bibcode:2006PhRvD..74a2006D. doi:10.1103/PhysRevD.74.012006.

[18] Asztalos, S. J.; Carosi, G.; Hagmann, C.; Kinion, D.; Van Bibber, K.; Hoskins, J.; Hwang, J.; Sikivie, P.; Tanner, D. B.; Hwang, J.; Sikivie, P.; Tanner, D. B.; Bradley, R.; Clarke, J.; ADMX Collaboration (2010). "SQUID-Based Microwave Cavity Search for Dark-Matter Axions". *Physical Review Letters* **104** (4): 41301. arXiv:0910.5914. Bibcode:2010PhRvL.104d1301A. doi:10.1103/PhysRevLett.104.041301.

22.5. REFERENCES

[19] "ADMX | Axion Dark Matter eXperiment". Phys.washington.edu. Retrieved 2014-05-10.

[20] Phase 1 Results, dated 2006-03-04

[21] Beck, Christian (December 2, 2013). "Possible Resonance Effect of Axionic Dark Matter in Josephson Junctions". *Physical Review Letters* **111** (23): 1801. arXiv:1309.3790. Bibcode:2013PhRvL.111w1801B. doi:10.1103/PhysRevLett.111.231801.

[22] Moskvitch, Katia. "Hints of cold dark matter pop up in 10-year-old circuit". New Scientist magazine (Reed Business Information). Retrieved 3 December 2013.

[23] "First axion results from the XENON100 experiment".*Phys. Rev. D 90, 062009*. 9 September 2014.Bibcode:2014PhRvD..90f. doi:10.1103/PhysRevD.90.062009.

[24] De Angelis, A.; Mansutti, O.; Roncadelli, M. (2007). "Evidence for a new light spin-zero boson from cosmological gamma-ray propagation?". *Physical Review D* **76** (12): 121301. arXiv:0707.4312. Bibcode:2007PhRvD..76l1301D. doi:10.1103/PhysR

[25] De Angelis, A.; Mansutti, O.; Persic, M.; Roncadelli, M. (2009). "Photon propagation and the very high energy gamma-ray spectra of blazars: How transparent is the Universe?". *Monthly Notices of the Royal Astronomical Society: Letters* **394**: L21–L25. arXiv:0807.4246. Bibcode:2009MNRAS.394L..21D. doi:10.1111/j.1745-3933.2008.00602.x.

[26] Chelouche, Doron; Rabadan, Raul; Pavlov, Sergey S.; Castejon, Francisco (2009). "Spectral Signatures of Photon-Particle Oscillations from Celestial Objects". *The Astrophysical Journal Supplement Series* **180**: 1–29. arXiv:0806.0411. Bibcode: doi:10.1088/0067-0049/180/1/1.

[27] Chelouche, Doron; Guendelman, Eduardo I. (2009). "COSMIC ANALOGS OF THE STERN-GERLACH EXPERIMENT AND THE DETECTION OF LIGHT BOSONS". *The Astrophysical Journal* **699**: L5–L8. arXiv:0810.3002. Bibcode: doi:10.1088/0004-637X/699/1/L5.

[28] The International Axion Observatory (IAXO)

[29] Sample, Ian. "Dark matter may have been detected – streaming from sun's core". *www,theguardian.com*. The Guardian. Retrieved 16 October 2014.

[30] Wilczek, Frank (1987-05-04). "Two applications of axion electrodynamics". *Physical Review Letters* **58** (18): 1799–1802. Bibcode:1987PhRvL..58.1799W. doi:10.1103/PhysRevLett.58.1799. PMID 10034541.

[31] Qi, Xiao-Liang; Taylor L. Hughes, Shou-Cheng Zhang; Zhang, Shou-Cheng (2008-11-24). "Topological field theory of time-reversal invariant insulators". *Physical Review B* **78** (19): 195424. arXiv:0802.3537. Bibcode:2008PhRvB..78s5424Q. doi:10.1103/PhysRevB.78.195424.

[32] Franz, Marcel (2008-11-24). "High-energy physics in a new guise". *Physics* **1**: 36. Bibcode:2008PhyOJ...1...36F. doi:10.1103/Physics.
[33] 1.36.

[34] Abe, Nobutaka, Takeo Moroi and Masahiro Yamaguchi; Moroi; Yamaguchi (2002). "Anomaly-Mediated Supersymmetry Breaking with Axion". *Journal of High Energy Physics* **1**: 10. arXiv:hep-ph/0111155.Bibcode:2002JHEP...01..010A.doi:10.1088/1126-6708/2002/01/010.

[35] Hooper, Dan and Lian-Tao Wang; Wang (2004). "Possible evidence for axino dark matter in the galactic bulge". *Physical Review D* **70** (6): 063506. arXiv:hep-ph/0402220. Bibcode:2004PhRvD..70f3506H. doi:10.1103/PhysRevD.70.063506.

[36] Redondo, J.; Raffelt, G.; Viaux Maira, N. (2012). "Journey at the axion meV mass frontier". *Journal of Physics: Conference Series*. **375** 022004. doi:10.1088/1742-6596/375/2/022004 (inactive 2015-03-30).

[37] P. Sikivie,*Dark matter axions*,arXiv.

[38] P. L. Jain, G. Singh, *Search for new particles decaying into electron pairs of mass below 100* MeV/c^2, J. Phys. G: Nucl. Part. Phys., **34**, 129–138, (2007); doi:10.1088/0954-3899/34/1/009, (possible early evidence of 7±1 and 19±1 MeV axions of less than 10^{-13} s lifetime).

[39] Alberto Salvio, Alessandro Strumia, Wei Xue, Alberto; Strumia, Alessandro; Xue, Wei (2014). "Thermal axion production". *Jcap 1401 (2014) 011* **2014**: 011. arXiv:1310.6982. Bibcode:2014JCAP...01..011S. doi:10.1088/1475-7516/2014/01/011.

[40] P. Sikivie, "Dark matter axions and caustic rings"

[41] P. Sikivie (personal website): pictures of alleged triangular structure in Milky Way; hypothetical flow diagram which could give rise to such a structure.

[42] The enigmatic Sun: a crucible for new physics

22.5.2 Journal entries

- Peccei, R. D.; Quinn, H. R. (1977). "*CP* Conservation in the Presence of Pseudoparticles". *Physical Review Letters* **38** (25): 1440–1443. Bibcode:1977PhRvL..38.1440P. doi:10.1103/PhysRevLett.38.1440.

- Peccei, R. D.; Quinn, H. R.; Quinn (1977). "Constraints imposed by *CP* conservation in the presence of pseudoparticles". *Physical Review* **D16** (6): 1791–1797. Bibcode:1977PhRvD..16.1791P. doi:10.1103/PhysRevD.16.1791.

- Weinberg, Steven(1978). "A New Light Boson?".*Physical Review Letters***40**(4): 223–226.Bibcode:1978PhRvL..40..223W. 10.1103/PhysRevLett.40.223.

- Wilczek, Frank (1978). "Problem of Strong *P* and *T* Invariance in the Presence of Instantons". *Physical Review Letters* **40** (5): 279–282. Bibcode:1978PhRvL..40..279W. doi:10.1103/PhysRevLett.40.279.

22.6 External links

- November 24, 2008 article in APS Physics
- January 28, 2007 news article by newscientist.com
- December 06, 2006 news article by physorg.com
- July 17, 2006 news article from Scientific American
- March 27, 2006 news article by PhysicsWeb.org
- November 24, 2004 news article by PhysicsWeb.org
- CAST Experiment
- CAST at MPI/MPE
- CAST at University of Technology Darmstadt
- ADMX at University of Washington

Chapter 23

Mixed dark matter

Mixed dark matter was a theory of dark matter which was promising up to the late 1990s.[1]

"Mixed" dark matter is also called hot + cold dark matter. Hot dark matter has only one known form – neutrinos, although other forms are speculated to exist. Cosmologically important dark matter is now believed to be pure cold dark matter. However, in the early 1990s, the power spectrum of fluctuations in the galaxy clustering did not agree entirely with the predictions for a standard cosmology built around pure cold dark matter. Mixed dark matter with a composition of about 80% cold and 20% hot (neutrinos) was investigated and found to agree better with observations. This model was made obsolete by the discovery in 1998 of the acceleration of universal expansion, which eventually led to the dark energy + dark matter paradigm of this decade.

23.1 References

[1] Andrew R Liddle, David H Lyth (1993). [astro-ph/9304017] Inflation and Mixed Dark Matter Models

23.2 Text and image sources, contributors, and licenses

23.2.1 Text

- **Dark matter** *Source:* https://en.wikipedia.org/wiki/Dark_matter?oldid=671959464 *Contributors:* AxelBoldt, Chenyu, Derek Ross, CYD, BF, Bryan Derksen, The Anome, Tarquin, Taw, XJaM, Arvindn, William Avery, Roadrunner, Mintguy, Bth, Stevertigo, Edward, Nealmcb, Boud, FrankH, Cprompt, DopefishJustin, Bobby D. Bryant, Ixfd64, SebastianHelm, Alfio, CesarB, Looxix~enwiki, Mkweise, William M. Connolley, JWSchmidt, Glenn, Mxn, Charles Matthews, Timwi, Fuzheado, Rednblu, Haukurth, DW40, Dragons flight, Furrykef, Saltine, Dogface, Populus, Jusjih, Finlay McWalter, Bearcat, Robbot, Zandperl, Korath, Nurg, Naddy, Arkuat, Gandalf61, Pingveno, Rursus, Rtfisher, Wereon, Diberri, Adam78, Aasim75, Marc Venot, Ancheta Wis, Giftlite, Graeme Bartlett, Laudaka, Barbara Shack, Herbee, Fropuff, Xerxes314, Dratman, Curps, Joconnor, Jdavidb, Unconcerned, Eequor, Bobblewik, Andycjp, Alexf, Geni, Antandrus, HorsePunchKid, Melikamp, PDH, Rdsmith4, Bosmon, Bbbl67, Icairns, Sam Hocevar, Cynical, Lumidek, Iantresman, Burschik, Joyous!, Adashiel, Urvabara, Discospinster, Rich Farmbrough, Oliver Lineham, Vsmith, Jpk, ArnoldReinhold, Murtasa, D-Notice, JPX7, KaiSeun, SpookyMulder, Bender235, Kjoonlee, Kaisershatner, Pk2000, PsychoDave, RJHall, Mr. Billion, El C, Bletch, PhilHibbs, Shanes, Frankenschulz, Art LaPella, RoyBoy, Themusicgod1, Bobo192, Smalljim, Shenme, Cmdrjameson, Reuben, Kmaguire, I9Q79oL78KiL0QTFHgyc, Zelda~enwiki, Mr. Brownstone, E is for Ian, Jumbuck, Storm Rider, Alansohn, Gary, Anthony Appleyard, Guy Harris, Eric Kvaalen, Arthena, Keenan Pepper, Kocio, Bart133, RPellessier, Benna, ClockworkSoul, Cal 1234, Count Iblis, Guthrie, H2g2bob, Bsadowski1, GabrielF, Pauli133, Leondz, DV8 2XL, Gene Nygaard, Feline1, Oleg Alexandrov, Brookie, Natalya, Flying fish, WilliamKF, Yeastbeast, Mindmatrix, RHaworth, Plek, BillC, JPFlip, Benbest, ^demon, WadeSimMiser, Gxojo, MONGO, Jwanders, Torqueing, ???????, Joke137, Wisq, Christopher Thomas, Palica, Mandarax, RedBLACKandBURN, Aarghdvaark, SqueakBox, Ashmoo, Graham87, Malangthon, Mamling, Jclemens, Drbogdan, Loris Bennett, Rjwilmsi, Lars T., Strait, Patrick Gill, Tangotango, Tawker, Smithfarm, Stevenscollege, Mike Peel, HappyCamper, SeanMack, ScottJ, Krash, Dermeister, Rangek, Madcat87, FlaBot, Ian Pitchford, PlatypeanArchcow, A scientist, Margosbot~enwiki, Gark, Nivix, Gparker, Pathoschild, Gurch, Stevenfruitsmaak, Goudzovski, Tomer Ish Shalom, Smithbrenon, Chobot, Moocha, DVdm, Gwernol, The Rambling Man, YurikBot, Wavelength, RobotE, Koveras, Hairy Dude, Huw Powell, Phmer, Hillman, RussBot, Michael Slone, Ohwilleke, Bhny, JabberWok, GLaDOS, DanMS, Zelmerszoetrop, Eleassar, Merick, Big Brother 1984, NawlinWiki, Alpertron, Długosz, Schlafly, FFLaguna, BlackAndy, Dbmag9, SCZenz, Haoie, Raven4x4x, Ospalh, Durval, Bota47, Supspirit, Pegship, Noosfractal, Charlie Wiederhold, WAS 4.250, Smoggyrob, Reyk, Tvaughan, Joedixon, Eric TF Bat, Emc2, Ilmari Karonen, Allens, Bernd in Japan, InsayneWrapper, Bclayabt, SmackBot, Cubs Fan, Ashill, IddoGenuth, Tomer yaffe, Stellea, InverseHypercube, KnowledgeOfSelf, Clpo13, Nickst, RedSpruce, Nightbat, Doc Strange, Herbm, Edgar181, HalfShadow, Flux.books, Dheerajkakar, Yamaguchi??, Richmeister, Gilliam, Folajimi, Oscarthecat, Skizzik, Kmarinas86, Chris the speller, SuperBuuBuu, Quinsareth, Persian Poet Gal, Sirex98, MalafayaBot, Silly rabbit, Sangrolu, Villarinho, DHN-bot~enwiki, Sbharris, Hongooi, Jdthood, CheerLeone, Gtkysor, Can't sleep, clown will eat me, Nick Levine, Tamfang, Kelvin Case, Vladislav, Vanished User 0001, Rrburke, Jgoulden, Auvii, Krich, Robma, Radagast83, Engwar, Nakon, VegaDark, John D. Croft, Alexander110, KimO, Adrigon, SpiderJon, Ultraexactzz, Zadignose, Tesseran, Byelf2007, L337p4wn, K7lim, SashatoBot, Mchavez, Swatjester, Leftydan6, Minaker, John, Ashoat, Scientizzle, Acitrano, Linnell, JoshuaZ, James.S, JorisvS, Coredesat, Goodnightmush, ICBB, Plunge, JHunterJ, Hypnosifl, Silverthorn, Descubes, Freederick, Dr.K., Vanished user, Iridescent, Darkerprojects, Astrobayes, Newone, MOBle, Igoldste, CapitalR, AGK, Courcelles, Tawkerbot2, Dlohcierekim, Chetvorno, Hammer Raccoon, Owen214, Eastlaw, Peledre, Pukkie, Anakata, Runningonbrains, DKOH, NickW557, Gregbard, MikeWren, Vttoth, Necessary Evil, Ryan, Viciouspiggy, Gogo Dodo, Anonymi, Xxanthippe, A Softer Answer, Odie5533, Tawkerbot4, DumbBOT, Robertinventor, Kozuch, Mtpaley, Philza85, Starship Trooper, UberScienceNerd, Crum375, Thijs!bot, Epbr123, Astroceltica, Passaggio, Barbarina, Mbell, Eugenespeed, N5iln, Mojo Hand, Carlif, Headbomb, Tonyle, Marek69, Lars Lindberg Christensen, OtterSmith, SusanLesch, Mmortal03, Hmrox, Hires an editor, AntiVandalBot, Seaphoto, Orionus, Opelio, Shirt58, Rehnn83, Joehodge, AaronY, Jj137, TTN, Dylan Lake, Chill doubt, Spencer, Yellowdesk, Sniktaw, CPitt76, Gökhan, Jcarter1, Res2216firestar, JAnDBot, Leuko, Husond, MER-C, CosineKitty, Plantsurfer, Mcorazao, Therealintellectual, Folkform, Balbers, 100110100, Autotheist, Wasell, Magioladitis, Bongwarrior, VoABot II, Timothy McVeigh, Charlesrkiss, AuburnPilot, Krkaiser, Mbarbier, Kaivosukeltaja, Foroa, Swpb, Stigmj, T a y l o s, Ekantik, Brusegadi, Bubba hotep, Fabrictramp, Catgut, Lilian.Kaufmann, Zhanghia, Acornwithwings, Vssun, LtHija, Whisky5, DerHexer, Prisca6023, PeteSF, Rickard Vogelberg, NatureA16, DancingPenguin, MartinBot, Schmloof, STBot, Pagw, Fs644, Nikpapag, Anaxial, CommonsDelinker, Jean-Pierre Petit~enwiki, PrestonH, WelshMatt, Chrishy man, Tgeairn, J.delanoy, Pharaoh of the Wizards, Trusilver, Adavidb, Kudpung, Rod57, Arion 3x3, PedEye1, McSly, Tarotcards, Davy p, HiLo48, NewEnglandYankee, Ohms law, Jorfer, Blckavnger, Potatoswatter, KylieTastic, Joshua Issac, Infiniteglitch, Remember the dot, Pitpif, Vanished user 39948282, Neekap, Natl1, Ldebain, BernardZ, SoCalSuperEagle, Squids and Chips, CardinalDan, Idioma-bot, Sheliak, Funandtrvl, Lights, VolkovBot, Craigheinke, Itsfullofstars, ColdCase, Jeff G., JohnBlackburne, Mocirne, AlnoktaBOT, Scikid, Grammarmonger, Leojohns, Larry R. Holmgren, Philip Trueman, TXiKiBoT, Docanton, Authorized User, Theophilus reed, Drestros power, Strichek, MarekMahut, Monkey Bounce, Lradrama, Sintaku, Carillonatreides, Martin451, Broadbot, Wikiisawesome, Mazarin07, Inductiveload, Knightshield, Telecineguy, Spiral5800, Kurowoofwoof111, Greswik, RobertFritzius, SwordSmurf, Falcon8765, Hellothere17, Enviroboy, Littlehollah, Wanchung Hu, Illumini85, SonOfMog Worf, Jazzman123, PGWG, 19merlin69, NHRHS2010, Neparis, Bfpage, S-n-ushakov, SieBot, Calliopejen1, Tresiden, Wibubba48, Tachyonics, Pallab1234, Paradoctor, KGyST, Bentogoa, Jimlester51, Oda Mari, Aaarnooo, Suomichris, Crowstar, PromX1, Lightmouse, Tombomp, Cyberplasm, Diego Grez, Spartan-James, Thinghy, Mygerardromance, Hamiltondaniel, Superbeecat, Denisarona, JL-Bot, Escape Orbit, Starcluster, Troy 07, Atif.t2, ArepoEn, Ak47gforce, Ratemonth, Sfan00 IMG, ClueBot, Phoenix-wiki, GorillaWarfare, The Thing That Should Not Be, ArdClose, Rodhullandemu, Cptmurdok, Drmies, Uncle Milty, Iuhkjhk87y678, MrBosnia, Bhaskarns, Andwor, Ktr101, Excirial, Dombom12, Cromescythe, Barbarinaz, FOARP, Brews ohare, Jotterbot, Iohannes Animosus, R.Andrae, Kentgen1, Ordovico, Mastertek, Rgoogin, Thehelpfulone, 1ForTheMoney, Versus22, Palmer666palmer, PCHS-NJROTC, Burner0718, Pillar of Babel, SoxBot III, Erodium, Vanished user uih38riiw4hjlsd, 1ofhissheep, TimothyRias, Arianewiki1, XLinkBot, DCCougar, Oldnoah, Rror, Gwark, Ost316, Avoided, Webmaster369, Gthomson, Tugrul irmak, Noctibus, Ploversegg, ZooFari, Parejkoj, Tayste, Addbot, Xp54321, Grayfell, Experimental Hobo Infiltration Droid, Willking1979, Some jerk on the Internet, Uruk2008, 04aeverington, DOI bot, Tcncv, Nohomers48, CharlesChandler, Gmeyerowitz, Haasfelix, Download, Proxima Centauri, Ashirgo, RTG, Redheylin, Glane23, Darkmatter654, SamatBot, Nanzilla, Lzkelley, Clone 209, Tassedethe, Numbo3-bot, Peridon, Chinchinthehun, Evildeathmath, Tide rolls, Lightbot, OlEnglish, Qemist, Gail, North Polaris, Legobot, Artichoke-Boy, Luckas-bot, Yobot, WikiDan61, Cosoce, Dov Henis, Aldebaran66, KillYourLove, CzechFalcon, Amble, Mmxx, CinchBug, Perusnarpk, IW.HG, Einstein vs Dark energys, Eric-Wester, Tempodivalse, Synchronism, AnomieBOT, Letuño, Girl Scout cookie, IRP, JackieBot, RBM 72, AdjustShift, Nicolaas Vroom, Henrykan-

23.2. TEXT AND IMAGE SOURCES, CONTRIBUTORS, AND LICENSES

drup, Iluziat, Materialscientist, Dendlai, ImperatorExercitus, The High Fin Sperm Whale, Citation bot, Ternity0127, Maxis ftw, Frankenpuppy, Quebec99, LilHelpa, Aksel89, Xqbot, Stlwebs, Random astronomer, Sionus, Cureden, Jradis1337, Capricorn42, Wperdue, Deleance, Raspw, Tomwsulcer, Magicxcian, AbigailAbernathy, Srich32977, NOrbeck, Artemis6234, Almabot, Abell 1367, Feldhaus, False vacuum, RibotBOT, Waleswatcher, Mikedr, Kongkokhaw, Rvnieuwe, Shadowjams, MeDrewNotYou, A. di M., Peter470, Sageman7, ??, Luminique, Captain-n00dle, Imyfujita, FrescoBot, Andyradke0, Ag allstar, Paine Ellsworth, Originalwana, Styxpaint, Mark Renier, VS6507, PhysicsExplorer, Dbirkhofer, Steve Quinn, Nestlefolife, Adrian Akau, 1414rwbt, SF88, Citation bot 1, Redrose64, DUUJEEGWEEM, Tyler6298, Pinethicket, I dream of horses, Grammarspellchecker, Danlof, 10metreh, Jonesey95, Tom.Reding, Pmokeefe, A8UDI, For.a.limited.time.only, Elentirno, TedderBot, Aknochel, SkyMachine, IVAN3MAN, Kgrad, Nieuwenh, Trappist the monk, Puzl bustr, Fama Clamosa, Domeinthebumhole, Michael9422, UrukHaiLoR, Allen4names, JLincoln, Jeffrd10, Lovemybluetooth, Diannaa, Fastilysock, Innotata, DrCrisp, Whisky drinker, Onel5969, RjwilmsiBot, 5mgoblue5, Blakelewis122, Þorri, Mathewsyriac, Leandro.lelas, Mserard313, Mdznr, Ultima821, EmausBot, Francophile124, Grrow, Super48paul, GoingBatty, Gimmetoo, Solarra, Jmencisom, Slightsmile, Tommy2010, Winner 42, SusanaMultidark, Gocows2, Wikipelli, Serketan, Krifferjel, Zurich Astro, Hhhippo, Mhatthei, Svolin, Micahqgecko, JSquish, Josve05a, Trojanmice, MithrandirAgain, Edwinkaren, Devilaza, Arbnos, Oraclan, Suslindisambiguator, AlbertusmagnusOP, Tolly4bolly, L1A1 FAL, Ancient Anomaly, L Kensington, Maj den, Corabilek, Donner60, Aldnonymous, Ihardlythinkso, RockMagnetist, Terra Novus, TYelliot, DASHBotAV, Kroupap, D Phoesheezey, Travies10, Jxraynor, TheTimesAreAChanging, ClueBot NG, Rich Smith, Afjvanraan, Crystal7878, Catinthehat93, Bped1985, Infinifold, Wiggit002, Jj1236, PapaMike, MonEyshOt42069210, Muon, Esdacosta, Asukite, Masssly, Ph.d Carl edenburgh, Widr, Gavin.perch, Helpful Pixie Bot, Curb Chain, Calabe1992, Bibcode Bot, BG19bot, Dualus, Kishanparekh, Stevenwilkins, NacowY, Cheeseray1, Cyberguy5, Darkmatter adam, Yomomma8102, Hza a 9, Rarelight, Cyberpower678, Cosmologist77, தமென்காரி சுப்பிரமணியன், Dahliamtl, Dodshe, Mark Arsten, Darkmatterotheruniverses, Cadiomals, Achowat, Rolandwilliamson, A2Die, Clint55555, Mgka79, NotEither, BattyBot, Ronin712, Babymushrooms, Davidmexican, Drphilmarshall, Dilaton, Quin71901, U-95, ChrisGualtieri, Npmay, Kvark92, Lukasz.astrus, Ducknish, JYBot, Davidlwinkler, Astrohap, Hunterf12, Caroline1981, Gravityking100, Junavia, Fredrikdn, Jcardazzi, Lugia2453, Wjs64, Andwor42, Frosty, Honneydewp243, Junjunone, DrHowzer73, JustAMuggle, WadiElNatrun, Reatlas, Rfassbind, Acetotyce, I am One of Many, DirkXcal, Melonkelon, Ybidzian, Gig9876, M.ashrafinia, Trolololman12, Ilikedeletingstufffromhere, DavidLeighEllis, Onecreation, Zenibus, Jernahthern, Hipposaregrey, Frinthruit, Stamptrader, Cyberalchemyst, Aaronknowsitall, FelixRosch, Darkmer, Doubleknockout, Monkbot, Wardinstrument, Leegrc, Vikas Rauniyar, Apipia, Upsalla, Jkvaternik, Lol kaptyn troll, Mohammedshukoor, Callum92, Stefania.deluca, Ashweigh, Oldstone James, Astezar, 39Debangshu, YoYoDude012, Anunaki truth, Pyrotle, Tetra quark, Carazmatic, God of matterrr, Silversparkcontributions, Isambard Kingdom, Rizi0909, Anand2202, Kbap2002, Kb2002, DN-boards1, Yohoona, KasparBot, I love trains sooo much, Id6040 and Anonymous: 1209

- **Dark energy** Source: https://en.wikipedia.org/wiki/Dark_energy?oldid=671909269 Contributors:The Anome, Dachshund, Roadrunner, Schew,St evertigo, Nealmcb, Michael Hardy, Tim Starling, FrankH, Bobby D. Bryant, SebastianHelm, Ahoerstemeier, Glenn, Tristanb, Reddi, Wik,DW40, Dragons flight, Anupamsr, Pierre Boreal, BenRG, Jeffq, Donarreiskoffer, Robbot, Zandperl, Korath, Scott McNay, Vespristiano,Peak, Gandalf 61, Rursus, Mlaine, UtherSRG, SC, Mattflaschen, Acm, Ancheta Wis, Giftlite, Graeme Bartlett, Awolf002, Jyril, Art Carl-son, Herbee, Perl, C urps, Henry Flower, Gzornenplatz, Manuel Anastácio, Andycjp, BruceR, LucasVB, Antandrus, Beland, Karol Langner,Kevin B12, Bbbl67, Urvab ara, JimJast, Discospinster, Rich Farmbrough, Pjacobi, Vsmith, D-Notice, Dbachmann, Bender235, Eric Forste,RJHall, JustinWick, Omnibus, El C, Lycurgus, Jomel, Kwamikagami, Frankenschulz, RoyBoy, Stesmo, Reuben, I9Q79oL78KiL0QTFHgyc,Diego Moya, Keenan Pepper, Slugmast er, Axl, Benna, Wtmitchell, RainbowOfLight, Mikeo, Vuo, Freyr, DV8 2XL, Kazvorpal, Falcorian,Velho, Batintherain, Hottscubbard, OwenX, Mindmatrix, FeanorStar7, Velvetsmog, Uncle G, Netdragon, Jeff3000, GregorB, Isnow, SDC,??????, Joke137, Abd, Christopher Thomas, Sn eakums, Dysepsion, BD2412, Doc Savage, Malangthon, RadioActive~enwiki, Drbogdan,Loris Bennett, Rjwilmsi, Strait, TheRingess, Salleman, H appyCamper, Sohmc, Ems57fcva, DonJuan~enwiki, BitterMan, Tomer Ish Shalom,Srleffler, Smithbrenon, CJLL Wright, Chobot, DVdm, Wavelen gth, RobotE, SamuelR, Diliff, Bhny, Stephenb, CambridgeBayWeather, Mer-ick, NawlinWiki, Msikma, FFLaguna, LiamE, SCZenz, FoolsWar, Bo ta47, Rwxrwxrwx, Daniel C, Enormousdude, 2over0, Helge Rosé, Pb30,Dr.alf, Joedixon, Rlove, Geoffrey.landis, Ilmari Karonen, Moonsleeper7, K ungfuadam, Bernd in Japan, GrinBot~enwiki, Treesmill, Smack-Bot, Ashill, Saravask, Bayardo, Tom Lougheed, InverseHypercube, Knowledge OfSelf, Melchoir, J.Sarfatti, Nickst, Silverhand, Edgar181,Vixus, Gilliam, Skizzik, Jlsilva, Andy M. Wang, Tyciol, Sirex98, Oli Filth, DHNbot~enwiki, Sbharris, Colonies Chris, Jdthood, Can'tsleep, clown will eat me, ThePromenader, PoiZaN, Chlewbot, Joema, Cybercobra, Lpgeffen , Rpf, Kendrick7, Byelf2007, Rory096, Boradis,Richard L. Peterson, Xerxesx18, Writtenonsand, JorisvS, Mgiganteus1, Ckatz, Hypnosifl, Mega ne~enwiki, Ryulong, Quaeler, Dan Gluck,Spebudmak, Paul venter, Cxat, UncleDouggie, Courcelles, Tawkerbot2, JRSpriggs, Atomobot, Trevor.to mbe, JForget, CRGreathouse, Lavat-eraguy, Nadyes, Mlsmith10, Arnavion, Logical2u, Rob Maguire, Cydebot, Stebbins, Gmusser, 879(CoDe), R racecarr, Soetermans, MichaelC Price, Chrislk02, Kozuch, Landroo, Thijs!bot, Headbomb, Marek69, Electron9, Second Quantization, Chris goule t, Davidhorman, Turelli,Dawnseeker2000, AntiVandalBot, Orionus, Gnixon, Fayenatic london, Tim Shuba, Empyrius, Archmagusrm, AstroPaul,B agster, JAnDbot,Carl1011, Davewho2, MER-C, CosineKitty, Rkomatsu, Michael Wood-Vasey, Felix116, Acroterion, Bongwarrior, VoABot II, Tripbeetle,LordCémOnur, Seleucus, Kevinwiatrowski, Ours18, DerHexer, Nevit, Simplizissimus, NatureA16, Johann1870, Jimmilu, ARC Gritt,Ni kpa-pag, TechnoFaye, Christian424, Tgeairn, J.delanoy, Trusilver, Maurice Carbonaro, Natty4bumpo, Komowkwa, OttoMäkelä, Jlechem, Tsuite,S JP, Videokunst~enwiki, Malerin, Jorfer, Potatoswatter, Cmichael, DorganBot, Jcmargeson, Ja 62, JHussein, Jjabellar, Sheliak, Johnassas-sin, Car ibbean H.Q., VolkovBot, ColdCase, JohnBlackburne, D A Patriarche, AlnoktaBOT, Fences and windows, Philip Trueman, Darren22,HowardFram pton, TXiKiBoT, Oshwah, Zanardn, Someguy1221, Oxfordwang, Jackfork, UnitedStatesian, Mazarin07, Venny85, Goaliemas-ter121, SwordSmur f, RayNorris, Fourthark, Wanchung Hu, Obsidianmile, Radical Robert, Noncompliant one, Donauland~enwiki, PlanetStar,TrulyBlue, Murad.Shibli, Likebox, Flyer22, Hotdiggity, Avidallred, Faradayplank, Poindexter Propellerhead, OKBot, Aquijex, Loren.wilton,Martarius, BillWilliam, ClueBo t, Dead10ck, The Thing That Should Not Be, Rodhullandemu, SuperHamster, Andwor, Tms9, Jusdafax, Darulz07, Barbarinaz, Kentgen1, Razor flame, Stevecrye, AC+79 3888, Pillar of Babel, TimothyRias, Gwark, Ost316, PL290, MikeSmith10,Parejkoj, Andreaprins, Dgirl1723, HexaCh ord, D.M. from Ukraine, Addbot, Gravitophoton, Uruk2008, DOI bot, Nernom, LaaknorBot, Ad-fellin, Glane23, Delaszk, ChenzwBot, Sophia889 1, Combatman~enwiki, Craigsjones, Arbitrarily0, Gurusoft2, Cosmos72, Luckas-bot, Yobot,Cosoce, Systemizer, Aldebaran66, Fulcanelli, Amble,A nomieBOT, Iluziat, Materialscientist, Citation bot, Icosmology, ArthurBot, Xqbot, S h i v a (Visnu), Sionus, Drilnoth, Wperdue, Tomwsulcer, BLP-ou trageous move logs, ProtectionTaggingBot, Mathonius, Shadowjams, Finncarey,PrimeMatter, FrescoBot, Tobby72, Sławomir Biały, Zero Thrust, Kvgyarmati, Woodingdean, Alpha plus (a+), Citation bot 1, Redrose64,Pinethicket, I dream of horses, Jonesey95, Three887, Tom.Reding,Sha hidur Rahman Sikder, Efficiency1101e, Casimir9999, Aknochel,IVAN3MAN, Meier99, BradTheBadWiki, TADEET, Jordgette, Heurisko,Mich ael9422, Adi4094, Earthandmoon, Wellsmax, RjwilmsiBot,Alph Bot, EmausBot, Grrow, Quantanew, RA0808, Slightsmile, Italia2006, NicatronTg , H3llBot, Suslindisambiguator, Paulstarpaulstar, Frigotoni, Colin.campbell.27, Iiar, HCPotter, Tunborough, RockMagnetist, Herk1955, Deathglass, DASHBotAV, Fire Vortex, Mjbmrbot, Yceren Loq, ClueBot NG, Ccalen, Chester Markel, Matias Pocobi, Jj1236, Frietjes, Helvitica Bold, Curb Chain, Bibcode Bot, BG19bot, Gordonben, Cheeseray1, FiveColourMap, Hippokrateszholdacskai, Yizlpku, Snow Blizzard, Gerhardtschmerhardt, Migrainus, Mcspaans, Szczurcq, Unclejoe0306, Akshay Lattimardi, CityOfUr, Jcardazzi, Wjs64, JustAMuggle, WorldWideJuan, Epicgenius, Yheyma, MiceEater, LindaYeah, DavidLeighEllis, Federicoturner, Babitaarora, Isateach, Onecreation, Prokaryotes, BerdanII, Anrnusna, Stamptrader, Monkbot, Mlsmith55, Haxxorz596, THemanRE$%S23, Jnojha007, Richard.drapeau, MF22, Larsyxa, EpicLX, Tibenas, ScrapIronIV, 39Debangshu, Anunaki truth, Tetra quark, Isambard Kingdom, Anand2202, Jman135, KasparBot, ShankZeTank, Tgorewic, Esadri21, Phseek, Buckbill10 and Anonymous: 512

- **Big Bang nucleosynthesis** *Source:* https://en.wikipedia.org/wiki/Big_Bang_nucleosynthesis?oldid=670687654 *Contributors:* Vicki Rosenzweig, AstroNomer~enwiki, Roadrunner, Space Cadet, PaulDSP, Bueller 007, LittleDan, Schneelocke, Reddi, Phil Boswell, Korath, Sanders muc, Peak, Rursus, Harp, Art Carlson, Herbee, Anville, Dmmaus, Eroica, JoJan, Karol Langner, Deglr6328, Pjacobi, Vsmith, SpookyMulder, Brian0918, RJHall, Pilatus, Art LaPella, Army1987, I9Q79oL78KiL0QTFHgyc, GeorgeStepanek, Jheald, Oleg Alexandrov, Camw, BlaiseFEgan, Wdanwatts, Joke137, Grundle, Qwertyus, Rjwilmsi, Oo64eva, Mishuletz, Goudzovski, Phoenix2~enwiki, Chobot, Amaurea, Rmbyoung, YurikBot, Sir48, Fobos~enwiki, Uber nemo, Enormousdude, Modify, Ilmari Karonen, Cmglee, SmackBot, Dauto, Bluebot, Kashami, Silly rabbit, Sbharris, Colonies Chris, Ligulembot, GodBlessTheNet, JorisvS, Stevebritgimp, Getjonas, Mssgill, George100, Vyznev Xnebara, Jsd, Gregbard, Thijs!bot, Markus Pössel, Headbomb, John254, Peter Gulutzan, DPdH, Uruiamme, Orionus, Nipisiquit, VoABot II, ThoHug, LorenzoB, DerHexer, Geboy, MartinBot, Pagw, Peter Chastain, Eliz81, BobEnyart, Vegasprof, Wesino, Biglovinb, Juliancolton, Sheliak, VolkovBot, TXiKiBoT, Calwiki, Thrawn562, OlavN, Broadbot, UnitedStatesian, BotKung, Pamputt, SwordSmurf, Newsaholic, Gdude95, Ashdabash, SieBot, Jdaloner, Escape Artist Swyer, ClueBot, CLCalver, ChandlerMapBot, NuclearWarfare, DumZiBoT, TimothyRias, Chanakal, Addbot, Shiba6, Uruk2008, DOI bot, Njaelkies Lea, Yobot, AnomieBOT, Tad Lincoln, Rainald62, Physdragon, Citation bot 1, Gil987, Dogaru Florin, Pinethicket, I dream of horses, Edderso, Jonesey95, Tom.Reding, Pmokeefe, Footwarrior, Double sharp, RobertMfromLI, RjwilmsiBot, DASHBot, XinaNicole, GenyAncalagon, Ad3l, Rcsprinter123, Zueignung, ClueBot NG, Kevin pirotto, Bibcode Bot, Krastanov, AvocatoBot, Zedshort, Mrt3366, Garuda0001, Wjs64, James floodhall, Frosty, Rsenk326, Jwratner1, HamiltonFromAbove, Anrnusna, Monkbot, Sofia Koutsouveli, ComicsAreJustAllRight, Tetra quark, Phseek, TychosElk and Anonymous: 103

- **Abundance of the chemical elements** *Source:* https://en.wikipedia.org/wiki/Abundance_of_the_chemical_elements?oldid=671667208 *Contributors:* Bryan Derksen, Timo Honkasalo, The Anome, Dcljr, Cherkash, Smack, SEWilco, Pakaran, Jni, Donarreiskoffer, Arkuat, Rursus, Al-khowarizmi, MisfitToys, Karol Langner, FT2, Pjacobi, Vsmith, Chad okere, Paul August, Blade Hirato~enwiki, ESkog, Art LaPella, Kjkolb, Pearle, Andrewpmk, Flying fish, Rodii, Woohookitty, Oliphaunt, Rend~enwiki, TotoBaggins, CharlesC, Magister Mathematicae, BD2412, Eteq, DePiep, Drbogdan, Saperaud~enwiki, Rjwilmsi, R.e.b., AndyKali, Alphachimp, Flcelloguy, Roboto de Ajvol, Wavelength, Russoc4, Rsrikanth05, Joel7687, Syrthiss, Brainwad, ASmartKid, NielsenGW, Ordinary Person, Cmglee, Itub, SmackBot, Hydrogen Iodide, Elminster Aumar, Onebravemonkey, Michbich, Bluebot, Sbharris, Darth Panda, Tamfang, Polonium, DMacks, Nishkid64, Titus III, JorisvS, JHunterJ, Hypnosifl, Geologyguy, Mdanziger, Dan Gluck, Wizard191, Iridescent, PavelCurtis, Vaughan Pratt, Thermochap, Runningonbrains, Korandder, Sopoforic, Xminivann, Pcu123456789, Headbomb, Marek69, Porqin, Paul from Michigan, Gdo01, Frankie816, Drollere, Talon Artaine, PaulTaylor, SBarnes, ChemNerd, J.delanoy, GoatGuy, AntiSpamBot, Warut, Nwbeeson, Rex07, Atropos235, VoidLurker, Joeinwap, 28bytes, Holme053, Philip Trueman, Zidonuke, McM.bot, Billinghurst, Falcon8765, Insanity Incarnate, Junkinbomb, Czmtzc, Minion87, Jauerback, Orthorhombic, Scorpion451, R0uge, Lightmouse, Tombomp, Nn123645, Nergaal, Denisarona, ClueBot, PipepBot, The Thing That Should Not Be, CounterVandalismBot, DragonBot, Sun Creator, LarryMorseDCOhio, SchreiberBike, Count Truthstein, DumZiBoT, Addbot, Foggynight, Roentgenium111, Substar, Marx01, CanadianLinuxUser, Favonian, LinkFA-Bot, Tide rolls, ⁇, Arbitrarily0, Luckasbot, Yobot, AnomieBOT, Floozybackloves, Materialscientist, Citation bot, Frankie0607, Riventree, Citation bot 1, Pinethicket, I dream of horses, Tom.Reding, RedBot, Mikespedia, Tim1357, Double sharp, Trappist the monk, Vrenator, Pbrower2a, DARTH SIDIOUS 2, EmausBot, Katherine, ScottyBerg, ZéroBot, MacHyver, Wayne Slam, ClueBot NG, KlappCK, Widr, Reify-tech, Helpful Pixie Bot, Bibcode Bot, Churchgoer251, Glacialfox, Wastednow, Dexbot, EvergreenFir, Abitslow, Trackteur, IiKkEe, Firetraner, Sleepneeder and Anonymous: 194

- **Observable universe** *Source:* https://en.wikipedia.org/wiki/Observable_universe?oldid=671891453 *Contributors:* Boud, Michael Hardy, Minesw, Marteau, Ciphergoth, Doradus, BenRG, Nurg, SoLando, Clementi, Giftlite, DocWatson42, DavidCary, 0x0077BE, Barbara Shack, No Guru, Kmote, Beland, Karol Langner, Latitude0116, Mschlindwein, TJSwoboda, Rich Farmbrough, Pjacobi, Pie4all88, R6144, Roodog2k, Dbachmann, Bender235, Ben Standeven, El C, Mytg8, CDN99, Väsk, Davidruben, Viriditas, Cmdrjameson, Ziggurat, I9Q79oL78KiL0QTFHgyc, Quaoar, Alansohn, Hadlock, ProhibitOnions, Mmxbass, WilliamKF, Zanaq, Fred Condo, Mindmatrix, LOL, Chris Mason, Duncan.france, Matt Mahoney, Jleon, Thruston, GregorB, Aarghdvaark, SqueakBox, Rnt20, Graham87, BD2412, Drbogdan, Rjwilmsi, Koavf, Zbxgscqf, Strait, Woodsja, Spott, Mike Peel, Drrngrvy, 01101001, Kolbasz, Choess, Mathrick, Fresheneesz, Goudzovski, Lord Patrick, Wavelength, Vedranf, Hairy Dude, Jimp, Wolfmankurd, RussBot, Xoloz, Mattgibson, Koffieyahoo, NawlinWiki, Trovatore, Sir48, Fulltruth, Dbfirs, Falcon9x5, Cambion, WAS 4.250, 2over0, Rpvdk, Endomion, Esprit15d, CWenger, Fram, Caco de vidro, Kungfuadam, Lengau, Serendipodous, Robertd, Sardanaphalus, SmackBot, Unschool, Ashill, Zazaban, Melchoir, McGeddon, Eskimbot, MQQ, Mad Bill, Papa November, Hibernian, Nbarth, Colonies Chris, Jdthood, JGXenite, Scwlong, Fotoguzzi, Vanished User 0001, EOZyo, Theanphibian, Radagast83, Kntrabssi, Trieste, Adrigon, Mostlyharmless, ArglebargleIV, Thanatosimii, Siva1979, JunCTionS, Jpagel, UberCryxic, JorisvS, RomanSpa, Ckatz, Booksworm, Mr Stephen, Hypnosifl, Quarty~enwiki, Autonova, Alan.ca, Ossipewsk, K, NEMT, Astrobayes, Paul venter, Newone, Twas Now, LethargicParasite, CapitalR, Richard75, JRSpriggs, CmdrObot, Olaf Davis, Ruslik0, Icarus of old, AndrewHowse, Treybien, Michael C Price, Tawkerbot4, Casliber, Thijs!bot, Barticus88, Jwt015, Keraunos, Headbomb, Pmrobert49, Afabbro, Elert, Anttilk, G Rose, Cdunn2001, Neitsa, Gumby600, VoABot II, Alienpeach, Harelx, Nyttend, Avicennasis, Sam Medany, JJ Harrison, Zepheriah, StuFifeScotland, Wikianon, Robin S, Flaming Ferrari, NatureA16, Rrostrom, Maurice Carbonaro, Natty4bumpo, NerdyNSK, Potatoswatter, Greatestrowerever, Inwind, Dorftrottel, Funandtrvl, VolkovBot, Haade, LeilaniLad, Aliento, DarkShroom, Kww, Anonymous Dissident, Macslacker, Wiendietry~enwiki, Martin451, Broadbot, Israeld, Robert1947, SheffieldSteel, SwordSmurf, Parsifal, Hellothere17, Kbenoit, DarthBotto, SieBot, Paradoctor, Hertz1888, Dawn Bard, Wing gundam, Oda Mari, Heikki m, Faradayplank, Beast of traal, Jdaloner, Lightmouse, Sunrise, Adamtester, Cosmo0, Anchor Link Bot, Roded86400, Gwpray, Tomahiv, Startswithj, Soporaeternus, Sfan00 IMG, Madang1965, Ronald12, Myqueminnetz, Agge1000, Chimesmonster, Excirial, PixelBot, Sun Creator, Brews ohare, NuclearWarfare, M.O.X, Pyrofork, BOTarate, Panos84, Versus22, Vanished User 1004, DumZiBoT, XLinkBot, MystBot, Infonation101, Maldek, Addbot, Mathbuddy8888, Darko.veberic, Samiswicked, Fieldday-sunday, Zarcadia, Download, SpBot, Romulocortezdepaula, Numbo3-bot, OlEnglish, Flash.starwalker, Luckas-bot, Justintan88, Aldebaran66, Azcolvin429, AnomieBOT, UnitarianUniversalism, Materialscientist, Citation bot, Rodhas, YouthoNation, Plastadity, Jsharpminor, Martnym,

Sirmc, Mlpearc, Gap9551, Jhbdel, False vacuum, Appple, Omnipaedista, Kyng, Dngnta, Mnmngb, FrescoBot, Paine Ellsworth, KTParadigm, Citation bot 1, Pinethicket, Tom.Reding, Full-date unlinking bot, SkyMachine, Orenburg1, Xeracles, Lam Kin Keung, OnesimusUnbound, Zachareth, Earthandmoon, Sideways713, RjwilmsiBot, WildBot, Troy wahl, EmausBot, Ge3lan, WikitanvirBot, Preceding easy, Treymix, Silverlight2010, Einkleinestier, Italia2006, Chasrob, Yiosie2356, Brandmeister, JanAson, CountMacula, ChuispastonBot, Just granpa, Mjbmrbot, Alcazar84, ClueBot NG, Natey7, RaptorHunter, Gilderien, Lepota, Bulldog73, Physics is all gnomes, Jj1236, Machina Lucis, Helpful Pixie Bot, Bibcode Bot, Astrofan7, BG19bot, Jwchong, Vagobot, Ugncreative Usergname, Physicssmart, Harizotoh9, BattyBot, NOWEASELWORDS, YFdyh-bot, Khazar2, Jimjohnson2222, Thehoopisonfire, MacGreenbear, Rogerstrolley, Alysonbloom, Reatlas, Jamesmcmahon0, Ruwshun, Unmismoobjetivo, Tango303, NorthBySouthBaranof, PirtleShell, Jwratner1, Johndric Valdez, Uclmaps, Anrnusna, Potterbaby, Sjzaslaw, Proref2, Monkbot, Sofia Koutsouveli, DSCrowned, SkyFlubbler, Gabe schulhof, Fimatic, Tetra quark and Anonymous: 238

- **Lambda-CDM model** *Source:* https://en.wikipedia.org/wiki/Lambda-CDM_model?oldid=670675110 *Contributors:* Bryan Derksen, The Anome, Roadrunner, Boud, Michael Hardy, Dcljr, Charles Matthews, Timwi, Forseti, Gandalf61, Wjhonson, Giftlite, Andycjp, Pjacobi, Vsmith, Jonathanischoice, AdamSolomon, Art LaPella, I9Q79oL78KiL0QTFHgyc, Diego Moya, Plumbago, Ceyockey, Joke137, Rnt20, Drbogdan, Zbxgscqf, Mike Peel, Bubba73, Phoenix2~enwiki, Karch, YurikBot, Vuvar1, Gadget850, CharlesHBennett, Caco de vidro, McGeddon, Bluebot, Jdthood, Hve, Yannick Copin, JorisvS, Beetstra, Hypnosifl, Spebudmak, Petr Matas, CmdrObot, Dr.enh, Thijs!bot, Headbomb, Vertium, Peter Gulutzan, Escarbot, Rico402, Huttarl, Drollere, Yobol, DAID, KylieTastic, STBotD, Sheliak, VolkovBot, RedAndr, MariAlexan, BotKung, SwordSmurf, Catdogqq, SieBot, Hertz1888, Droog Andrey, BartekChom, IlkkaP, Sunrise, Coldcreation, Duae Quartunciae, Astrohou, DragonBot, Telekenesis, Brews ohare, Cenarium, Scog, Ich42, Parejkoj, Addbot, LaaknorBot, Yobot, Ptbotgourou, Amirobot, Amble, Azcolvin429, AnomieBOT, Citation bot, Xqbot, Dendropithecus, Omnipaedista, Mnmngb, FrescoBot, Craig Pemberton, SF88, Tom.Reding, Pmokeefe, Puzl bustr, Sehatfield, Dr. Salvia, Earthandmoon, RjwilmsiBot, Ripchip Bot, Italia2006, Suslindisambiguator, Brandmeister, One.Ouch.Zero, Senator2029, Milk Coffee, Fire Vortex, Jj1236, Bibcode Bot, Technical 13, BG19bot, Flekkie, Harizotoh9, SoylentPurple, Khazar2, Wjs64, Junjunone, Thewarriltonsiegedoc, Prokaryotes, Orrerysky, Sjzaslaw, Monkbot, Unatnas1986, Sofia Koutsouveli, Verdana Bold, Mof-tan, Tetra quark, TychosElk and Anonymous: 67

- **Mass–energy equivalence** *Source:* https://en.wikipedia.org/wiki/Mass%E2%80%93energy_equivalence?oldid=671240907 *Contributors:* Tarquin, Stevertigo, Edward, Ubiquity, Patrick, Michael Hardy, SebastianHelm, Ahoerstemeier, Darkwind, Julesd, Charles Matthews, Stone, Kbk, Andrewman327, Evgeni Sergeev, Doradus, Tpbradbury, Dragons flight, McKay, AnonMoos, Eugene van der Pijll, BenRG, Twang, Robbot, Owain, ZimZalaBim, Gandalf61, Postdlf, Tobias Bergemann, Enochlau, Jimpaz, Giftlite, Muzzle, C2357, Kpalion, Jackol, ConradPino, Antandrus, OverlordQ, Thorwald, Mike Rosoft, Discospinster, Rich Farmbrough, FT2, Pjacobi, Vsmith, Ponder, SpookyMulder, Chadlupkes, Bender235, ESkog, ZeroOne, JustinWick, Ben Webber, El C, Carlon, Lycurgus, Haxwell, Causa sui, Bobo192, Longhair, Savvo, Jojit fb, Kjkolb, Officiallyover, Yalbik, Landroni, Alansohn, Gary, Anthony Appleyard, Tek022, Riana, Ashley Pomeroy, Scarecroe, Mysdaao, Stillnotelf, Bart133, Melaen, Clubmarx, Danhash, Count Iblis, RainbowOfLight, Dirac1933, Mikeo, H2g2bob, Bsadowski1, GabrielF, Gene Nygaard, Ron Ritzman, Zntrip, Mindmatrix, StradivariusTV, Kzollman, Robert K S, ^demon, WadeSimMiser, Qwertyman~enwiki, GregorB, Zzyzx11, Mandarax, Jclemens, Rjwilmsi, Koavf, Jake Wartenberg, Vary, Bob A, SMC, Bfigura, Yamamoto Ichiro, FayssalF, Wikiliki, Eubot, RobertG, Gurch, Fresheneesz, Chobot, DVdm, Bgwhite, McGinnis, Vyroglyph, Wavelength, Hairy Dude, Huw Powell, Sarranduin, Zafiroblue05, Ericorbit, Bhny, Gaius Cornelius, CambridgeBayWeather, Bovineone, Thane, NawlinWiki, Arichnad, JoeBruno, RazorICE, Dureo, RUL3R, Dbfirs, BOT-Superzerocool, Woscafrench, Poochy, WAS 4.250, Enormousdude, 21655, 2over0, Zzuuzz, Dspradau, Staxringold, RG2, NeilN, Finell, Sardanaphalus, SmackBot, Unschool, Ashill, InverseHypercube, Hydrogen Iodide, McGeddon, Shoy, Frasor, ASarnat, Canthusus, Gilliam, Skizzik, GwydionM, Chris the speller, Bluebot, JCSantos, Rakela, Oli Filth, Miquonranger03, Silly rabbit, Sbharris, Colonies Chris, Darth Panda, Derekt75, Can't sleep, clown will eat me, Timothy Clemans, Onorem, Rrburke, Addshore, Stevenmitchell, Jmnbatista, Robma, Cybercobra, Nakon, Andrew c, DMacks, Acdx, Aftertheend, Ohconfucius, Angela26, SashatoBot, Lambiam, ArglebargleIV, Kuru, Mgiganteus1, Zarniwoot, Aleenf1, Ben Moore, 16@r, Smith609, Slakr, MarcAurel, Dicklyon, Waggers, Spiel496, Cbuckley, Dan Gluck, Iridescent, Joseph Solis in Australia, R~enwiki, Blehfu, Courcelles, Achoo5000, JForget, Sakurambo, CmdrObot, Ninetyone, Editorius, Green caterpillar, Gdbiederman, Cydebot, Kanags, MC10, Subwoofer, Gogo Dodo, Yuzz, Tkynerd, Edgerck, Capedia, Christian75, DumbBOT, DarkLink, JSal, Malleus Fatuorum, Epbr123, Barticus88, Biruitorul, Pajz, Headbomb, Marek69, Davidlawrence, John254, NorwegianBlue, James086, X201, Davidhorman, Thljcl, D.H, Klausness, Ellid021, Mentifisto, Majorly, Seaphoto, Orionus, Elmoosecapitan, Smittycity42, Edokter, TimVickers, Joe Schmedley, Naveen Sankar, Farosdaughter, Spencer, Spartaz, Gökhan, Res2216firestar, DOS-Guy, JAnDbot, Aheyfromhome, MER-C, CosineKitty, Txomin, Thenub314, Andonic, Hut 8.5, Cvkline, Casmith 789, Magioladitis, Puellanivis, Pedro, Bongwarrior, VoABot II, Sekfetenmet, Sikory, Rimibchatterjee, Jatkins, Twsx, DAGwyn, Tristan Horn, Zanibas, Fabrictramp, Catgut, Indon, Crunchy Numbers, JJ Harrison, 28421u2232nfenfcenc, DerHexer, JNF Tveit, An Sealgair, G.A.S, MartinBot, Ariel., Lelandrb, Sm8900, Keith D, R'n'B, Dgcaste, CommonsDelinker, Onixz100, Jaredroussel, J.delanoy, Pharaoh of the Wizards, GoatGuy, C. Trifle, Maurice Carbonaro, AngleWyrm, DD2K, Lantonov, Ajmint, Dispenser, Nsigniacorp, Uranium grenade, Greater mind, NewEnglandYankee, Rominandreu, Zojj, DavidCBryant, Remember the dot, JohnOdhner, Barraki, WillPF, Scott Illini, JavierMC, Useight, Halmstad, SoCalSuperEagle, Funandtrvl, X!, VolkovBot, Jeff G., AlnoktaBOT, TXiKiBoT, Oshwah, Drhtl, NPrice, Mieszko the first, Sintaku, Graham Wellington, Martin451, Solo1234, Jackfork, UnitedStatesian, Madhero88, Enigmaman, Jacobandrew2012, Antixt, Enviroboy, Insanity Incarnate, Dufo, Why Not A Duck, HiDrNick, Logan, Tvinh, Vegardo, Xgllo, SieBot, Ivan Štambuk, Madman, Timb66, Euryalus, Ziolkovsky, VVVBot, Arpose, Triwbe, Jason Patton, RatnimSnave, AvengedSevenfold00, Keilana, Happysailor, Likebox, Flyer22, Qst, Le Pied-bot~enwiki, Aly89, Oxymoron83, Robertfreemanfund, Harry~enwiki, Beast of traal, GaryColemanFan, Steven Zhang, Lightmouse, Wackedout, Techman224, OKBot, Onopearls, Coldcreation, Sapoty, Susan118, Ascidian, Dolphin51, Denisarona, Bschaeffer~enwiki, Loren.wilton, Elassint, ClueBot, Orangedolphin, Chalmersss, Binksternet, Jackollie, The Thing That Should Not Be, NunchuckJack, VQuakr, Mild Bill Hiccup, MathGeek123, Polyamorph, Chwilliamson, Boing! said Zebedee, Yamakiri, Unitfreak, Adamslattery54, EricTN, Estevoaei, Blanchardb, NakedEye71, Agge1000, Dazzafar, Oxnard27, Hi777, Paulcmnt, Excirial, -Midorihana-, Jusdafax, Leonard^Bloom, Sgroupace, Brews ohare, NuclearWarfare, LongLiveRock72, Animalality, Lumpy27, Cartledge555, Noosentaal, Dekisugi, GluonBall, Mikaey, Melkijad, La Pianista, Fernandinho1000, Thingg, Zeekyb00gydoog wrongwhom7, Versus22, LieAfterLie, Djk3, AC+79 3888, Vanished user uih38riiw4hjlsd, Ashish16328, Bgeelhoed, BendersGame, XLinkBot, Hotcrocodile, Jovianeye, Rror, Dthomsen8, Avoided, WikHead, Thewho65, Alexius08, Kodster, Tayste, Xp54321, Cxz111, JPINFV, Willking1979, DOI bot, Tcncv, Captain Ref Desk, Friginator, Valejo10005, DougsTech, Ymath, Ronhjones, CanadianLinuxUser, Jaeger123, NjardarBot, TomTyldesley, OliverTwisted, Yujie1, Download, Brett37, LaaknorBot, The yoster1, Epik phale, Glane23, Favonian, Playerace5, Jaf24, LinkFA-Bot, Jasper Deng, 5 albert square, Amoskowitz, Agathor222, Squandermania, Botbotkins, Slushieeater, Dayewalker, Bigzteve, Tide rolls, Lightbot, Cesiumfrog, Teles, Gail, Stevek 85, CountryBot, StarLight,

Angrysockhop, Luckas-bot, Yobot, Fraggle81, JakeH07, Paepaok, Spysdudeqazwsx, THEN WHO WAS PHONE?, Gunnar Hendrich, Myktk, Prometheusindisguise, Backslash Forwardslash, AnomieBOT, VanishedUser sdu9aya9fasdsopa, FatAndSassy4, Joule36e5, Jim1138, IRP, Piano non troppo, A09fa2, Kingpin13, ConsciousUniverse, Fredd2374, Visiting Guest, Materialscientist, Are you ready for IPv6?, Citation bot, Tonytony9, ArthurBot, Monkeybutts93, Pyrodude431, Gravityforce, Rittigai, Xqbot, Aman2007007, Ertebatbama, Drilnoth, Acebulf, Dandelion Jane, Pvkeller, DSisyphBot, Nitrxgen, Jeffwang, Runaway9995, Laa Careon, Srich32977, Coretheapple, Domjm, FrånKlarhetTillKlarhet, UNCLE ROCKA, Pmlineditor, Loosah, Omnipaedista, Anhydrobiosis, Elbigger1, RibotBOT, Amaury, Ace111, Paul maul123, Aaron35510, IShadowed, Mikesoc28, DanielDisastrous, N419BH, Shadowjams, A. di M., Databytecorp, CES1596, Hegaldi, FrescoBot, Paine Ellsworth, Ryryrules100, Dogposter, Maxamilliona, Yaser soleimani, Alex-c-johnson, Rymmen, Gfjohnsn, Evalowyn, PirateSmackK, Masked Turk, Citation bot 1, Aditya narain srivastava, Dterp, Pinethicket, I dream of horses, 10metreh, Jonesey95, A412, Tom.Reding, Mekeretrig, Triplestop, Random editor, RedBot, Artem Korzhimanov, Rausch, Reconsider the static, IVAN3MAN, Meier99, Odenjr, Nishant1997nishu, Vrenator, Anti-Nationalist,Mr.98, Reaper Eternal, Xobekil, Linguisticgeek, Tbhotch, Lolmanz, DARTH SIDIOUS 2, Whisky drinker, Harryking177, Mineadwaly, RjwilmsiBot, Skipstar7, Rajettan, WildBot, Wintonian, Slon02, Spamking93, EmausBot, Williamthomasandrews, Immunize, Gfoley4, Mikey12348, Britannic124, RA0808, Twisindia, Lja514, Havoc606 Wakka-Pakka, Blink'em, Youncej, Zeusiscute, Bjwill13, Wikipelli, Sepguilherme, JSquish, ZéroBot, Cogiati, Daonguyen95, 1howardsr1, SomDood, A930913, Quondum, WinstonsDomain, SporkBot, ThatBird, Wayne Slam, Tercerista, The Talking Toaster, Maschen, Donner60, Carmichael, Dmyasish, RockMagnetist, TYelliot, Llightex, Scooter12345, MFDMICRO, Xanchester, ClueBot NG, Tpain1776, BriiGarcia, Kenny90655, Caute AF, Mkoconnor, SusikMkr, Carbon editor, Cntras, Nikhileditor, Widr, Antiqueight, Eimeardoneapoo, NO FUSE HERE, Helpful Pixie Bot, Martin Berka, Smoothieking7, Bibcode Bot, BG19bot, Albert012101, Xtfcr7, Hallows AG, Who.was.phone, Mark Arsten, Zachaysan, Ivor Ludlam, F=q(E+v^B), Eelkeher, ███ █████, Harizotoh9, Tremere2, Lee Kyle Jay, Shawn Worthington Laser Plasma, SirTobiasII, Zeegeorge, Mitch H. Waylee, RudolfRed, BattyBot, N lasters, Timothy Gu, NatalieAvigailL, Smartdonkey, Khazar2, EuroCarGT, Dexbot, Physicsmaster 1 3, Jayster294, Josepht404, Thebaconhawk, The Nuke, Numbermaniac, Lugia2453, Mihir John, Cobalt174, Sui docuit, Adwait.a.raste, Reatlas, Thebaconhawk69, WorldWideJuan, Liugaila, CsDix, Curtis P.... Heimberg, Eyesnore, PhantomTech, Zelliej, Dr DonZi, NeapleBerlina, Nigellwh, The Herald, Surfscoter, Ginsuloft, 8i347g8gl, DavRosen, Physikerwelt, Punit chaudhary, Frinthruit, Tssbender, Monkbot, Johnnyideal, Anuvarshanw, Gianluca Di Fiore, Lomtucas, Akifumii, CoolOppo, Asdklf;, TranquilHope, JMP EAX, Narky Blert, Samangivian, Arisht Aveiro, Beckzilla178, Infogamer12345, Videogamefreak43rv, Knaveknight, Karlswag, Brobroswagens, Alleballeeeeeeeeee, Corsairio, Tetra quark, Samfart20, Swaglord908199920088, SWAGlordLOLZ, Praneeth Sarvade, Pac6mon9, Zarrus.rasaili94, Kdkddkkdkeksksjdj, Adeptussoratis stormlord1, SayanChakraborty1234567890, Brekkestewart, Amrit kushwaha and Anonymous: 1030

- **Gravitational lens** *Source*:https://en.wikipedia.org/wiki/Gravitational_lens?oldid=666089405 *Contributors*:Bryan Derksen, AstroNomer~enwi,--April, Michael Hardy, EddEdmondson, Arpingstone, Looxix~enwiki, Glenn, Nikola Smolenski, Gamma~enwiki, Reddi, Sbwoodside, Top-banana,Raul654, Shantavira, Northgrove, Zandperl, Jheise, Meanos~enwiki, Xanzzibar, David Gerard, Giftlite, Michael Devore, Hugh2414,Python eg gs,Urhixidur, Zro, PRiis, Rich Farmbrough, Spoon!, .:Ajvol:., La goutte de pluie, Daniel Arteaga~enwiki, Enirac Sum, Kjetil,Wricardoh~en wiki,Capecodeph, Falcorian, Uncle G, Vinter, Ulcph, Crucis, Joke137, Aarghdvaark, Rnt20, Kbdank71, Drbogdan, Rjwilmsi,Urbane Legend, Mike Peel, Ems57fcva, Marozols, Margosbot~enwiki, Dimator, BradBeattie, Chobot, CaseKid, The Rambling Man, Yurik-Bot, Hillman, Salsb, ErkDemon, Joel7687, GENtLe, Bobak, Tony1, Gadget850, Open2universe, Spacebirdy, Alain r, Geoffrey.landis, Vidarlo,Sardanaphalus, SmackB ot, C.Fred, Ifnord, Delldot, Athaler, Hbackman, Cuddlyopedia, Chris the speller, Jprg1966, Imaginaryoctopus, Zy-MOS, Hve, OrphanBot, Lou Scheffer,JMO, Andi47,JorisvS, Zzzzzzzzzzz, Togamoos, Az1568, Cryptic C62, Geremia, Myrrhlin, Van helsing,Friendlystar, Mb.bret, WillowW, Besieged,Thijs!bot,Markus Pössel, EdJohnston, Escarbot, AntiVandalBot, Uvaphdman, Sbattersby, Arch-magusrm, Myanw, CosineKitty, .anacondabot, JNW, J mcandrews,EagleFan, Gwern, Middlenamefrank, WhatUpPeopole, CommonsDelinker,Erockrph, HEL, AstroHurricane001, Numbo3, Athaenara, Zedmelon,Plasticup, Entropy, STBotD, Idioma-bot, Sheliak, Maghnus, VasilievVV,TXiKiBoT, UnitedStatesian, BotKung, Brachiopod, AlleborgoBot, Phe-bot,Likebox, Paolo.dL, Overpet, ClueBot, Agge1000, Polina Ivanova,RichardMassey, Wnt, Gumbosea, ErgoSum88, A ddbot, Iemaster77, DOI bot, Smitten912, Auspex1729, Wseaman, Wikicorona, TallJimbo,Lightbot, Zorrobot, Luckas-bot, Yobot, Amirobot, Jre hmeyer, Groaznic, The High Fin SpermWhale, Citation bot, ArthurBot, Xqbot, Tasfhkl,NOrbeck, SassoBot, Hauganm, Mnmngb, Grav-universe, LucienBOT, Originalwana, Citation bot1, Alipson, Jonesey95, Tom.Reding, Lithiumcyanide, Hardikkatyarmal1234, TobeBot, Puzl bustr, عقىل كاشف, Earthandmoon, ElPeste, Beyond My Ken, EmausBot, Primefac, Rename-dUser01302013, Jmencisom, Slawekb, Karthikndr, L Kensington, ClueBot NG, MIKHEIL, Polstar83, Braincricket, Tr00rle, Helpful PixieBot, Bibcode Bot, Bcxfu75k, BattyBot, Mdann52, Kevinfrancis1067, Frosty, Yuedongfang, Et2brute, ✩, Frinthruit, Seabuckthorn, Monkbot,Kcw19, Tetra quark, KasparBot and Anonymous: 116

- **Physics beyond the Standard Model** *Source*: https://en.wikipedia.org/wiki/Physics_beyond_the_Standard_Model?oldid=670089290 *Contributors*: David spector, Ewen, Michael Hardy, Andrewman327, Donarreiskoffer, Nurg, Rursus, David Gerard, Alison, David Schaich, RJHall, El C, Kwamikagami, I9Q79oL78KiL0QTFHgyc, Jeodesic, 4v4l0n42, Alinor, Count Iblis, Rjwilmsi, Strait, Eyu100, HappyCamper, Lmatt, BradBeattie, Ohwilleke, Bhny, SCZenz, CecilWard, Karl Andrews, Nlu, Dna-webmaster, Pawyilee, 2over0, Caco de vidro, Jaysbro, SmackBot, Mdj, Nickst, Chris the speller, Bluebot, Scwlong, QFT, Pepsidrinka, Jgwacker, Yevgeny Kats, Doug Bell, John, Dspitzle, RandomCritic, JarahE, Kurtan~enwiki, Headbomb, Peter Gulutzan, N shaji, Lenny Kaufman, VoABot II, Email4mobile, Maliz, R'n'B, HEL, Natsirtguy, Rod57, Tarotcards, DadaNeem, Goop Goop, Fences and windows, Michael H 34, Venny85, Wing gundam, Beast of traal, Bhuna71, Mild Bill Hiccup, Djr32, Excirial, RCalabraro, Brews ohare, Mastertek, TimothyRias, Truthnlove, Addbot, Luckas-bot, Zhitelew, Yobot, AnomieBOT, Citation bot, LilHelpa, Smk65536, Stevebow, Omnipaedista, Seeleschneider, A. di M., Kenneth Dawson, Steve Quinn, Citation bot 2, Aturen, Tom.Reding, ErgSlider, Physics therapist, Gistmass, Bj norge, Vstarsky, Serketan, ZéroBot, Galaktiker, Arbnos, Suslindisambiguator, Wiggles007, Smtchahal, ClaudeDes, Braincricket, Widr, Helpful Pixie Bot, Mike9110, DryRun, Bibcode Bot, BG19bot, Brainssturm, Qtom.masters, ThePeriodicTable123, M0532062613, Andyhowlett, Cinaro, I am One of Many, Kowtje, CtrlAltBackspace, 22merlin, Monkbot, Delbert7, Tetra quark, TQuentin, MauiPhoenix and Anonymous: 73

- **Structure formation** *Source*: https://en.wikipedia.org/wiki/Structure_formation?oldid=670660277 *Contributors*: Zundark, Edward, Rtfisher, Everyking, Grm wnr, FT2, Vsmith, Nabla, El C, Cmdrjameson, I9Q79oL78KiL0QTFHgyc, RHaworth, Mu301, Jeff3000, Joke137, Drbogdan, Rjwilmsi, R Lee E, Wavelength, Gaius Cornelius, Redgolpe, That Guy, From That Show!, SmackBot, Ashill, MalafayaBot, Droll, Colonies Chris, Myasuda, BobQQ, Corpx, Thijs!bot, Headbomb, Magioladitis, R'n'B, Warut, Sheliak, AlnoktaBOT, SwordSmurf, Lamro, SieBot, Shahidur Rahman, Jonmtkisco, Panos84, Dana boomer, Tailedkupo, DumZiBoT, RexxS, DCCougar, Hess88, Addbot, DOI bot, AkhtaBot, Samiswicked, Glane23, Yobot, Amirobot, AnomieBOT, Christopher.Gordon3, Citation bot, Icosmology, Kikuyu3, Citation bot 1, Berkeleyjess, Tom.Reding, RockSolidCosmo, Puzl bustr, JLincoln, RjwilmsiBot, TjBot, John of Reading, Italia2006, RaptureBot, Brownie Charles,

23.2. TEXT AND IMAGE SOURCES, CONTRIBUTORS, AND LICENSES

EdoBot, Machina Lucis, Lincoln Josh, Bibcode Bot, Khazar2, Tzymne, Penitence, PirtleShell, Jwratner1, Kogge, Xibalban Alchemist, Cosmic connection, Monkbot, Tetra quark, TychosElk and Anonymous: 15

- **Gravitational binding energy** *Source:* https://en.wikipedia.org/wiki/Gravitational_binding_energy?oldid=662128528 *Contributors:* Bryan Derksen, Olivier, Patrick, Julesd, Robertb-dc, BenRG, Wwoods, Gracefool, Karol Langner, Vsmith, Harley peters, Hooperbloob, Allen McC.~enwiki, Gene Nygaard, Kbdank71, Rjwilmsi, Ems57fcva, Mordecai, Fresheneesz, Wavelength, Gaius Cornelius, Moomoomoo, Smack-Bot, BeteNoir, Bbhart, Sbharris, Just plain Bill, Zarniwoot, Ckatz, Iridescent, Moreschi, Hans Dunkelberg, Katalaveno, Deep Atlantic Blue, Gillyweed, KyleOwens, Trang Oul, Explicit, Bobathon71, The Thing That Should Not Be, Kbdankbot, PV=nRT, Legobot, Yobot, AnomieBOT, Nickkid5, Mackem222, Achim1999, Akshit Goyal, Octaazacubane, Zueignung, MerlIwBot, Bibcode Bot, Kondephy, E8xE8, Rwh2100 and Anonymous: 26

- **Galaxy formation and evolution** *Source:* https://en.wikipedia.org/wiki/Galaxy_formation_and_evolution?oldid=668646767 *Contributors:* AxelBoldt, Bryan Derksen, AstroNomer~enwiki, -- April, Roadrunner, FlorianMarquardt, Olivier, Boud, Arpingstone, Looxix~enwiki, Ahoerstemeier, Nikai, Charles Matthews, Grendelkhan, Chris 73, Gandalf61, Merovingian, Sverdrup, Fuelbottle, Stirling Newberry, Giftlite, Jyril, Zigger, Crag, Andycjp, Karol Langner, RetiredUser2, Tothebarricades.tk, Icairns, Iantresman, Trevor MacInnis, Noisy, FT2, Vsmith, R6144, RJHall, MJT1331, Megaton~enwiki, Viriditas, Reuben, I9Q79oL78KiL0QTFHgyc, Eric Kvaalen, Keflavich, Evil Monkey, Nuno Tavares, Bacteria, Palica, Aarghdvaark, SqueakBox, Drbogdan, FlaBot, Gurch, YurikBot, Wavelength, Rodasmith, CambridgeBayWeather, Grafen, Tetracube, Hebb l, Arthur Rubin, Wsiegmund, ColinFrayn, Ilmari Karonen, Shp0ng1e, GorgonzolaCheese, SmackBot, GwydionM, Dark jedi requiem, MalafayaBot, Colonies Chris, Modest Genius, Vanished User 0001, Robma, John D. Croft, Hurricane Floyd, Jitterro, Muadd, Dl2000, Hetar, Richard Nowell, Ruslik0, Michael C Price, DumbBOT, Ward3001, Abtract, Headbomb, CharlotteWebb, AntiVandalBot, Dr. Submillimeter, Farosdaughter, Caper13, Alastair Haines, .anacondabot, VoABot II, Cgingold, Bruin69, HKL47, Rod57, Hubie59, Warut, Sheliak, Lights, VolkovBot, AlnoktaBOT, Vanished user ikijeirw34iuaeolaseriffic, Luckypengu07, Jahter, Doc Perel, Pika ten10, Harry~enwiki, Jdaloner, OKBot, Cosmo0, PipepBot, DragonBot, Scog, Astrotwitch, DumZiBoT, XLinkBot, LikeHolyWater, Ronhjones, TutterMouse, LaaknorBot, Kris1284x, Jorichoma, Tide rolls, QuadrivialMind, Legobot, Luckas-bot, Yobot, Viking59, Azcolvin429, Citation bot, Obersachsebot, Xqbot, Celiviel, FrescoBot, Originalwana, Charles Edwin Shipp, Oashi, Citation bot 1, Kot Barsik, Tom.Reding, Rushbugled13, RedBot, MastiBot, BlackHades, Barras, IVAN3MAN, DixonDBot, UrukHaiLoR, RjwilmsiBot, Immunize, Jmencisom, ZéroBot, Pringl123, Paymanpayman, Donner60, Just granpa, ClueBot NG, Bibcode Bot, Eric M. Jones, Saiph8, Jason from nyc, Samwalton9, RickV88, Benstrick, Khazar2, Lonesomehenry, Jwratner1, Abitslow, Ajb31, Monkbot, Sofia Koutsouveli, Rtjmca, Tetra quark and Anonymous: 104

- **Anisotropy** *Source:* https://en.wikipedia.org/wiki/Anisotropy?oldid=667840017 *Contributors:* Bryan Derksen, SimonP, DrBob, Michael Hardy, Charles Matthews, Wik, Book~enwiki, Furrykef, Rogper~enwiki, Robbot, Fredrik, Cyrius, DavidCary, Etune, Zigger, Bensaccount, Frencheigh, Karol Langner, Atemperman, Iantresman, WpZurp, M1ss1ontomars2k4, Eep^2, Slipstream, RJHall, Kwamikagami, RoyBoy, Wisdom89, Slicky, Kjkolb, Helix84, Alansohn, JYolkowski, Keflavich, Wtmitchell, Stephan Leeds, Uncle G, JohnJohn, Palica, V8rik, BD2412, Josh Parris, Drbogdan, Isaac Rabinovitch, Lockley, Amhaun01, Patrickr, Rrcoulter, Srleffler, YurikBot, Bergsten, Tubantia, DragonHawk, WAS 4.250, Femmina, Sardanaphalus, KnightRider~enwiki, SmackBot, RDBury, Ze miguel, Eskimbot, Duke Ganote, Bluebot, Berland, Fmindlin, Ixnayonthetimmay, Fitzhugh, Jfitzger, Spiritia, SashatoBot, Dl2000, JoeBot, George100, Propower, VoxLuna, James pic, Equendil, Cydebot, Davidrforrest, Hebrides, Thijs!bot, Mbell, Nonagonal Spider, M0ffx, Escarbot, Luna Santin, Chepyle, JAnDbot, Deflective, Hamsterlopithecus, Lopkiol, Americanhero, N.Nahber, DerHexer, WvEngen, Theron110, James-W, Terrek, Inwind, Vlmastra, Cremepuff222, Ebarbero, BotMultichill, VVVBot, Water and Land, 9eyedeel, Adam Cuerden, Anchor Link Bot, Karlhendrikse, Blackbat, Estirabot, NuclearWarfare, DumZiBoT, Farmerstam, Ngebbett, AgadaUrbanit, Pietrow, Zorrobot, Legobot, Luckas-bot, AnomieBOT, 1exec1, Essin, Materialscientist, Christopher Blubaugh, ArthurBot, LilHelpa, Tasudrty, GrouchoBot, Ianromanick, Sono53, Racingstripes, Matthieu.berthome, Reaper Eternal, Marie Poise, StephenNewby, EmausBot, Immunize, ZéroBot, Prayerfortheworld, Brandmeister, RockMagnetist, ClueBot NG, Jhsttshj, Satellizer, T. Packham, Helpful Pixie Bot, Riley Huntley, Eoktay, Xprofj, SlipperyDongDumpster, Polkmnqwaszx, KasparBot, Proutyr and Anonymous: 114

- **Baryon** *Source:* https://en.wikipedia.org/wiki/Baryon?oldid=671472233 *Contributors:* AxelBoldt, Tobias Hoevekamp, Bryan Derksen, BenZin~enwiki, Heron, Tim Starling, Alan Peakall, Paul A, Salsa Shark, Glenn, Mxn, Charles Matthews, The Anomebot, ElusiveByte, Phys, Bevo, Traroth, Donarreiskoffer, Robbot, Korath, Kristof vt, Merovingian, Ojigiri~enwiki, Sunray, Wikibot, Giftlite, DocWatson42, ShaunMacPherson, Herbee, Xerxes314, Dratman, DÅ‚ugosz, Kaldari, OwenBlacker, Icairns, JohnArmagh, Rich Farmbrough, Guanabot, Mani1, E2m, Tompw, El C, Bobo192, I9Q79oL78KiL0QTFHgyc, Giraffedata, Physicistjedi, Jumbuck, Gary, ABCD, Oleg Alexandrov, Woohookitty, Tevatron~enwiki, BD2412, Kbdank71, Nightscream, Ae77, MZMcBride, Chekaz, R.e.b., Erkcan, Maxim Razin, Oo64eva, Chobot, Roboto de Ajvol, YurikBot, Bambaiah, Jimp, Salsb, Ergzay, DragonHawk, SCZenz, E2mb0t~enwiki, Bota47, Simen, Sbyrnes321, Lainagier, Timotheus Canens, Bluebot, Colonies Chris, Kingdon, Shadow1, Bigmantonyd, Drphilharmonic, Kseferovic, Wierdw123, Physicsdog, Torrazzo, Verdy p, Michael C Price, Thijs!bot, Headbomb, Hcobb, Orionus, QuiteUnusual, Spartaz, Plantsurfer, Amateria1121, Diamond2, Swpb, BatteryIncluded, Hveziris, Saxophlute, Gwern, Ben MacDui, R'n'B, Ash, Tgeairn, Maurice Carbonaro, STBotD, VolkovBot, GimmeBot, NoiseEHC, Tearmeapart, BotKung, BrianADesmond, Antixt, AlleborgoBot, Lou427, SieBot, VVVBot, Gerakibot, LeadSongDog, Keilana, Paolo.dL, Doctorfluffy, TrufflesTheLamb, OKBot, Hamiltondaniel, TubularWorld, ClueBot, Artichoker, ChandlerMapBot, CalumH93, Addbot, LaaknorBot, CarsracBot, Jonhstone12, Legobot, Luckas-bot, Bugbrain 04, AnomieBOT, JackieBot, Materialscientist, Citation bot, ArthurBot, Xqbot, Omnipaedista, SassoBot, Spellage, WaysToEscape, FrescoBot, Citation bot 1, FoxBot, Noommos, EmausBot, John of Reading, JSquish, ZéroBot, StringTheory11, Stibu, Ethaniel, Markinvancouver, ClueBot NG, Koornti, Kasirbot, Rezabot, Bibcode Bot, Atomician, Zedshort, Marioedesouza, ChrisGualtieri, WorldWideJuan, CoolHandLouis, Monkbot, KasparBot and Anonymous: 105

- **Weakly interacting massive particles** *Source:* https://en.wikipedia.org/wiki/Weakly_interacting_massive_particles?oldid=667005504 *Contributors:* Damian Yerrick, Bryan Derksen, The Anome, Andre Engels, Rgamble, Arvindn, Roadrunner, Hashar, Timwi, Dysprosia, Maximus Rex, Robbot, Jakohn, Fredrik, Kizor, Cyrius, Awolf002, Sj, Dratman, Waltpohl, Eequor, Gadfium, Xenoglossophobe, LucasVB, Deglr6328, Pjacobi, AdamSolomon, Mr. Billion, El C, Fischej, Reuben, Riana, Snowolf, Count Iblis, Martian, Ciroa, Rjwilmsi, Strait, Bubba73, FlaBot, Eubot, Mindloss, Tedder, Tone, YurikBot, RussBot, Hellbus, Salsb, Zunaid, Mig21bp, KnightRider~enwiki, SmackBot, Incnis Mrsi, Rentier, Nickst, Fueled~enwiki, Skizzik, Dauto, Kmarinas86, Chris the speller, Bluebot, Sbharris, Vladislav, Tasc, Brandizzi, RekishiEJ, Tawkerbot2, Memetics, Lentower, MaxEnt, Difluoroethene, Thijs!bot, Irigi, Headbomb, Denverjeffrey, Autotheist, Hypershock, Soulbot, Vanished user ty12kl89jq10, Jarod997, Scrawfo, 0-Jenny-0, MarkBoulay, Antixt, StAnselm, Martarius, Niceguyedc, NuclearWarfare, Frongle, Wnt, Doug80, Ost316, Stephen Poppitt, Addbot, LaaknorBot, PoizonMyst, Csmallw, AnomieBOT, Citation bot, LilHelpa, Cydelin, Tomwsulcer,

J04n, Trongphu, RibotBOT, Ace111, CRea80, A. di M., FrescoBot, Styxpaint, Tom.Reding, RedBot, MastiBot, IVAN3MAN, Trappist the monk, Puzl bustr, Michael9422, EmausBot, Tommy2010, Zurich Astro, Mhatthei, Ebrambot, Wikfr, Timetraveler3.14, Nerd bzh, Ulflund, Jwyates, Esdacosta, Helpful Pixie Bot, Bibcode Bot, BG19bot, Luizpuodzius, Gorthian, Glevum, Gregaus, Gwickwire, Haraujo, Penguinstorm300, WebTV3, BattyBot, Gshahali, C.einstein1, Aszewci, Lmartin78, Monkeybuffer and Anonymous: 85

- **Cold dark matter** *Source:* https://en.wikipedia.org/wiki/Cold_dark_matter?oldid=671153639 *Contributors:* Bryan Derksen, Roadrunner, Tim Starling, Llywrch, Looxix~enwiki, Schneelocke, Doradus, Rho~enwiki, Eequor, Christopherlin, Keith Edkins, Balcer, Karl Dickman, Rich Farmbrough, StephanKetz, I9Q79oL78KiL0QTFHgyc, Fwb22, Lysdexia, Uris, Rjwilmsi, Chobot, 2over0, Kungfuadam, KnightRider~enwiki, Nickst, Nightbat, MalafayaBot, Cmanser, Farseer, OhioFred, John, Robofish, Eridani, Joeyfox10, Alaibot, Mbell, Headbomb, CosineKitty, Magioladitis, Ryan WMD, Idioma-bot, 0-Jenny-0, Michael H 34, BotKung, Mazarin07, Paradoctor, Niceguyedc, Auntof6, Addbot, Lightbot, Sebas310, Yobot, Systemizer, AnomieBOT, GrouchoBot, Waleswatcher, IAP Astro, Erik9bot, Kikuyu3, Sae1962, Citation bot 4, Pinethicket, Jonesey95, Tom.Reding, Trappist the monk, Gwyneth99, Zurich Astro, AvicAWB, SporkBot, AThinkingScientist, ClueBot NG, CocuBot, BBCDM, Helpful Pixie Bot, Bibcode Bot, BG19bot, Dualus, Winston Trechane, Willoakley, Prokaryotes, Sjzaslaw, Monkbot, Sofia Koutsouveli, Astronome de Meudon, Stefania.deluca, TychosElk and Anonymous: 45

- **Warm dark matter** *Source:* https://en.wikipedia.org/wiki/Warm_dark_matter?oldid=630369519 *Contributors:* Kwamikagami, Uogl, Rjwilmsi, Marasama, Profero, Nickst, Folajimi, Tawkerbot2, Foice, CosineKitty, Polytropoi74, VolkovBot, Addbot, Gaothird, Fluffernutter, Omnipaedista, Finncarey, Kikuyu3, Zurich Astro, ZéroBot, ClueBot NG, Helpful Pixie Bot, SilentPlanet22, ?? and Anonymous: 9

- **Hot dark matter** *Source:* https://en.wikipedia.org/wiki/Hot_dark_matter?oldid=667020437 *Contributors:* Mav, Lexor, Cassini~enwiki, Looxix~enwiki, Theresa knott, AnthonyQBachler, Korath, Naddy, Eequor, DragonflySixtyseven, Sam Hocevar, Burschik, Kate, Tom, Mac Davis, RJFJR, Rjwilmsi, Strait, Marasama, Zaak, Rangek, Margosbot~enwiki, Chobot, DVdm, DanMS, Salsb, Kungfuadam, Nickst, MalafayaBot, Lenoxus, Alaibot, OEYoung, Idioma-bot, BotKung, AlleborgoBot, MystBot, Addbot, LaaknorBot, Yobot, Romul~enwiki, AnomieBOT, Kikuyu3, LittleWink, Tom.Reding, Zurich Astro, Omyojj, Helpful Pixie Bot, PuruMuthal and Anonymous: 12

- **Baryonic dark matter** *Source:* https://en.wikipedia.org/wiki/Baryonic_dark_matter?oldid=665124216 *Contributors:* Roadrunner, SimonP, Heron, Yann, Ahoerstemeier, Laudaka, JeffBobFrank, Eequor, Gzornenplatz, ESkog, Dapete, Reuben, DV8 2XL, Koavf, Tone, Dancing Meerkat, SmackBot, Ashill, Nickst, Sbharris, Hemlock Martinis, Mentifisto, Wvaughan, STBot, UnitedStatesian, Mazarin07, ClueBot, Djr32, Vanished User 1004, Addbot, Rakesh sharma ujjain, Sebas310, GrouchoBot, False vacuum, UMinnAstro, Carlog3, FrescoBot, Bibcode Bot, Caroline1981 and Anonymous: 12

- **Massive compact halo object** *Source:* https://en.wikipedia.org/wiki/Massive_compact_halo_object?oldid=665790925 *Contributors:* Bryan Derksen, Roadrunner, Zadcat, EddEdmondson, Looxix~enwiki, Docu, Mark Foskey, Robbot, Jheise, Curps, Eequor, RetiredUser2, Sam Hocevar, Avriette, Gadykozma, Jonathanischoice, Reuben, Eleland, Idont Havaname, Eteq, Nanite, Pyb, Gary Brown, FlaBot, RussBot, Incnis Mrsi, Nickst, Aaron of Mpls, Skizzik, Kmarinas86, Rogermw, JorisvS, Younesmaia, V111P, OS2Warp, Lentower, Thijs!bot, Headbomb, Madhava 1947, Idioma-bot, Leebo, Mazarin07, Akroli000001, SieBot, Starcluster, The Thing That Should Not Be, DumZiBoT, Doug80, Addbot, LaaknorBot, Rob2funky, Yobot, AnomieBOT, Djxerox, GrouchoBot, FrescoBot, DaLeBu, Michael9422, Tommy2010, Suslindisambiguator, Widr, Saylors and Anonymous: 32

- **Neutralino** *Source:* https://en.wikipedia.org/wiki/Neutralino?oldid=671536470 *Contributors:* Angela, Julesd, Schneelocke, Charles Matthews, Saltine, Donarreiskoffer, Robbot, Rursus, Moink, Awolf002, Herbee, Xerxes314, Waltpohl, Pharotic, Eequor, Icairns, Urvabara, Pjacobi, Slicky, Physicistjedi, JPFlip, SDC, Theofilatos, Kbdank71, FlaBot, Roboto de Ajvol, YurikBot, Conscious, SCZenz, AndrewWTaylor, SmackBot, Stepa, Nickst, Vladislav, Pulu, Lester, Newone, Wadoli Itse, Thijs!bot, Barticus88, Headbomb, Stannered, Squantmuts, NicZ~enwiki, Choihei, Antixt, ClueBot, SkyLined, Addbot, Prim Ethics, Yobot, AnomieBOT, Citation bot, ArthurBot, Br77rino, Ernsts, A. di M., Erik9bot, Tom.Reding, ZéroBot, David C Bailey, Helpful Pixie Bot, Bibcode Bot, Halfb1t, Manar al Zraiy, Makecat-bot, Phseek and Anonymous: 24

- **Axion** *Source:* https://en.wikipedia.org/wiki/Axion?oldid=666181988 *Contributors:* Bryan Derksen, Roadrunner, Maury Markowitz, Heron, Ste vertigo, Edward, Ahoerstemeier, J'raxis, Jengod, Dragons flight, Saltine, Phys, Thue, BenRG, Seglea, Pmineault, Rursus, Herbee, Xerxes314, Jeff BobFrank, Eequor, Pjacobi, David Schaich, Bender235, JustinWick, Themusicgod1, John Vandenberg, .:Ajvol:., Reuben, I9Q79oL78KiL0QTPh ysicistjedi, Fwb22, Guy Harris, Axl, Mac Davis, Wiccan Quagga, RJFJR, Count Iblis, Poseidon^3, ????, Gene Nygaard, Betsythedevine, Kb dank71, Rjwilmsi, Captmondo, Mike Peel, AndyKali, Erkcan, Itinerant1, Lmatt, Peri~enwiki, Ggb667, Chobot, Peter Grey, Uriah923, Yurik Bot, Spacepotato, Bambaiah, RussBot, Salsb, Oni Lukos, Buster79, Długosz, Dogcow, Ravedave, 2over0, A13ean, SmackBot, TomLougheed, Nickst, Colonies Chris, Martin Blank, Vladislav, OrphanBot, Pwjb, A5b, Lambiam, JorisvS, Anescient, Ossipewsk, Dsspiegel, Danlev, Friendly Neighbour, Foice, Runningonbrains, Phatom87, Thijs!bot, Epbr123, Wikid77, Headbomb, Davidhorman, Escarbot, OreoPriest, Orionus, Billdad, Bubsir, Dricherby, Mwzappe, Qev, Swpb, Homunq, Otivaeey, Maliz, Alsee, HEL, William H. Kinney, 0-Jenny-0, Adamwang, Markhealey, Voorlandt, Antixt, SieBot, Senor Cuete, Martarius, ArdClose, Uxorion, FOARP, Kcren, SkyLined, Addbot, Uruk2008, DOI bot, Howard Landman, Zorrobot, KusterM, Luckas-bot, Yobot, Ptbotgourou, AnomieBOT, RoundPanda, JackieBot, Cita-tion bot, Xqbot, Reality006, Omnipaedista, FrescoBot, Chuli2802, Citation bot 1, Citation bot 4, Three887, Governus, Trappist the monk, Dinamik-bot, Earthandmoon, Sideways713, Marie Poise, StringTheory11, Suslindisambiguator, Zhskyy, Timetraveler3.14, ClueBot NG, Alessandro.de.angelis, Koornti, Asstrak, Bibcode Bot, BG19bot, Trevayne08, Cerabot~enwiki, Wipark, Andyhowlett, Ajbilan, Monkbot, Ter-ryAlex, PS83, BradNorton1979 and Anonymous: 83

- **Mixed dark matter** *Source:* https://en.wikipedia.org/wiki/Mixed_dark_matter?oldid=645246581 *Contributors:* Bbbl67, TJSwoboda, Stemonitis, SmackBot, Nickst, JRSpriggs, Leuko, Appraiser, Kiwizoid, Rickn27, Addbot, Tom.Reding, Quondum, ChrisGualtieri and Jocelyndurrey

23.2.2 Images

- **File:080998_Universe_Content_240.jpg** *Source:* https://upload.wikimedia.org/wikipedia/commons/a/a5/080998_Universe_Content_240.jpg *License:* Public domain *Contributors:* http://map.gsfc.nasa.gov/media/080998/index.html *Original artist:* Credit: NASA / WMAP Science Team

23.2. TEXT AND IMAGE SOURCES, CONTRIBUTORS, AND LICENSES

- **File:080998_Universe_Content_240_after_Planck.jpg** *Source:* https://upload.wikimedia.org/wikipedia/commons/b/b6/080998_Universe_Content_240_after_Planck.jpg *License:* Public domain *Contributors:* http://map.gsfc.nasa.gov/media/080998/index.html updated data from http://www.nasa.gov/mission_pages/planck/news/planck20130321.html *Original artist:* NASA, Modified by User:??

- **File:14-23000-Sparky-MassiveGalaxyFormation-20140827.jpg***Source:*https://upload.wikimedia.org/wikipedia/commons/2/23/14-23000-Sparky-MassiveGalaxyFormation-20140827.
jpg *License:* Public domain *Contributors:* http://www.nasa.gov/sites/default/files/14-230_0.jpg *Original artist:* NASA, Z. Levay, G. Bacon (STScI)

- **File:14-283-Abell2744-DistantGalaxies-20141016.jpg***Source:*https://upload.wikimedia.org/wikipedia/commons/d/d2/14-283-Abell2744-DistantGalaxies-20141016.
jpg *License:* Public domain *Contributors:* http://www.nasa.gov/sites/default/files/14-283_0.jpg *Original artist:* NASA, J. Lotz, (STScI)

- **File:1e0657_scale.jpg** *Source:* https://upload.wikimedia.org/wikipedia/commons/a/a8/1e0657_scale.jpg *License:* Public domain *Contributors:* Chandra X-Ray Observatory: 1E 0657-56 *Original artist:* NASA/CXC/M. Weiss

- **File:2MASS_LSS_chart-NEW_Nasa.jpg** *Source:* https://upload.wikimedia.org/wikipedia/commons/7/7d/2MASS_LSS_chart-NEW_Nasa.jpg *License:* Public domain *Contributors:* "Large Scale Structure in the Local Universe: The 2MASS Galaxy Catalog", Jarrett, T.H. 2004, PASA, 21, 396 *Original artist:* IPAC/Caltech, by Thomas Jarrett

- **File:2dfdtfe.gif** *Source:* https://upload.wikimedia.org/wikipedia/commons/b/b3/2dfdtfe.gif *License:* CC-BY-SA-3.0 *Contributors:* http://en.wikipedia.org/wiki/Sloan_Great_Wall *Original artist:* Willem Schaap

- **File:Abell_1689.jpg** *Source:* https://upload.wikimedia.org/wikipedia/commons/9/9a/Abell_1689.jpg *License:* CC BY 3.0 *Contributors:* http://www.spacetelescope.org/images/heic1014a/ *Original artist:* NASA, ESA, E. Jullo (JPL/LAM), P. Natarajan (Yale) and J-P. Kneib (LAM).

- **File:Abell_S740,_cropped_to_ESO_325-G004.jpg** *Source:* https://upload.wikimedia.org/wikipedia/commons/d/d3/Abell_S740%2C_cropped_to_ESO_325-G004.jpg *License:* Public domain *Contributors:* http://hubblesite.org/newscenter/archive/releases/galaxy/elliptical/2007/08/image/a/warn/ *Original artist:* J. Blakeslee (Washington State University)

- **File:Antennae_galaxies_xl.jpg** *Source:* https://upload.wikimedia.org/wikipedia/commons/f/f6/Antennae_galaxies_xl.jpg *License:* Public domain *Contributors:*

- http://www.spacetelescope.org/images/html/heic0615a.html *Original artist:* NASA, ESA, and the Hubble Heritage Team (STScI/AURA)-ESA/Hubble Collaboration

- **File:Artist'{}s_impression_of_gravitational_lensing_of_a_distant_merger.ogg** *Source:* https://upload.wikimedia.org/wikipedia/commons/4/49/Artist%27s_impression_of_gravitational_lensing_of_a_distant_merger.ogg *License:* CC BY 4.0 *Contributors:* ESO *Original artist:* ESO/M. Kornmesser

- **File:Artist's_impression_of_a_gamma-ray_burst_shining_through_two_young_galaxies_in_the_early_Universe.jpg** *Source:* https://upload.wikimedia.org/wikipedia/commons/3/3a/Artist%E2%80%99s_impression_of_a_gamma-ray_burst_shining_through_two_young_galaxies_in_the_early_Universe.jpg *License:* CC BY 4.0 *Contributors:* ESO *Original artist:* ESO/L. Calçada

- **File:Artist's_impression_of_the_expected_dark_matter_distribution_around_the_Milky_Way.ogv** *Source:* https://upload.wikimedia.org/wikipedia/commons/0/03/Artist%E2%80%99s_impression_of_the_expected_dark_matter_distribution_around_the_Milky_Way.ogv *License:* CC BY 4.0 *Contributors:* ESO *Original artist:* ESO/L. Calçada

- **File:Baryon-decuplet-small.svg** *Source:* https://upload.wikimedia.org/wikipedia/commons/7/78/Baryon-decuplet-small.svg *License:* Public domain *Contributors:* Own work *Original artist:* Trassiorf

- **File:Baryon-octet-small.svg** *Source:* https://upload.wikimedia.org/wikipedia/commons/b/b5/Baryon-octet-small.svg *License:* Public domain *Contributors:* Own work *Original artist:* Trassiorf

- **File:Black_hole_lensing_web.gif** *Source:* https://upload.wikimedia.org/wikipedia/commons/0/03/Black_hole_lensing_web.gif *License:* CC-BY-SA-3.0 *Contributors:* en:Image:BlackHole_Lensing_2.gif *Original artist:* Urbane Legend (optimised for web use by Alain r)

- **File:CDMS_parameter_space_2004.png** *Source:* https://upload.wikimedia.org/wikipedia/commons/7/7f/CDMS_parameter_space_2004.png *License:* Public domain *Contributors:* From 2004 CDMS [1] presentation by Richard Schnee and was created with dark matter plot generator [2]. *Original artist:* PNG created with Inkscape and Photoshop by Kieff from a JPEG data plot.

- **File:CERN_LHC_Tunnel1.jpg** *Source:* https://upload.wikimedia.org/wikipedia/commons/f/fc/CERN_LHC_Tunnel1.jpg *License:* CC BY-SA 3.0 *Contributors:* Own work *Original artist:* Julian Herzog (website)

- **File:CL0024+17.jpg** *Source:* https://upload.wikimedia.org/wikipedia/commons/c/c3/CL0024%2B17.jpg *License:* Public domain *Contributors:* http://hubblesite.org/newscenter/archive/releases/2007/17/image/a/ (direct link) *Original artist:* NASA, ESA, M.J. Jee and H. Ford (Johns Hopkins University)

- **File:COSMOS_3D_dark_matter_map.jpg** *Source:* https://upload.wikimedia.org/wikipedia/commons/3/38/COSMOS_3D_dark_matter_map.jpg *License:* Public domain *Contributors:* Originally uploaded at en.wikipedia as File:COSMOS 3D dark matter map.jpg by User:RichardMassey. (Transfered by User:Quibik.)

Also available at http://spacetelescope.org/images/heic0701b/

Original artist: NASA/ESA/Richard Massey (California Institute of Technology)

- **File:Capital_Lambda.svg** *Source:* https://upload.wikimedia.org/wikipedia/commons/1/1d/Capital_Lambda.svg *License:* Public domain *Contributors:* DarkEvil *Original artist:* DarkEvil

- **File:Collage_of_six_cluster_collisions_with_dark_matter_maps.jpg** *Source:* https://upload.wikimedia.org/wikipedia/commons/0/03/Collage_of_six_cluster_collisions_with_dark_matter_maps.jpg *License:* ? *Contributors:* http://www.spacetelescope.org/images/heic1506a/ *Original artist:* NASA, ESA, D. Harvey (École Polytechnique Fédérale de Lausanne, Switzerland), R. Massey (Durham University, UK), the Hubble SM4 ERO Team, ST-ECF, ESO, D. Coe (STScI), J. Merten (Heidelberg/Bologna), HST Frontier Fields, Harald Ebeling(University of Hawaii at Manoa), Jean-Paul Kneib (LAM)and Johan Richard (Caltech, USA)

- **File:Commons-logo.svg** *Source:* https://upload.wikimedia.org/wikipedia/en/4/4a/Commons-logo.svg *License:* ? *Contributors:* ? *Original artist:* ?

- **File:Crab_Nebula.jpg** *Source:* https://upload.wikimedia.org/wikipedia/commons/0/00/Crab_Nebula.jpg *License:* Public domain *Contributors:* HubbleSite: gallery, release. *Original artist:* NASA, ESA, J. Hester and A. Loll (Arizona State University)

- **File:DMPie_2013.svg** *Source:* https://upload.wikimedia.org/wikipedia/commons/1/1f/DMPie_2013.svg *License:* CC BY-SA 3.0 *Contributors:* Own work *Original artist:* Szczureq

- **File:Dark_Energy.jpg** *Source:* https://upload.wikimedia.org/wikipedia/commons/c/ce/Dark_Energy.jpg *License:* Public domain *Contributors:*http://hubblesite.org/newscenter/archive/releases/2001/09/image/g/ORhttp://science.nasa.gov/astrophysics/focus-areas/what-is-dark-energy/*Original artist:*Ann Feild (STScI)

- **File:E=mc²-explication.svg** *Source:* https://upload.wikimedia.org/wikipedia/commons/c/c9/E%3Dmc%C2%B2-explication.svg *License:* CC BY-SA 3.0 *Contributors:* Own work *Original artist:* JTBarnabas

- **File:E_equals_m_plus_c_square_at_Taipei101.jpg** *Source:* https://upload.wikimedia.org/wikipedia/commons/6/62/E_equals_m_plus_c_square_at_Taipei101.jpg *License:* CC-BY-SA-3.0 *Contributors:* zh:File:E_equals_m_plus_c_square_at_Taipei101.jpg *Original artist:* SElefant

- **File:Earth'{}s_Location_in_the_Universe_(JPEG).jpg***Source:*https://upload.wikimedia.org/wikipedia/commons/b/b6/Earth%27s_Location_in_the_Universe_%28JPEG%29.jpg *License:* CC BY-SA 3.0 *Contributors:* Own work *Original artist:* Andrew Z. Colvin

- **File:Edit-clear.svg** *Source:* https://upload.wikimedia.org/wikipedia/en/f/f2/Edit-clear.svg *License:* Public domain *Contributors:* The *Tango! Desktop Project*. *Original artist:*

    The people from the Tango! project. And according to the meta-data in the file, specifically: "Andreas Nilsson, and Jakub Steiner (although minimally)."

- **File:Einstein_-_Time_Magazine_-_July_1,_1946.jpg** *Source:* https://upload.wikimedia.org/wikipedia/en/5/57/Einstein_-_Time_Magazine_-_July_1%2C_1946.jpg *License:* ? *Contributors:*

    www.time.com *Original artist:*

    Artist: Ernest Hamlin Baker (1889-1975)
    Time Magazine

- **File:Einstein_cross.jpg** *Source:* https://upload.wikimedia.org/wikipedia/commons/c/c8/Einstein_cross.jpg *License:* Public domain *Contributors:* http://hubblesite.org/newscenter/archive/releases/1990/20/image/a/ *Original artist:* NASA, ESA, and STScI

- **File:Elemental_abundances.svg** *Source:* https://upload.wikimedia.org/wikipedia/commons/0/09/Elemental_abundances.svg *License:* Public domain *Contributors:* http://pubs.usgs.gov/fs/2002/fs087-02/ *Original artist:* Gordon B. Haxel, Sara Boore, and Susan Mayfield from USGS; vectorized by User:michbich

- **File:Eso1516a.jpg** *Source:* https://upload.wikimedia.org/wikipedia/commons/0/0b/Eso1516a.jpg *License:* CC BY 4.0 *Contributors:* ESO website *Original artist:* ESO

- **File:Fermi_Observations_of_Dwarf_Galaxies_Provide_New_Insights_on_Dark_Matter.ogv***Source:*https://upload.mons/a/a9/Fermi_Observations_of_Dwarf_Galaxies_Provide_New_Insights_on_Dark_Matter.ogv*License:*Public domain*Contributors:*Goddard Multimedia*Originalartist:*NASA/Goddard Space Flight Center

- **File:Folder_Hexagonal_Icon.svg** *Source:* https://upload.wikimedia.org/wikipedia/en/4/48/Folder_Hexagonal_Icon.svg *License:* Cc-by-sa-3.0 *Contributors:* ? *Original artist:* ?

- **File:GalacticRotation2.svg** *Source:* https://upload.wikimedia.org/wikipedia/commons/b/b9/GalacticRotation2.svg *License:* CC-BY-SA-3.0 *Contributors:* Own work in Inkscape 0.42 *Original artist:* PhilHibbs

- **File:Gravitational_lens-full.jpg** *Source:* https://upload.wikimedia.org/wikipedia/commons/0/02/Gravitational_lens-full.jpg *License:* Public domain *Contributors:* ? *Original artist:* ?

- **File:Gravitational_lensing_of_distant_star-forming_galaxies_(schematic)_-vid-.webm** *Source:* https://upload.wikimedia.org/wikipedia/commons/0/02/Gravitational_lensing_of_distant_star-forming_galaxies_%28schematic%29_-vid-.webm *License:* CC BY 4.0 *Contributors:* ESO *Original artist:* ALMA (NRAO/ESO/NAOJ)/Luis Calçada (ESO)

- **File:Gravitational_lensing_of_distant_star-forming_galaxies_(schematic)_2.webm***Source:*https://upload.wikimedia.org/wikipedia/coons/e/e7/Gravitational_lensing_of_distant_star-forming_galaxies_%28schematic%29_2.webm *License:* CC BY 4.0 *Contributors:* ESO *Original artist:* ALMA (NRAO/ESO/NAOJ)/Luis Calçada (ESO)

- **File:Gravitationally-lensed_distant_star-forming_galaxy.jpg***Source:*https://upload.wikimedia.org/wikipedia/commons/b/b9/Gravitationally-lensed_distant_star-forming_galaxy.jpg*License:*CC BY 4.0*Contributors:*http://www.eso.org/public/images/eso1313a/*Original artist:*ALMA (ESO/NRAO/NAOJ),J. Vieira et al.

- **File:Gravitationell-lins-4.jpg** *Source:* https://upload.wikimedia.org/wikipedia/commons/0/0b/Gravitationell-lins-4.jpg *License:* Public domain *Contributors:* http://hubblesite.org/newscenter/newsdesk/archive/releases/2003/01/image/a *Original artist:* NASA, N. Benitez (JHU), T. Broadhurst (Racah Institute of Physics/The Hebrew University), H. Ford (JHU), M. Clampin (STScI),G. Hartig (STScI), G. Illingworth (UCO/Lick Observatory), the ACS Science Team and ESA

- **File:Half-logarithm_graph.jpg** *Source:* https://upload.wikimedia.org/wikipedia/commons/0/0b/Half-logarithm_graph.jpg *License:* CC BY-SA 4.0 *Contributors:* Own work *Original artist:* Alexey Alekseenko

23.2. TEXT AND IMAGE SOURCES, CONTRIBUTORS, AND LICENSES

- **File:Horizonte_inflacionario.svg** *Source:* https://upload.wikimedia.org/wikipedia/commons/b/b4/Horizonte_inflacionario.svg *License:* CC-BY-SA-3.0 *Contributors:* Transferred from en.wikipedia to Commons.; original: *I created this work in Adobe Illustrator*. *Original artist:* Joke137 at English Wikipedia
- **File:HubbleTuningFork.jpg** *Source:* https://upload.wikimedia.org/wikipedia/commons/2/21/HubbleTuningFork.jpg *License:* Public domain *Contributors:* Transferred from en.wikipedia *Original artist:* Original uploader was Cosmo0 at en.wikipedia
- **File:HubbleUltraDeepFieldwithScaleComparison.jpg***Source:*https://upload.wikimedia.org/wikipedia/commons/e/e4/HubbleUltraDearison.jpg *License:* Public domain *Contributors:* http://hubblesite.org/newscenter/archive/releases/2004/07/image/a/warn/ *Original artist:* NASA and the European Space Agency. Edited by Autonova
- **File:Hubble_image_of_the_galaxy_cluster_Abell_3827.jpg** *Source:* https://upload.wikimedia.org/wikipedia/commons/5/51/Hubble_image_of_the_galaxy_cluster_Abell_3827.jpg *License:* CC BY 4.0 *Contributors:* http://www.eso.org/public/images/eso1514a/ *Original artist:* ESO
- **File:Ilc_9yr_moll4096.png** *Source:* https://upload.wikimedia.org/wikipedia/commons/3/3c/Ilc_9yr_moll4096.png *License:* Public domain *Contributors:* http://map.gsfc.nasa.gov/media/121238/ilc_9yr_moll4096.png *Original artist:* NASA / WMAP Science Team
- **File:Incorrect_plaque_at_the_Rose_Center_for_Earth_and_Space,_April_2011.jpg***Source:*https://upload.wikimedia.org/wikipedia/2/24/Incorrect_plaque_at_the_Rose_Center_for_Earth_and_Space%2C_April_2011.jpg *License:* CC BY-SA 3.0 *Contributors:* Picture taken at the Rose Center for Earth and Space, New York, New York, during a visit. *Original artist:* Rogerstrolley
- **File:LCDM.jpg** *Source:* https://upload.wikimedia.org/wikipedia/commons/7/7d/LCDM.jpg *License:* CC BY-SA 3.0 *Contributors:* Own work *Original artist:* Michael L. Umbricht
- **File:Lambda-Cold_Dark_Matter,_Accelerated_Expansion_of_the_Universe,_Big_Bang-Inflation.jpg** *Source:* https://upload.wikimedia.org/wikipedia/commons/c/c2/Lambda-Cold_Dark_Matter%2C_Accelerated_Expansion_of_the_Universe%2C_Big_Bang-Inflation.jpg *License:* CC BY-SA 3.0 *Contributors:* Own work *Original artist:* User:Coldcreation
- **File:Large-scale_structure_of_light_distribution_in_the_universe.jpg** *Source:* https://upload.wikimedia.org/wikipedia/commons/6/6d/Large-scale_structure_of_light_distribution_in_the_universe.jpg *License:* CC BY 2.0 *Contributors:* https://www.flickr.com/photos/uclmaps/15051460475/ *Original artist:* Andrew Pontzen and Fabio Governato
- **File:M101_hires_STScI-PRC2006-10a.jpg** *Source:* https://upload.wikimedia.org/wikipedia/commons/c/c5/M101_hires_STScI-PRC2006-10a.jpg *License:* CC BY 3.0 *Contributors:* http://www.spacetelescope.org/news/html/heic0602.html (direct link) *Original artist:*

  Credit:

  Image: European Space Agency & NASA

  Acknowledgements:

  Project Investigators for the original Hubble data: K.D. Kuntz (GSFC), F. Bresolin (University of Hawaii), J. Trauger (JPL), J. Mould (NOAO), and Y.-H. Chu (University of Illinois, Urbana)

  Image processing: Davide De Martin (ESA/Hubble)

  CFHT image: Canada-France-Hawaii Telescope/J.-C. Cuillandre/Coelum

  NOAO image: George Jacoby, Bruce Bohannan, Mark Hanna/NOAO/AURA/NSF
- **File:MACS_J1206.jpg** *Source:* https://upload.wikimedia.org/wikipedia/commons/5/56/MACS_J1206.jpg *License:* ? *Contributors:* http://www.spacetelescope.org/images/heic1115a/ *Original artist:* NASA, ESA, M. Postman (STScI) and the CLASH Team
- **File:NGC4676.jpg** *Source:* https://upload.wikimedia.org/wikipedia/commons/d/db/NGC4676.jpg *License:* Public domain *Contributors:* APOD 2004-06-12 *Original artist:* NASA, H. Ford (JHU), G. Illingworth (UCSC/LO), M.Clampin (STScI), G. Hartig (STScI), the ACS Science Team, and ESA
- **File:NGC891.jpg** *Source:* https://upload.wikimedia.org/wikipedia/commons/2/2e/NGC891.jpg *License:* CC BY-SA 2.5 *Contributors:* http://www.martin-x.de/astro/astro.html *Original artist:* Martin Baessgen
- **File:Nucleosynthesis_periodic_table.svg** *Source:* https://upload.wikimedia.org/wikipedia/commons/3/31/Nucleosynthesis_periodic_table.svg *License:* CC BY-SA 3.0 *Contributors:* Own work *Original artist:* Cmglee
- **File:Observable_Universe_with_Measurements_01.png***Source:*https://upload.wikimedia.org/wikipedia/commons/9/98/Observable_Universe_with_Measurements_01.png *License:* CC BY-SA 3.0 *Contributors:* Own work *Original artist:* Azcolvin429
- **File:Observable_universe_logarithmic_illustration.png***Source:*https://upload.wikimedia.org/wikipedia/commons/e/e7/Observable_universe_logarithmic_illustration.png *License:* CC BY-SA 3.0 *Contributors:* Own work *Original artist:* Unmismoobjetivo
- **File:Office-book.svg** *Source:* https://upload.wikimedia.org/wikipedia/commons/a/a8/Office-book.svg *License:* Public domain *Contributors:* This and myself. *Original artist:* Chris Down/Tango project
- **File:Plasma-lamp_2.jpg** *Source:* https://upload.wikimedia.org/wikipedia/commons/2/26/Plasma-lamp_2.jpg *License:* CC-BY-SA-3.0 *Contributors:*
- own work www.lucnix.be *Original artist:* Luc Viatour
- **File:Portal-puzzle.svg** *Source:* https://upload.wikimedia.org/wikipedia/en/f/fd/Portal-puzzle.svg *License:* Public domain *Contributors:* ? *Original artist:* ?
- **File:Question_book-new.svg** *Source:* https://upload.wikimedia.org/wikipedia/en/9/99/Question_book-new.svg *License:* Cc-by-sa-3.0 *Contributors:*

  Created from scratch in Adobe Illustrator. Based on Image:Question book.png created by User:Equazcion *Original artist:* Tkgd2007

- **File:Relativity3_Walk_of_Ideas_Berlin.JPG** *Source:* https://upload.wikimedia.org/wikipedia/commons/4/49/Relativity3_Walk_of_Ideas_Berlin.JPG *License:* CC BY 2.5 *Contributors:* Own work *Original artist:* Lienhard Schulz
- **File:SN1994D.jpg** *Source:* https://upload.wikimedia.org/wikipedia/commons/a/a2/SN1994D.jpg *License:* CC BY 3.0 *Contributors:* http://www.spacetelescope.org/images/html/opo9919i.html *Original artist:* NASA/ESA, The Hubble Key Project Team and The High-Z Supernova Search Team
- **File:Scheme_of_nuclear_reaction_chains_for_Big_Bang_nucleosynthesis.svg** *Source:* https://upload.wikimedia.org/wikipedia/commons/5/5f/Scheme_of_nuclear_reaction_chains_for_Big_Bang_nucleosynthesis.svg *License:* CC BY-SA 4.0 *Contributors:* Own work ; vectorisation de The main nuclear reaction chains for Big Bang nucleosynthesis.jpg *Original artist:* Pamputt
- **File:Science.jpg** *Source:* https://upload.wikimedia.org/wikipedia/commons/5/54/Science.jpg *License:* Public domain *Contributors:* ? *Original artist:* ?
- **File:SolarSystemAbundances.png** *Source:* https://upload.wikimedia.org/wikipedia/commons/e/e6/SolarSystemAbundances.png *License:* CC BY-SA 3.0 *Contributors:* Transferred from en.wikipedia *Original artist:* Original uploader was 28bytes at en.wikipedia
- **File:Spacer.gif** *Source:* https://upload.wikimedia.org/wikipedia/commons/5/52/Spacer.gif *License:* Public domain *Contributors:* ? *Original artist:* ?
- **File:Spacetime_curvature.png** *Source:* https://upload.wikimedia.org/wikipedia/commons/2/22/Spacetime_curvature.png *License:* CC-BY-SA-3.0 *Contributors:* ? *Original artist:* ?
- **File:Standard_Model_of_Elementary_Particles.svg** *Source:* https://upload.wikimedia.org/wikipedia/commons/0/00/Standard_Model_of_Elementary_Particles.svg *License:* CC BY 3.0 *Contributors:* Own work by uploader, PBS NOVA [1], Fermilab, Office of Science, United States Department of Energy, Particle Data Group *Original artist:* MissMJ
- **File:Structure_mode_history.svg** *Source:* https://upload.wikimedia.org/wikipedia/commons/e/ef/Structure_mode_history.svg *License:* CC-BY-SA-3.0 *Contributors:* This is my work, calculated using a code I wrote myself, plotted in Gnuplot and edited in Adobe Illustrator. *Original artist:* Joke137 at English Wikipedia
- **File:Stylised_Lithium_Atom.svg** *Source:* https://upload.wikimedia.org/wikipedia/commons/e/e1/Stylised_Lithium_Atom.svg *License:* CC-BY-SA-3.0 *Contributors:* based off of Image:Stylised Lithium Atom.png by Halfdan. *Original artist:* SVG by Indolences. Recoloring and ironing out some glitches done by Rainer Klute.
- **File:Symbol_book_class2.svg** *Source:* https://upload.wikimedia.org/wikipedia/commons/8/89/Symbol_book_class2.svg *License:* CC BY-SA 2.5 *Contributors:* Mad by Lokal_Profil by combining: *Original artist:* Lokal_Profil
- **File:TaskForce_One.jpg** *Source:* https://upload.wikimedia.org/wikipedia/commons/4/41/TaskForce_One.jpg *License:* Public domain *Contributors:* Official US Navy photograph [1] available on Navsource.org [2] *Original artist:* USN
- **File:The_incomplete_circle_of_everything.svg** *Source:* https://upload.wikimedia.org/wikipedia/commons/0/0d/The_incomplete_circle_of_everything.svg *License:* CC BY 3.0 *Contributors:* Own work *Original artist:* Zhitelew
- **File:Turning_Black_Holes_into_Dark_Matter_Labs.webm** *Source:* https://upload.wikimedia.org/wikipedia/commons/d/d0/Turning_Black_Holes_into_Dark_Matter_Labs.webm *License:* Public domain *Contributors:* NASA's Goddard Space Flight Center *Original artist:* NASA'sGoddard Space Flight Center
- **File:WMAP_2010.png** *Source:* https://upload.wikimedia.org/wikipedia/commons/2/2d/WMAP_2010.png *License:* Public domain *Contributors:* http://wmap.gsfc.nasa.gov/media/101080 *Original artist:* ?
- **File:Warped_galaxy.jpg** *Source:* https://upload.wikimedia.org/wikipedia/commons/0/06/Warped_galaxy.jpg *License:* Public domain *Contributors:* http://hubblesite.org/newscenter/archive/releases/2001/23/image/a (direct link) *Original artist:* NASA and The Hubble Heritage Team (STScI/AURA)
- **File:Wikiquote-logo.svg** *Source:* https://upload.wikimedia.org/wikipedia/commons/f/fa/Wikiquote-logo.svg *License:* Public domain *Contributors:* ? *Original artist:* ?
- **File:Wikisource-logo.svg** *Source:* https://upload.wikimedia.org/wikipedia/commons/4/4c/Wikisource-logo.svg *License:* CC BY-SA 3.0 *Contributors:* Rei-artur *Original artist:* Nicholas Moreau
- **File:Wiktionary-logo-en.svg** *Source:* https://upload.wikimedia.org/wikipedia/commons/f/f8/Wiktionary-logo-en.svg *License:* Public domain *Contributors:* Vector version of Image:Wiktionary-logo-en.png. *Original artist:* Vectorized by Fvasconcellos (talk · contribs), based on original logo tossed together by Brion Vibber
- **File:World_line.svg** *Source:* https://upload.wikimedia.org/wikipedia/commons/1/16/World_line.svg *License:* CC-BY-SA-3.0 *Contributors:* Transferred from en.wikipedia.
Original artist: SVG version: K. Aainsqatsi at en.wikipedia
- **File:Wz-z.jpg** *Source:* https://upload.wikimedia.org/wikipedia/commons/f/f0/Wz-z.jpg *License:* CC BY-SA 4.0 *Contributors:* Own work *Original artist:* Esadri21

23.2.3 Content license

- Creative Commons Attribution-Share Alike 3.0

Printed in Great Britain
by Amazon